CHEMICAL SIGNALS
Vertebrates and Aquatic Invertebrates

CHEMICAL SIGNALS
Vertebrates and Aquatic Invertebrates

Edited by
Dietland Müller-Schwarze
and
Robert M. Silverstein

State University of New York
Syracuse, New York

PLENUM PRESS · NEW YORK AND LONDON

Library of Congress Cataloging in Publication Data

Symposium on Chemical Signals in Vertebrates and Aquatic Animals, 2d, State University of New York, 1979.
  Chemical signals.

  Includes index.
  1. Vertebrates — Behavior — Congresses. 2. Aquatic invertebrates — Behavior — Congresses. 3. Animal communication — Congresses. 4. Pheromones — Congresses. I. Müller—Schwarze, Dietland. II. Silverstein, Robert Milton, 1916-   III. Title.
  QL776 S94 1979   591.59   80-184
  ISBN-13: 978-1-4684-1029-7   e-ISBN-13: 978-1-4684-1027-3
  DOI: 10.1007/978-1-4684-1027-3

Proceedings of the Symposium on Chemical Signals, held at
State University of New York College of Environmental Science
and Forestry, Syracuse, New York, May 30—June 2, 1979.

© 1980 Plenum Press, New York
Softcover reprint of the hardcover 1st edition 1980
A Divison of Plenum Publishing Corporation
227 West 17th Street, New York, N.Y. 10011

Preface

Research on chemical communication in animals is in a very active and exciting phase; more species are studied, data are accumulating, concepts are changing, and practical application seems feasible.

While most of the work on chemical ecology and chemical signals deals with insects, vertebrate communication provides a formidable challenge and progress has been slow. Joint efforts and frequent direct contacts of ecologists, behaviorists, psychologists, physiologists, histologists and chemists are required. Such an interdisciplinary exchange of information took place on the occasion of the Symposium on Chemical Signals in Vertebrates and Aquatic Animals in Syracuse, New York, from May 31 to June 2, 1979. More than one hundred investigators from seven countries participated, and the papers presented comprise this volume.

Since the first Symposium on Vertebrate Chemical Signals at Saratoga Springs in 1976, considerable progress has been made with field studies, the physiology of the vomeronasal organ, and its role in reproductive behavior. The behavioral functions and chemical nature of priming pheromones are better understood. Efforts to isolate and identify mammalian pheromones are gaining ground, and the bioassays are becoming more sophisticated.

In addition to formal presentations, one evening of the Symposium was devoted to round-table discussions of particular topics. The selected themes indicate the "growing points" of chemical communication research: priming pheromones, vomeronasal organ, bioassay, and practical applications.

In the discussion of priming pheromones in rodents, it was suggested that inhibition of maturation in females may be the most important effect, and it was recommended that the research be extended to other species. The chemical considerations centered on molecular weight, volatility, and bonding of primer pheromone molecules.

The vomeronasal system's essential independence of the olfactory pathways was stressed, although there exist complex interactions. More information is needed on the electrophysiological responses to compounds over a range of volatility. Taxonomically, the vomeronasal system seems to emerge at the level of fishes; it is functional in primates and is present in human infants.

In the bioassay, behavior tests can and should be augmented by measuring physiological functions, such as cardiac and respiration rates, although by themselves, these indicators may often be too general. Both complex odor mixtures and complex responses seem to be the rule in vertebrates. In addition, changes of odors and in responsiveness over time further complicate bioassays. Parallels between bioassays in pollution studies and pheromone studies were pointed out. A consensus developed that a sound bioassay must start with a thorough knowledge of the life cycle, ecology, and behavior of the animal in its natural habitat. Only then can a chemically released or modulated response be selected for study so that collaboration with chemists may commence.

Applications of social, interspecific, or environmental odors for animal management are still in their infancy. Important potential manipulations of vertebrate behavior include food aversions, used to discourage predators or crop pests, imprinting of fish on the odor of their home streams, harvesting fish with chemical signals, use of chemicals in trapping, monitoring of rut in game animals by odors or secretions, short-term use of alarm odors as repellents, and utilization of macrosmatic mammals, such as dogs, for odor detection. Currently, the only practical application of mammalian pheromones is the use of a synthesized pheromone (a steroid emitted in the boar's saliva) to stimulate sows in artificial insemination programs.

We thank Miss Cathy Northway for typing the entire camera-ready manuscript, Mrs. Christine Müller-Schwarze for preparing the index, and Steve Dyer and Ms. Pat Vann of Plenum for their efficient help in the speedy publication of this volume.

Syracuse, New York   D. Müller-Schwarze
October 1979   R.M. Silverstein

Contents

Part Three: Food Responses

Part Four: Learning

Part Five: Priming

Part Six: Vomeronasal Organ

SOME RESPONSES OF A FREE LIVING COMMUNITY OF RODENTS TO THE ODORS OF PREDATORS

D. Michael Stoddart

Dept. of Zoology, University of London King's College

Strand, London WC2R 2LS England

INTRODUCTION

When the forces of Natural Selection created mammalian predators which feed upon mammalian prey, they incidentally created a
paradox, at least in the nostrils of a human observer. Mammals
generally have a very good olfactory sense; small rodents particularly so (Stoddart 1974). Paradoxically their predators are
characterized by their ability to produce copious quantities of
extraordinarily pungent secretions, as anyone who has ever kept
ferrets or encountered a skunk that has been run over on the road
will aver. Are rodent prey animals unable to smell their predators?
The long term stability and persistence of predator-prey interactions indicates that small mammals, even with their excellent
noses, still get caught and eaten in sufficient numbers to support
substantial predator populations. Clearly they do not <u>always</u> react
to the odor of their predators, but do they ever? In answering
this question it would be helpful to know whether there is any odor
quality fundamentally associated with predators - a sort of olfactory
Leitmotiv harking back to some ancient evolutionary contact between
predator and prey. As Marlitt says in her study Das Eulenhaus:
"Nichts in der Welt macht Vergangenes so lebendig wie der Geruch".
('Nothing on earth brings the past so alive as odor').

The object of this paper is to analyze the effect of predator
odor on a free-living community of small rodents and while so doing
to appraise the validity of a field based approach to problems of
interspecific scent communication.

1

REVIEW OF LITERATURE

　　All reported studies on the behavioural effect of the presentation of predator odors have been performed under conditions of
captivity and most actually in the laboratory. The first studies,
conducted over half a century ago (Griffith 1920), established
that laboratory rats responded to the whole body odor of a cat by
"freezing". When a glass bell jar was placed over the cat such
that the rats could no longer perceive its odor but could still
see it, the rats went about their business apparently quite unper-
turbed by the sight of the cat. That it was the odor of the cat
which brought about the change in rat behavior was demonstrated
again more recently by Mollenauer, Plotnik and Snyder (1974). The
normal response to cat odor (i.e. almost total suppression of
drinking and long periods of "freezing" interspersed by brief bouts
of high speed activity) could be significantly impaired by the
surgical removal of the olfactory bulbs. A problem facing experimenters in this type of study is that there is no fully adequate
control procedure that can be applied. Mollenauer et al. lifted
and replaced the nasal bones of control subjects, to mimic approximately the experimental subjects operation trauma, but even having
done that there is no way of knowing whether the subsequent behavioral indifference to the previously upsetting stimulus in
experimental animals is caused by the lack of olfactory ability per
se, or by some other effect of olfactory bulbectomy on emotionality.
Furthermore, although the point is probably not relevant here,
olfactory bulbectomy will render the olfactory mucosa and Jacobson's
organ ineffective (this latter being connected via vomeronasal
nerves to the accessory olfactory bulb which is closely appressed
to the olfactory bulb and would be removed with it), but will have
no effect upon the septal organ of Rudolfo-Masera nor upon the
endings of the terminalis and trigeminal nerves, all of which constitute part of the olfactory receptive system (Bojsen-Møller 1975).

　　Returning to Griffith's study, the interpretation presented
makes it clear that the odorous stimuli emanating from the cat were
very largely, if not exclusively, responsible for the fear response
triggered in the rat. Griffith made no attempt to ascertain if any
odor would have been effective and how easy it would have been to
daub some novel odor, not associated with cats, onto the outside
of the jar! Courtney, Reid and Wasden (1968) attempted to resolve
this omission by offering a commercial deodorant as a novel odor
in an experiment designed to examine the role of cat odor in the
time taken by a thirsty rat trained to run along a gangway to a
water bottle. Cat odor sharply increased the time taken but the
deodorant resulted in times no different from those achieved in
odorless controls. Müller-Schwarze (1972) demonstrated that black-
tailed deer (Odocoileus hemionus) fawns were deterred from feeding
by the odors of various predators mixed with their food. He

indicated that the response of the fawns was more intense to the odor of sympatric species of predator (coyote Canis latrans, mountain lion Felis concolor) than to the odor of allopatric species (lion Panthera leo, tiger Panthera tigris, snow leopard Panthera uncia) but his only control was odorless food. As his test subjects had never had the opportunity to learn to avoid the odors of predators he concluded that the ability to respond to predator odor was innate.

The main problem surrounding the interpretation of results from laboratory or captivity studies is that of adequate controls. The experimenter does not know how the effects of stress will manifest themselves at the behavioral level nor does he know that an observed response is truly the result of a particular set of experimental conditions. If he were working with insects he could be more confident but with mammals with their elaborate and versatile sensory systems, he must exercise extreme caution.

Field studies are potentially able to yield different sorts of results than laboratory studies because the test subjects, living their lives under natural conditions, are free from the unquantifiable influences created by captivity. The adverb 'potentially' is used advisedly since field studies bring with them their own problems and two of the most irritating in small rodent studies are "trap-shyness" and "trap-proneness" (Leslie, Chitty and Chitty 1953). Not all members of a population are equally trappable so the sample cannot be considered to be a representative one. Since the behavior towards traps is not predictable for all the population's members, conclusions drawn from experimental manipulation of trappability through the application of specific odorants cannot be applied to the whole population. Further problems beyond the experimenter's control include the prevailing climatic conditions. Sudden heavy rain or intense sunshine may result in animals dying in the traps and so it is necessary for the experimenter to exercise extreme vigilance. And finally the population may be inadequate in size for the particular study the experimenter has in mind because of a hard winter and a late spring. Notwithstanding these drawbacks, field studies, if conducted intensively over fairly short time periods, may yield data whose great value lies in the fact that they have been gathered free from laboratory constraints.

METHODS

As a result of these unknown aspects of the natural environment my own work on predator odor has had to utilize two different experimental designs. In the first part of the study (six experiments) conducted between August 19 and September 3, 1975 the rodent community inhabiting an 'old-field' grassland community at King's College Field Station at Rogate in Sussex, southern England, was large and it was possible statistically to compare the rodents'

response to either tainted or clean traps set simultaneously. In
the second part of the study (five experiments) conducted between
July 10 and August 6, 1978 the rodent community in the same area
was much smaller and could not be divided up for similar statistical
analyses to be made. The grassland community was otherwise used
only for undergraduate teaching purposes and was neither burned
nor grazed. Live-trapping of rodents was conducted on a 6 x 6
point grid in 1975 and a 7 x 7 point grid in 1978. The spacing
between trap lines was constant at 10 m. Excluding the effect traps
at the edge have on individuals living outside the grid, the larger
of these grids encompassed 3,600 m^2. Longworth live traps were
used, baited with mixed grain and supplied with carrot and hay. A
pair of traps was placed at each intersection of the grid. Prior
to the start of each experiment the traps were washed in hot water
and domestic detergent, rinsed and dried. Fresh bait and bedding
were supplied after each washing. In the experiments utilizing
weasel Mustela nivalis anal gland secretions (experiments numbers
2 and 5), the odor extracted in ethanol, ether and water was applied
to the inside of the trap with a fine artist's paintbrush. Only
one of each pair of traps was so tainted; the other had a similar
sized application of solvent alone. Each time a rodent was captured
the position of the two traps was reversed. (For other details of
the method see Stoddart 1976). In the experiments using jaguar
and rabbit urine (nos 8 and 10) small squares of filter paper
measuring approximately 7 mm x 7 mm and saturated with the odorant
were placed in both members of each pair of traps. Although the
squares were handled with forceps, no other special precautions
were taken to prevent the experimenter's hand odor from tainting
the traps. In every experiment the traps were visited and cleared
twice each day for four days, with the exception of experiment 10
when only seven clearances occurred. The rodents caught were
identified by a coded series of fur clip marks made with scissors.
The processing of captives was as speedy as possible to reduce handling trauma to a minimum.

RESULTS

     The results of the study are summarized in Tables 1, 2 and 3.
Experiments numbered 1 and 4, 2 and 5, and 3 and 6 are run together
but analyzed separately so that differences in response between
woodmice Apodemus sylvaticus and short-tailed voles Microtus
agrestis can be clarified. There is one very obvious difference
and that is that the number of individual M. agrestis caught
during experiment 5 when weasel odor was used was substantially,
though not statistically significantly, lower than in the control
experiments 4 and 6. The rate at which individuals that did enter
the traps were caught, however, was not significantly different
from the capture rate during the controls. A. sylvaticus by contrast showed a significant increase during experiment 2 over

Table 1. The effect of certain predator odors on the number of individual rodents captured and their individual capture rate.

Species	Experiment number	Experiment	N° of trap clearances	N° of individuals marked during experiment	Tot N° captures of all individuals caught during experiment	Individual capture* rate
Apodemus sylvaticus	1	1st control	8	5	13	0.325
	2	Weasel anal gland	8	13	32	0.308
	3	2nd control	8	14	33	0.294
Microtus agrestis	4	1st control	8	19	55	0.361
	5	Weasel anal gland	8	13	34	0.326
	6	2nd control	8	21	61	0.365
	7	1st control	8	20	52	0.325
	8	Jaguar urine	8	14	37	0.330
Microtus agrestis	9	2nd control	8	17	56	0.412
	10	Rabbit urine	7	17	58	0.470
	11	3rd control	8	20xx	61+	

*Mean number of captures per individual per trap clearance.

xxTotal increased by three to account for three trap deaths on final day of previous experiment.

+Total increased by mean total capture rate to account for three trap deaths on final day of previous experiment.

Table 2. The differential effect of weasel odor on captures of
          <u>Apodemus sylvaticus</u> and <u>Microtus agrestis</u>.

	Total number of captures	Number in tainted traps	Number in clean traps
A. sylvaticus	32	14	18
		$x^2 = 0.50$ P = > 0.05	
M. agrestis	34	11	23
		$x^2 = 4.2353$ P = < 0.05; > 0.02	

(From Stoddart 1976, reproduced by kind permission of Springer
  Verlag).

experiment 1 (x^2 P = 0.05) but this could have been caused by a
slow pattern of acceptance of the traps or even an invasion of
the grassland habitat. The number caught during experiment 3
suggests that the freakish figure is that observed during experiment 1, not experiment 2. Further consideration of trap choice
exhibited by these two rodents (Table 2) reveals that <u>A.</u>
<u>sylvaticus</u> was undeterred by weasel odor but that <u>M. agrestis</u>
chose to enter clean traps significantly more frequently than
tainted traps. This apparent difference is puzzling, since both
rodent species are predated upon by weasels (Day 1968; Walker 1972)
and each is as likely as the other to benefit by avoiding anything
that smells like a predator.

    In an attempt to examine this difference in more detail a
further series of experiments (numbers 7-11 Table 1) were designed
but when trapping started it was at once apparent that the number
of <u>A. sylvaticus</u> was too small to be of any value. The consequence
was that the situation regarding the response of the vole was
examined in detail but further examination of woodmouse response
must await further study. The question asked was whether <u>M.</u>
<u>agrestis</u> responded to some quality of the weasel odor which said
"weasel" or just "predator" and whether the response could be
induced by any novel odor occurring in an unusual place. The urine

Table 3. The sex ratio of <u>Microtus agrestis</u> caught in experiments
          7 through 11.

Experiment	Number of males	Number of females	Totals
Control – no odor experiment 7	10	10	20
Jaguar urine experiment 8	9	5	14
Control – no odor experiment 9	11	6	17
Rabbit urine experiment 10	10	8	18
Control – no odor experiment 11	8	12	20
Totals	48	40	88

$x^2 = 14.17$ $P < 0.01$, > 0.001

of a jaguar (<u>Panthera onca</u>) was chosen as a test odorant because
it came from a species completely unfamiliar to the voles, and from
a predatory species at that. The choice of a novel odor but one
which would not trigger a neophobic response was much more difficult,
since a novel odor ceases to be of value as a control if it produces
a response similar to that elicited by the experimental odor. It
was decided to use the urine of New Zealand white rabbits which had
been laboratory bred and fed on a standard laboratory diet. It was
hoped that it would be sufficiently different from that produced by
the wild rabbits which shared the grass field with the rodents to
be novel. In any case, a degree of novelty was created by making
the rabbit odor emanate from a point source inside the trap. Table
1 shows that the individual capture rate of voles rose slightly but

steadily from experiments 7 through 10, with a slight fall in experiment 11. There is no evidence that either jaguar or rabbit odor severely influenced individual capture rate. Jaguar odor did, however, act like weasel odor in reducing the number of individuals caught although the reduction is not statistically significant. Although the small sample sizes are acknowledged it appears that the odor of rabbits does not act to deter entry into the traps in the way that predator odor does. This suggests that the voles in experiment 8 reacted to a signal which said "predator" and not just to any novel odor.

A detailed examination of the pattern of entry of the voles, however, suggests there is a sex related response to the varying experimental conditions. Table 3 shows the numbers of each sex caught during experiments 7 to 11 inclusive. The χ^2 value indicates that the heterogeneity in the data is too great to have been caused by chance. Use of jaguar odor caused a sudden decrease in the number of females caught but not males. While the number of males remained at a fairly constant level throughout the series, that of females climbed steadily from the nadir in experiment 8 to overtake that of males in the final control. It is not known why the number of females appeared to take so long to return after the jaguar odor experiment; clearly this is an aspect of the investigation which demands close study.

DISCUSSION

Leaving aside the problem of the apparent differential response to predator odor between the woodmouse and the short-tailed vole which cannot be resolved until the natural population conditions become more favorable, the following conclusions can be drawn regarding the response of vole populations to predator odor:

1. The presence of the odor of weasel brings about a sharp reduction in the numbers of short-tailed voles which enter traps.

2. The frequency of capture of those individuals which do enter.traps, whether the traps be tainted or clean, is no different to the frequency of capture revealed during control periods.

3. When presented with a pair of traps, one clean, the other tainted with weasel odor, significantly fewer individuals choose to enter tainted traps than clean traps.

4. When presented with a non-predator odor deemed to be novel, no change in the number of individuals caught or their individual capture rate is seen.

5. When presented with a novel odor taken from an unfamiliar predator species, a sharp though statistically insignificant drop in numbers of individuals caught is observed. Their individual capture rate remains unaffected. Less females than males enter. It is assumed that the predator odor elicits a species-specific avoidance reaction, though why more females than males respond in this way is not known.

Little is known of the ontogeny of response to predatory odors in mammals, although Henessy and Owings (1978) have demonstrated that ground squirrels Spermophilus beecheyi react more intensely to the odor of rattlesnakes Crotalus viridus and gopher snakes Pituophis melanoleucus than to the sight of them. The response is innate, as naive squirrels show it as strongly as squirrels which have encountered snakes or been in burrows along which snakes have traveled. The young fawns used by Muller-Schwarze had likewise never been exposed to any predator odor and thus had no opportunity to relate odor characteristics with the predatory habit. Bolles (1970) argues that the occasions for avoidance learning by prey animals are very limited indeed and that the prey uses its innate species-specific defense reaction whenever it encounters any novel environmental stimulus. This puts the idea of a novel odor as a control into a rather precarious position. If the odor truly is novel, a rejection of it should occur. If no behavioral response to the odor is observed, either the odor is not novel or Bolles' argument is not wholly correct. If, on the other hand, the population responds to a novel odor by rejecting it at first but later accepting it when it is clear it is not associated with any other unpleasant stimulus, then its use is vindicated. Particularly so if the response to the predator odor continues for as long as it is present, as was the case in the described experiments. The temporal base of field studies is very important.

The methodology for research into olfactory communication is still in a state of flux - the problem of comparing studies conducted in two or more laboratories is known to us all. In conducting field studies we must tread warily so as not to obscure or impede the expression of a behavioral response, the likes of which are impossible to reproduce in the laboratory.

Acknowledgements

My grateful thanks are due to Dr. M. Brambell of the Zoological Society of London for the supply of jaguar urine and to Dr. Gillian Sales for a valuable criticism of a draft of this paper.

REFERENCES

Bojsen-Møller, F. 1975. Demonstration of terminalis, olfactory trigeminal and perivascular nerves in the rat nasal septum. J. Comp. Neurol., 159 245-256.

Bolles, R.C. 1970. Species-specific defense reactions and avoidance learning. Psychol. Rev. 77 32-48.

Courtney, R.J., L.D. Reid and R.E. Wasden. 1968. Suppression of running times by olfactory stimuli. Psychronom. Sci. 12 315-316.

Day, M.G. 1968. Food habits of British stoats (Mustela erminea) and weasels (Mustela nivalis). J. Zool. 155 485-497.

Griffith, C.J. 1920. The behaviour of white rats in the presence of cats. Psychobiol. 2 19-28.

Henessy, D.F. and D.H. Owings. 1978. Snake species discrimination and the role of olfactory cues in the snake-directed behaviour of the California ground squirrel. Behaviour 65 116-124.

Leslie, P.H., D. Chitty and H. Chitty. 1953. The estimation of population parameters from data obtained by means of the capture-recapture method. III. An example of the practical application of the method. Biometrika 40 137-169.

Mollenauer, S., R. Plotnik and E. Snyder. 1974. Effects of olfactory bulb removal on fear responses and passive avoidance in the rat. Physiology and Behavior 12 141-144.

Müller-Schwarze, D. 1972. Responses of young black-tailed deer to predator odours. J. Mammal. 53 393-394.

Stoddart, D.M. 1974. The role of odor in the social biology of small mammals. in "Pheromones" ed. M.C. Birch 297-315. North Holland Publishing Co., Amsterdam.

Stoddart, D.M. 1976. Effect of the odour of weasels (Mustela nivalis L.) on trapped samples of their prey. Oecologia 22 439-441.

Walker, D.R.G. 1972. Observations on a collection of weasels (Mustela nivalis) from estates in south-west Hertfordshire. J. Zool. 166 474-480.

THE URINE MARKING BEHAVIOR AND MOVEMENT PATTERNS OF RED FOXES (VULPES VULPES) DURING A BREEDING AND POST-BREEDING PERIOD

J. David Henry

Department of Biology
University of Calgary
Calgary, Alberta, Canada

INTRODUCTION

This study is part of a series of field research projects on the behavior of free-ranging red foxes in Prince Albert National Park, Saskatchewan, Canada (Henry, 1976, 1977). Consequently it has several interrelated purposes.

First, the urine marking behaviors of male and female red foxes are documented during a breeding season and compared to the same behaviors during a post-breeding period. Changes in several parameters of these scent marking behaviors are statistically tested ($P < 0.05$) for significant seasonal and sexual differences.

Second, Henry (1977) suggested that urine marking may function as a type of "bookkeeping system" during the fox's scavenging behavior. Specifically, the hypothesis was advanced that red foxes urine mark inedible food remnants, and when the same or a different fox re-investigates this marked item, the urine mark signals "no food" and the food remnant is investigated for only a short period of time. It was suggested that this use of scent marking might increase the efficiency of the red fox's scavenging behavior whereby more food items are found per hour of scavenging. Henry (1977) carried out three field experiments which tested and supported various aspects of this hypothesis. This present study examines whether inedible food remnants are also urine marked by foxes during the breeding and post-breeding periods.

Third, the influence of snow texture on the spatial distribution and movement patterns of red foxes in the boreal forest during

winter is examined. The relationship of these movement patterns to
the distribution of red fox scent marks during the breeding and post-
breeding periods is discussed.

METHODS

Free-ranging red foxes were studied in the central portion of
Prince Albert National Park (see Henry, 1976). The park was
established in 1927, and red foxes in the park have been protected
from hunting and trapping since that date. The park is 3,980 km^2
in area and located at the approximate geographic center of the
province (54°00'N and 106°00'W). It is located in the mixed woods
section of the boreal forest (Rowe, 1972). Swan and Dix (1966)
and Jeglum (1972, 1973) offer descriptions of the physical environment, climate and vegetation of this region.

The winter behavior of red foxes was studied by following
their tracks across snow. A tracking technique has been used to
study the behavioral ecology of red foxes by several other investigators including Murie (1936), Schofield (1960), Kruuk (1964),
Tinbergen (1965), Jorgenson et al. (1978) and Whitten (in press).
Peters and Mech (1975) and Whitten (in press) have shown that a
tracking technique can successfully be used to study the scent
marking behavior of free-ranging canids.

In this study fox tracks were identified by criteria described
by Murie (1964). This identification was simplified by the fact
that there is only one species of fox in the park, no feral dogs
(Canis familiaris), and coyotes (Canis latrans) are relatively
uncommon on the study area. Furthermore, field observations suggest
that coyote and fox tracks in this area can be separated by the
following criteria: the greater body weight of coyotes as compared
to foxes (Henry, 1976), the width of the forefoot which in foxes is
approximately 5 cm as compared to 7 cm or more in coyotes, and the
different shape of the heel pad and greater amount of hair on the
feet of foxes as compared to coyotes (Murie, 1964).

In the field, each fox was tracked as far as possible,
and the following information was collected in each instance:
total distance the fox was tracked (measured by pacing), number of
scent marks observed, stimuli associated with each scent mark,
distance travelled in each habitat type and on the various types
of snow surfaces.

The age of each urine mark could be judged from its color,
which is straw-yellow when fresh but changes to an orange-brown
after several hours (Montague, 1975; Jorgenson et al., 1978).
Since the observer followed only fresh fox tracks, it was judged

that urine marks recently deposited could be separated from older
ones. At each fresh urine mark the observer carefully removed the
snow that surrounded it in order to uncover any older scent mark
buried in the snow that had been previously deposited at that
location by the same or a different fox.

Ables (1975) and Storm et al. (1976) state that red foxes in
North America may breed from December through April, but that most
matings occur during January and early February (see also Sheldon,
1949; Richards and Hine, 1953; Hoffman and Kirkpatrick, 1954;
Layne and McKeon, 1956). In Prince Albert Park, observations of
courting foxes travelling together and breeding behaviors directly
observed suggest that the breeding season among these foxes usually
begins in early January and is terminated by mid-February. Most
fox kits in this area are estimated to be born during late March
or early April (Henry, unpublished data), which based on a 52 day
gestation period (Storm et al., 1976) also suggests that most foxes
breed during late January or early February. Consequently, in this
study foxes were tracked in several natural areas of the park
during two periods: 29 December 1978 to 17 February 1979 which is
considered to be the breeding period, and 18 March to 8 April 1979
which is considered to be the post-breeding portion of the study
period.

The tracking technique used in this study has the disadvantage
that the foxes were not individually recognized. However, recent
telemetry studies of red foxes have shown them to be spatially
distributed with individual families occupying well defined, non-
overlapping, contiguous territories (Sargeant, 1972; Ables, 1975;
Storm et al., 1976, Macdonald, 1977). The fact that the foxes
were tracked in the same areas during the two study periods suggests
that similar foxes may have been studied during the two periods.
Nevertheless, the lack of individual recognition of the foxes is
an inherent limitation to the study which should be kept in mind
and limits the interpretation of the data.

RESULTS

During the breeding period of the study (29 December 1978
to 17 February 1979), 19 foxes were tracked in six different areas
of the park for a total of 20,072 m. The average length of a
tracking record during this period was 1056 m (range 194 m to
2433 m).

During the post-breeding period (18 March to 8 April 1979),
the presence of a sun crust on the snow and more frequent winds
made the tracking of foxes slightly more difficult. During this
time, 19 foxes were tracked in the same six areas for a total of

16,894 m. The average length of a tracking record during this
period was 889 m (range 217 m to 5546 m).

Identification of the Sex of Individual Foxes

    The sex of each fox could frequently be inferred from an
accumulation of information contained in the tracks. Tembrock
(1957), Henry (1977) and Macdonald (in press) describe the different
urine marking postures shown by male and female red foxes. Jorgenson et al. (1978: 797) recently studied red foxes by tracking and
state: "The sex of the individual was inferred from the nature of
the urine marks (Montague, 1975). Male urine is projected laterally
and may be found 20 to 25 cm above the hind paw mark, and obviously
one leg has been lifted... In contrast, urine deposited by females
is found between and behind the rear paw marks, and greater volumes
may be observed. Although females may raise one leg, there is no
lateral projection of urine; and, as a rule, females do not urinate
on raised objects."

    Field observations collected during the present study agree
with Jorgenson's conclusions with one important exception. Male
red foxes in Prince Albert Park have frequently been observed to
lower the inguinal region towards the ground and urine mark from
a squat posture (for example, see Figure 2 of Henry, 1977). For
example, during October 1971, a subadult male red fox was observed
to urine mark 104 times, 84 times using the squat posture. During
October 1972, another male red fox was observed to urine mark 72
times, 40 times using a squat posture. This squat posture was
also observed in the tracks of foxes identified as males during
the present study.

    Consequently the criteria used to identify the sex of red
foxes from tracks should be modified to include this posture.
Field observations from this study support that a male red fox
deposits urine in front of his hind paw marks, either directly in
front if a squat posture is used, or laterally in front if a
raised leg posture is used. The study agrees with Jorgenson's
statement that a female red fox deposits urine either between
or behind the hind paw marks.

    For each fox tracked in this study, an inference about its
sex was made after several scent marks were encountered. Additional information collected while tracking this fox either supported the inference, or, if it did not, then it was judged that
the sex of the fox could not be reliably determined. If a fox was
tracked for only a short distance, it was judged that not enough
information had been collected to reliably determine its sex.

During the breeding period of the study, ten of the foxes were inferred to be males, eight females, and the sex of three foxes could not be reliably determined. In two instances, a pair of foxes travelling together were tracked, and in both cases quantitative data was collected only for the male of the courting pair. Thus for this period quantitative data was collected for a total of 19 out of 21 foxes.

During the post-breeding period, seven of the foxes were inferred to be males, six females, and the sex of six foxes could not be reliably determined. No pairs of foxes travelling together were observed during this time. Thus quantitative data was collected for all 19 foxes tracked during this period.

Comparison of Red Fox Urine Marking Behavior during the Breeding and Post-breeding Periods

Several researchers have advanced hypotheses concerning how the scent marking behavior of red foxes changes during the breeding season. Jorgenson et al. (1978) suggest that the characteristic "skunky" odor of the urine marks intensifies during the breeding season. Fox (1975) and Macdonald (in press) suggest that the incidence of scent marking increases during the breeding season. Tembrock (1957) suggests that the incidence of urine marking previously scent marked locations also increases at this time.

These hypotheses have been advanced mainly on observations of captive or hand-reared red foxes. This study attempts to collect observations on free-ranging red foxes in order to test the validity of these hypotheses for foxes in a near natural environment.

Intensification of the Odor of Red Fox Urine Marks during the Breeding Season

The intensification of the odor of urine marks was studied only in a preliminary manner. Nevertheless, these descriptive observations suggest that the odor of the urine marks of both sexes abruptly increases during the early breeding season.

The intensity of the odor was determined by measuring the distance at which the researcher could first detect the odor of the scent mark. For this analysis only single fresh urine marks were examined. Of the 43 urine marks examined during December, the odor was detected normally at a distance of 30 to 60 cm. No difference was detected in the odor of male urine marks as compared to those of females.

The odor of the urine marks, however, abruptly changed during
early January. A urine mark with a strong musky odor was first
encountered on 5 January 1979. By January 11 a majority of the
fresh urine marks exhibited this strong odor. At that time the
observer detected the odor of the scent marks at a distance of
3 to 5 m, and it was a common experience that the urine marks were
detected olfactorily before they were seen (see also Montague, 1975).

No difference could be detected between the male and female
urine marks; both exhibited a strong musky odor. However, sex-
related differences in the biochemistry of the red fox urine do
appear to exist during the breeding season. Jorgenson et al. (1978)
using gas chromatography-mass spectrometry methods found that the
strong "skunky" odor characteristic of red fox urine is due to two
sulfur-containing compounds, Δ^3-isopentenyl methyl sulfide and
2-phenylethyl methyl sulfide. They also found that two terpene-
derived ketones, 6-methyl-Δ-hepten-2-one and geranylacetone, appeared
to be more characteristic of the female urine than the male urine
and that 2-methylquinoline (quinaldine) was present in male red
fox urine, but not detectable in female urine. The pheromonal
activity of these chemical constituents remains unknown at the
present time.

On the Prince Albert study area, pairs of courting foxes were
observed travelling together until 10 February 1979. However, the
intensity of the foxes' urine marks began to decrease during late
January. While the onset of the strong musky odor at the beginning
of the breeding season was abrupt, the intensity of the odor
diminished only gradually. The odor level decreased slowly over
the next two months until it approached a pre-breeding condition.
During late March and early April, the observer again measured
the distances at which fresh urine marks could be detected, and
for 57 urine marks examined the detection distance normally ranged
from 25 to 45 cm.

Consequently, the picture that emerges is that the odor of
red fox urine marks abruptly intensifies at the beginning of the
breeding season, stays at this intensity for approximately three
weeks and then gradually diminishes over the next two months. The
intensity of the odor of both male and female scent marks appears
to follow this pattern. The method used in this study to describe
these changes in odor intensity is admittedly preliminary and
inaccurate. Proper gas chromatography-mass spectrometry techniques
have already been used to analyze the urine of red foxes during
the breeding season (Jorgenson et al., 1978; Whitten, in press).
Hopefully, these same techniques will be used to document the rapid
onset of the odors at the beginning of the fox's breeding season
and their gradual diminishing during the post-breeding period.

Incidence of Urine Marking during the Breeding Season

The second parameter examined is whether either sex of fox shows an increased frequency of urine marking during the breeding season as compared to the post-breeding season. Because the tracking records were of different lengths, the rate of urine marks per 100 m of tracking record was calculated for each case. These data were then averaged. Table 1 reports the average rate of urine marking and its standard deviation for both sexes of fox during the breeding and post-breeding periods.

The data were subjected to an Analysis of Variance (Sokal and Rohlf, 1969) in order to test if significant differences ($P < 0.05$) exist among these groups of foxes regarding the incidence of urine marking. The ANOVA table is presented in Table 2.

Table 1. Average incidence of urine marking per 100 meters and its standard deviation for red foxes during a breeding and post-breeding period.

	Sample Size	Breeding Period	Sample Size	Post-breeding Period
Male Foxes	10	0.94 ± 0.55	7	0.74 ± 0.38
Female Foxes	6	0.99 ± 0.28	6	0.68 ± 0.31
Foxes of Undetermined Sex	3	0.46 ± 0.52	6	0.56 ± 0.37
Foxes of Both Sexes	19	0.88 ± 0.49	19	0.66 ± 0.34

Table 2. ANOVA table for the incidence of urine marking among red foxes during a breeding and post-breeding period.

Source of Variance	d.f.	S.S.	M.S.	F_s
Among Groups	5	1.190	0.238	1.325
Within Groups	32	5.748	0.179	
Total	37	6.938		

The critical $F_{0.05}$ (5,32) = 2.51. Therefore the analysis concludes that there is no significant (P > 0.05) added variance component among these groups of foxes for incidence of urine marking. The data support the conclusion that both sexes of foxes urine marked at approximately the same rate during the breeding period as compared to the post-breeding period.

Percentage of Urine Marks Deposited at Previously Scent Marked Locations

The third parameter examined involves the percentage of urine marks that each fox deposited at locations previously scent marked by the same or a different fox. Tembrock (1957) suggests that red foxes show an increased incidence of this type of marking behavior during the breeding season.

To examine this point, the snow around each fresh urine mark was removed and examined for other scent marks, either urine or feces. The percentage of urine marks deposited at previously scent marked locations was then calculated for each tracking record.

These data were then averaged. Table 3 reports the average percentage and standard deviations of urine marks placed at previously scent marked locations for each sex of fox during the breeding period as compared to the post-breeding period.

Table 3. Average percentage of urine marks and its standard deviation that were deposited at locations which were previously scent marked by red foxes.

	Sample Size	Breeding Period	Sample Size	Post-breeding Period
Male Foxes	10	38.5 ± 13.4%	7	31.6 ± 29.9%
Female Foxes	6	40.3 ± 35.1	6	40.0 ± 25.8
Foxes of Undetermined Sex	3	25.0 ± 25.0	6	13.9 ± 22.2
Foxes of Both Sex	19	36.9 ± 22.9	9	28.6 ± 27.2

The data were subjected to an Analysis of Variance (Sokal and Rohlf, 1969) in order to test if significant differences ($P < 0.05$) exist among these groups of foxes. The ANOVA table is presented in Table 4.

Table 4. ANOVA table for the percentage of urine marks deposited at locations previously scent marked by red foxes.

Source of Variance	d.f.	S.S.	M.S.	F_s
Among Groups	5	33.415	6.683	1.0677
Within Groups	32	200.317	6.259	
Total	37	233.732		

The critical value again is $F_{0.05}$ (5, 32) = 2.51. Therefore this analysis does not reject the null hypothesis and concludes that there is no significant ($P > 0.05$) added variance component among the groups for the fraction of urine marks placed at previously marked locations. The data support the conclusion that both sexes of foxes urine marked previously scent marked locations at approximately the same rate during the breeding period as compared to the post-breeding period.

Scent Marking of Inedible Food Remnants

As mentioned earlier, Henry (1977) advanced the hypothesis that red foxes urine mark inedible food remnants and that this type of marking system may make their scavenging behavior more efficient. From field observations collected between 1971 and 1974, he was able to verify that red foxes on the Prince Albert study area urine mark inedible food remnants in every month of the year from April to December. However, he was unable to state if this type of scent marking behavior occurs during the breeding season. The present study was able to collect information on this point.

During the breeding and post-breeding periods, both male and female foxes were observed to urine mark inedible food remnants. These remnants included bones of mammals, birds and fish; bird wings; ungulate scats; the remains of ungulates killed by wolves (Canis lupus); burrow entrances of prey such as snowshoe hare

(Lepus americanus), red squirrel (Tamiasciurus hudsonicus) and mink
(Mustela vison); and dug out caching holes. These types of stimuli
were associated with 46 of 258 (18%) urine marks observed during
this study, and they were approximately equally associated with
male and female foxes during both the breeding and post-breeding
periods.

Consequently, it appears that red foxes do urine mark inedible
food remnants during the breeding and post-breeding periods and that
this type of marking system may serve a function in their scavenging
behavior throughout the year (Henry, 1977).

Movements of Red Foxes in Relation to Snow Texture

Each tracking record documented the total distance that each
fox travelled on the various types of snow surface found in its
environment. During the analysis of these data, it became apparent
that the snow surfaces used by the foxes could be classified into
four broad categories (Table 5).

Table 5. Distance (meters) travelled by red foxes on various
snow surfaces during a breeding and post-breeding
period. Number in parentheses is percentage of
total distance travelled.

Category	Breeding Period	Post-breeding Period
I. Wind blown surfaces	2,669 (13.3%)	3,024 (17.9%)
II. Unpacked, soft snow	3,493 (17.4)	980 (5.8)
III. Packed trails	13,910 (69.3)	3,497 (20.7)
IV. Sun crusted surfaces	0 (0.0)	9,393 (55.6)
Total	20,072 (100%)	16,894 (100%)

The first category involved any snow surface that developed a
wind crust, and on the Prince Albert study area these surfaces were
usually associated with the large lakes. This category includes
the open areas of the lakes, the wind drift ridges along the shores
and the open tracts of forest adjacent to the lakes.

The second category included undisturbed, unpacked snow surfaces which during early and mid-winter were found throughout the

forested portion of the study area as well as in meadows and along
creeks. These were snow surfaces that have not been trodden upon
and do not exhibit either a wind or sun crust. It should be pointed
out that because of the constant low ambient temperatures (-20°C to
-40°C) and low amount of incoming solar radiation that characterize
the boreal region during the winter, the snow in this region remains extremely soft and yielding. Pruitt (1960, 1970, 1978) discusses the physical characteristics of snow in the boreal forest and
shows that it frequently exhibits only 5% the specific gravity of
water. Depending upon the weather, this extremely light and fluffy
snow is normally maintained until temperatures begin to ameliorate
during March. This typical boreal snow supports very little weight
and frequently restricts the movement of animals within the boreal
forest (Pruitt, 1957, 1960, 1978; Formozov, 1961).

The third category of snow surfaces included trails packed by
a number of wildlife species including wolves, elk (Cervus canaden-
sis), deer (Odocoileus hemionus and O. virginianus), moose (Alces
alces), snowshoe hares, red squirrels, other foxes, as well as
humans on snowshoes and cross country skis. Foxes also used plowed
roads for movement and frequently hunted along the edges of these
roadways (Henry, 1976). All these packed and plowed surfaces were
included in the third category.

The fourth category included any snow surface that developed
a sun crust. During early and mid-winter (November through
February), this type of snow surface was rarely encountered on the
study area. Approaching the equinox, however, the warmer temperatures and increased amount and incidence of solar radiation created
a sun crust on much of the snow of the study area. From mid-March
onwards this snow surface was frequently encountered along creeks
and rivers, on open lakes, in muskegs, and in open upland forests.

A cursory survey of the study area suggested that the undisturbed soft snow surfaces (category II) and wind crusted surfaces
(category I) represented a large percentage of the total area and
that the packed trails (category III) represented only a small
percentage of the total area. The sun crusted snow surfaces
(category IV), while they were extremely rare on the study area
during early and mid-winter, formed a large percentage of that
area from mid-March through early April.

Table 5 examines the movement of the foxes in relation to
these various types of snow surfaces. The table shows that during the breeding period over two thirds of the foxes' movements
occurred along trails packed by humans and other animals. Wind
crusted snow surfaces associated with the lakes were also
occasionally used by foxes and accounted for 13.3% of their movements.

Only 17.4% of the foxes' movements occurred on the fresh, unpacked snow surfaces. This soft, undisturbed snow did not support
the weight of a fox, and foxes were observed to wallow or lunge
through it. These types of movements appeared energetically more
demanding and also appeared to disrupt stalking and pursuit-types
of hunting behaviors exhibited by red foxes (Henry, 1976). The
movements that were observed on this type of snow were usually of
two types: (1) short movements through soft snow from one packed
surface to another, and (2) more extensive movements in closed
canopy conifer stands where the branches intercept much of the snow-
fall. In these dense conifer stands the accumulated snow on the
forest floor was only 16 to 20 cm deep.

In summary, the movements of red foxes during the breeding
period were quite restricted: 82.6% of their movements occurred
on trails packed by humans or other animals or on wind crusted
snow surfaces usually surrounding large lakes. After a heavy snow-
fall the movements of foxes became even more restricted until
animals such as elk, moose, and snowshoe hare re-established packed
trails in various areas which the foxes then made use of for more
extensive movements on their home range.

After mid-March a sun crust began to form on many parts of the
study area, and this greatly influenced the movement of the foxes.
During the post-breeding period 55.6% of the total distance travelled
by the foxes was on sun crusted snow surfaces (Table 5). Because
foxes were no longer restricted to packed trails and wind crusted
surfaces, they began to move into areas of their home range where
fox tracks had not been observed since early winter.

Table 5 shows that during the post-breeding period foxes used
wind crusted surfaces as often as during the breeding period, and
that trails packed by other animals were still used (20.7%) but to
a significantly lesser extent. Movement in soft, unpacked snow
decreased to 5.8% of the foxes' movement during the post-breeding
period.

Consequently, fox movements appeared to be less restricted
during late winter. Because of the sun crusted snow, the movements of the fox became more wide-ranging on its home range at
this time, and it is assumed that additional food resources may
have become available to the foxes because of this greater freedom
of movement.

DISCUSSION

Parameters of Urine Marking Behavior

Of the four parameters examined concerning the red fox's urine marking behavior, the only one to show a significant change during the breeding season was the intensification of the odor of the urine marks. It was found that the odor of the urine of both sexes of fox increased abruptly at the onset of breeding, stayed intense for approximately three weeks, and then gradually diminished. The pheromonal effects of this odor intensification of the urine on the reproductive physiology and behavior of red foxes should be investigated.

On the other hand, neither male or female red foxes were observed to show a significant change in the frequency of (1) urine marking, (2) urine marking previously scent marked locations, or (3) urine marking of inedible food remnants. These parameters stayed essentially the same during the breeding period as compared to the post-breeding period.

The Effects of Snow Texture on the Distribution of the Urine Marks

It was found in this study that the extremely light snow texture characteristic of the boreal forest greatly restricts the movement of foxes: 82% of their movements during the breeding period were on packed trails or wind crusted snow surfaces. This restricted movement also appears to greatly influence the distribution and density of scent marks on the fox's home range. Because fox movements in the boreal forest are restricted for most of the winter, these animals may encounter a higher density of scent marks during this time than at any other time of the year. The effects of this high density of scent marks on their breeding physiology and behavior should be investigated. In biomes such as tundra, grasslands or eastern deciduous forest, where the foxes' movements are not restricted by soft snow, red foxes may show an increased incidence of urine marking during the breeding season (Macdonald, in press).

This point illustrates the importance of considering such environmental parameters as snow texture when attempting to understand the winter movements of the red fox, the spatial distribution of its scent marks, or when attempting to construct a simulated computer model of red fox communication (Montgomery, 1974; Storm and Montgomery, 1975).

An important principle in understanding the movements of the red fox during winter involves the depth of snow the fox can tolerate while still maintaining its normal gait. Field observa-

tions suggest that when a red fox sinks into snow over 20 cm, its
gait changes from the normal trot to either a walk, wallowing
through the snow, or an energy-expensive loping gallop. Field
observations suggest that foxes avoid snow surfaces where they sink
in over 20 cm and alternately select snow surfaces that will support them above this threshold. For example, in this study the
foxes made extensive use of packed trails and wind crusted surfaces.
The use of trails created by other wildlife species is an interesting commensal interspecific relationship that the red foxes tend
to show in the boreal forest: in this study approximately two
thirds of their movements from December to mid-February depend upon
trails mainly created by other wildlife species.

Scent Marking of Inedible Food Remnants

Henry (1977) showed that a urine mark on a food remnant
elicited a short investigation time not only from the fox that
originally marked the remnant but also from other foxes that investigate it. He suggested that red foxes appear to use their own
as well as each other's urine marks in order to increase the
efficiency of their scavenging behavior.

It is interesting to examine this point in light of several
recent investigations concerning the social organization of red
foxes. Telemetry data from several studies suggest that red foxes
are spatially distributed with individual families occupying well-
defined, non-overlapping, contiguous territories (Sargeant, 1972;
Ables, 1975; Storm et al., 1976). Macdonald (1977) presents information which suggests that these family groups of foxes in some
habitats may stay together and occupy the same territory for years.

Assuming that family territories are the normal social organization of red foxes and that non-family interlopers on the territory
are rare, the mutual use of scent marks during scavenging behavior
could be favored by kin selection. The close index of relationship
(Dawkins, 1976) among the foxes on the territory may favor such
forms of social behavior in order to make the scavenging behavior of
each member of the family more efficient. This hypothesis is
admittedly speculative and needs further field experimentation;
nevertheless, it suggests that the social interactions of members
of a solitary species, such as the red fox, are often complex and
deserve detailed sociobiological analysis.

Consequently, this study agrees with Leyhausen (1965),
Seidensticker et al. (1973), Barash (1974), Barrette (1975), and
Kruuk (1976) when they say that highly gregarious species do not
necessarily represent a pinnacle of evolution and that solitary
does not mean 'asocial', that is, without social behavior or with
a reduced social repertoire. The more logical prediction is

that solitary species are likely to show a different kind but equally as complex a social repertoire as highly gregarious species (see Henry and Herrero, 1974; Barrette, 1975). Certainly the social adaptations and ecological advantages of solitary species deserve to be studied in greater detail.

REFERENCES

Ables, E.D. 1975. Ecology of the red fox in North America. In The Wild Canids (M.W. Fox, Ed.), pp. 216-236, Van Nostrand Reinhold Co., New York.

Barash, D.P. 1974. Neighbor recognition in two "solitary" carnivores: the raccoon (Procyon lotor) and the red fox (Vulpes fulva). Science (New York) 185: 794-796.

Barrette, C. 1975. Social behaviour of muntjac. Ph.D. Thesis (unpublished), University of Calgary, Canada.

Dawkins, R. 1976. The Selfish Gene. Oxford University Press, Oxford.

Formozov, A.N. 1961. On the significance of the structure of snow cover in the ecology and geography of mammals and birds. In Role of the Snow Cover in Natural Processes (M.I. Iveronova, Ed.), pp. 112-139, Academy of Sciences of the U.S.S.R., Moscow.

Fox, M.W. 1975. Evolution of social behavior in canids. In The Wild Canids (M.W. Fox, Ed.), pp. 429-459, Van Nostrand Reinhold Co., New York.

Henry, J.D. 1976. Adaptive strategies in the behaviour of the red fox, Vulpes vulpes L. Ph.D. Thesis (unpublished), University of Calgary, Canada.

Henry, J.D. 1977. The use of urine marking in the scavenging behavior of the red fox (Vulpes vulpes). Behaviour 61: 82-106.

Henry, J.D. and Herrero, S.M. 1974. Social play in the American black bear: its similarity to canid social play and an examination of its identifying characteristics. Amer. Zool. 14: 371-390.

Hoffman, R.A. and Kirkpatrick, C.M. 1954. Red fox weights and reproduction in Tippecanoe County, Indiana. J. Mammal. 35: 504-509.

Jeglum, J.K. 1972. Boreal forest wetlands near Candle Lake, central Saskatchewan, Part I. Vegetation. Musk-ox 11: 41-58.

Jeglum, J.K. 1973. Boreal forest wetlands near Candle Lake, central Saskatchewan, Part II. Relationships of vegetational variation to major environmental gradients. Musk-ox 12: 32-48.

Jorgenson, J.W., Novotny, M., Carmack, M., Copland, G.B., Wilson, S.R., Katona, S. and Whitten, W.K. 1978. Chemical scent constituents in the urine of the red fox (Vulpes vulpes L.) during the winter season. Science (New York) 199: 796-798.

Kruuk, H. 1964. Predators and anti-predator behaviours of the
    black-headed gull (Larus ridibundus L.). Behaviour Suppl.
    11: 1-130.
Kruuk, H. 1976. Review of E.O. Wilson's Sociobiology.   Anim.
    Behav. 24: 710-711.
Layne, J.N. and McKeon, W.H. 1956. Some aspects of red fox and
    gray fox reproduction in New York. N.Y. Fish Game J. 3: 44-74.
Leyhausen, P. 1965. The communal organization of solitary animals.
    Symp. Zool. Soc. London 14: 249-263.
Macdonald, D.W. 1977. The behavioural ecology of the red fox
    (Vulpes vulpes): a study of social organization and resource
    exploitation. D. Phil. Thesis (unpublished), Oxford University,
    Oxford.
Macdonald, D.W. (in press). Some observations and field experi-
    ments on the urine marking behaviour of the red fox, Vulpes
    vulpes L. Z. Tierpsychol.
Montague, F.H. 1975. The ecology and recreational value of the
    red fox in Indiana. Ph.D. Thesis (unpublished), Purdue
    University, Indiana.
Montgomery, G.G. 1974. Communication in red fox dyads: a
    computer simulation study. Smithsonian Contribution to
    Zoology 187: 1-30.
Murie, A. 1936. Following fox trails. Misc. Publ. Zool. Univ.
    Michigan 32: 1-45.
Murie, O.J. 1964. A Field Guide to Animal Tracks. Houghton
    Mifflin Co., Boston.
Peters, R.P. and Mech, L.D. 1975. Scent marking in wolves. Amer.
    Scientist 63: 628-637.
Pruitt, Jr., W.O. 1957. Observations on the bioclimate of some
    taiga mammals. Arctic 10: 131-138.
Pruitt, Jr., W.O. 1960. Animals in the snow. Sci. Amer. 202: 60-
    68.
Pruitt, Jr., W.O. 1970. Some ecological aspects of snow. Ecol.
    and Conservation 1: 83-99.
Pruitt, Jr., W.O. 1978. Boreal Ecology. University Park Press,
    Baltimore.
Richards, S.H. and Hine, R.L. 1953. Wisconsin fox populations.
    Wisc. Cons. Dept. Tech. Wildl. Bull. 6: 1-78.
Rowe, J.S. 1972. Forest regions of Canada. Canada Dept. of
    Fisheries and the Environment. Publication No. 1300,
    Ottawa, 171 p.
Sargeant, A.B. 1972. Red fox spatial characteristics in relation
    to waterfowl predation. J. Wildl. Manage. 36: 225-248.
Schofield, R.D. 1960. A thousand miles of fox trails in Michigan's
    ruffed grouse range. J. Wildl. Manage. 24: 432-434.
Seidensticker, J.C., Hornocker, M.G., Wiles, W.V. and Messick,
    J.P. 1973. Mountain lion social organization in the Idaho
    Primitive Area. Wildl. Monogr. 35: 1-60.
Sheldon, W.G. 1949. Reproductive behavior of foxes in New York
    State. J. Mammal. 30: 236-246.

Sokal, R.R. and Rohlf, F.J. 1969. Biometry· W.H. Freeman and
    Co., San Francisco.

Storm, G.L. and Montgomery, G.G. 1975. Dispersal and social
    contact among red foxes: results from telemetry and computer
    simulation. In The Wild Canids (M.W. Fox, Ed.), pp. 237-246,
    Van Nostrand Reinhold Co., New York.

Storm, G.L., Andrews, R.D., Phillips, R.L., Bishop, R.A., Siniff,
    D.B. and Tester, J.R. 1976. Morphology, reproduction, dis-
    persal, and mortality of midwestern red fox populations.
    Wildl. Monogr. 49: 1-82.

Swan, J.M.A. and Dix, R.L. 1966. The phytosociological structure
    of upland forest at Candle Lake, Saskatchewan. J. Ecology
    54: 13-40.

Tembrock, G. 1957. Zur Ethologie des Rotfuchses (Vulpes vulpes
    L.) unter besonderer Berücksichtigung der Fortpflanzung. Der
    Zoologische Garten 23: 289-532.

Tinbergen, N. 1965. Von den Vorratskammern des Rotfusches (Vulpes
    vulpes L.). Z. Tierspychol. 22: 119-149.

Whitten, W.K., Wilson, M.C., Wilson, S.R., Jorgenson, J.W., Novotny,
    M. and Carmack, M. (In press). Induction of marking behavior
    in wild red foxes (Vulpes vulpes L.) by synthetic urinary
    constituents. J. Chem. Ecol.

The author would like to thank M. Henry and D. Hamer for their
helpful reviews of this study.

MARKING BEHAVIOR IN WILD RED FOXES IN RESPONSE TO SYNTHETIC VOLATILE URINARY COMPOUNDS

M.C. Wilson[1], W.K. Whitten[1], S.R. Wilson[2], J.W. Jorgenson[2], M. Novotny[2], and M. Carmack[2]

[1]The Jackson Laboratory, Bar Harbor, ME 04609
[2]Department of Chemistry, Indiana University, Bloomington, Indiana 47401

Communication through the olfactory sense in wild canids may convey information relating to several variables including food, territory, dominance, pairbonding, individual or group recognition, and reproductive state. For purposes of discussion, the olfactory cues providing this information may be considered as arising from two major sources: the environment (including food, and other plant and animal odors), or other animals of the same species. Various environmental olfactory cues (including food and a few chemical substances noted by Heimburger (1959) to cause marking in wild animals), have been evaluated by Henry (1977). The second category of olfactory cues in foxes, i.e., other red foxes, is examined here.

This investigation took advantage of a natural situation in which wild red foxes travelled along a particular route, urine marking this route at intervals, particularly at sites which they had previously marked. This marking was observed during the breeding months of January through March and probably represented an intensification of behavior occurring with less frequency during the rest of the year (Montague, 1975; Fox, 1971; Tembrock, 1957). Henry (1977),for example, reported that only 12% of the markings he observed during the autumn, were not related to food, and therefore, were potentially of this type.

The function of trail marking by red foxes is in doubt. Henry (1977) theorized that "non-food related markings" might serve as a "social register." Alternatively, the trail markings of red foxes might be territorial warnings to neighboring foxes or attempts to make the trail smell "foxy" and more similar to "home." If the marking were done for this latter purpose, rather than for the identification of individual foxes, as might occur in a "social register" or in territorial marking, then it would seem likely that the urinary compound(s) required for the marking or overmarking would be one(s) characteristic of red foxes, and thus might properly be considered as potential pheromone(s).

Since red foxes and red fox urine do appear to have a very characteristic "skunk-like odor", the potential for identifying a species specific recognition substance, a pheromone, seemed quite likely. In addition, it seemed quite logical that it might be relatively easy to isolate and identify because of its characteristic odor. A synopsis of the work that led to the chemical identification of the characteristic scent of red fox urine and the results of subsequent field testing of this scent on wild red foxes is presented here.

IDENTIFICATION OF VOLATILE COMPOUNDS IN RED FOX URINE

Recently several volatile compounds in red fox urine have been identified (Jorgenson et al., 1978). During January through March, the mating and breeding season for red foxes in the northern hemisphere, the tracks made by red foxes in the fresh snow were followed in and around Acadia National Park, Mt. Desert Island, Maine. By carefully following the same fox trail, the pattern of urination could be identified as male or female depending upon whether there were several raised leg urinations (males) or whether there were urinations from the squatting position (female). In this manner, fourteen separate urine samples were collected and characterized as being derived from 3 female foxes and 4 male foxes. They were collected together with a small amount of adhering snow and shipped on dry ice to be analyzed by gas chromatography and mass spectrometry at the Chemistry Department at Indiana University. Volatile compounds in the 100–300 molecular weight range were separated as is shown in Figure 1 and Table 1. Section (a) of Figure 1 shows a gas chromatographic analysis of 8 different volatiles found in male red fox urine while Section (b) shows the comparable analysis for female red fox urine. Major differences between male and female fox urine were found in compound 2, Δ^3-isopentenyl methyl sulfide which is present in much higher quantities in male urine and compound 7, 2-methylquinoline which appears to be altogether missing from female urine. Compound 3 and 4, on the other hand, 6 methyl Δ^5-heptene-2-one and geranylacetone were more prevalent in female urine. The characteristic odor of fox urine, as perceived by the human olfactory sense, appeared to reside in the two sulfur containing compounds, Δ^3-isopentenyl methyl sulfide and 2-phenylethyl methyl sulfide. Of these two, the Δ^3-isopentenyl methyl sulfide clearly predominated in amount. It is unique to the red fox and has recently been synthesized (Wilson et al., 1978).

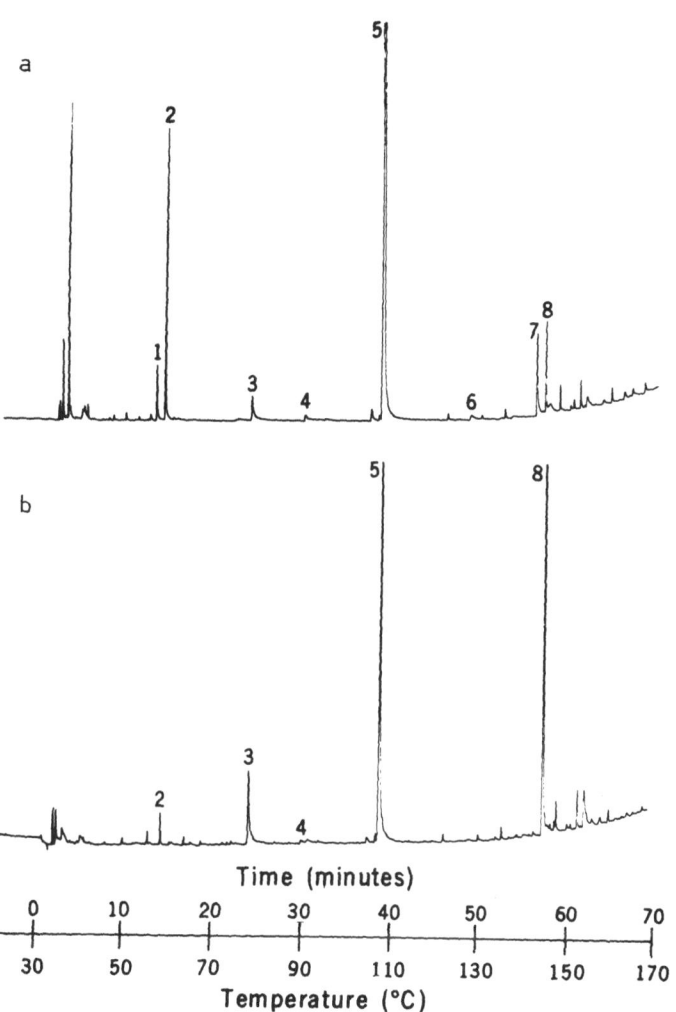

Figure 1. Gas chromatographic analysis in glass capillary columns
           of urinary volatile compounds in the red fox (Vulpes
           vulpes L.). (a) Male; (b) female. The constituents
           are identified by number in Table 1. (Jorgenson et al.,
           1978). Copyright 1978 by the American Association for
           the Advancement of Science.

Table 1.   Volatile compounds found in red fox urine as identified
           by gas chromatography and mass spectroscopy (Jorgenson
           et al., 1978). Copyright 1978 by the American Associa-
           tion for the Advancement of Science.

Peak No.	Mol. Wt.	Compound	Presence in Fox	
			Male	Female
1	114	4-Heptanone	+	+
2	116	Δ^3-Isopentenyl methyl sulfide	++	+
3	126	6-Methyl-Δ^5-hepten-2-one	+	++
4	106	Benzaldehyde	+	+
5	120	Acetophenone	++	++
6	152	2-Phenylethyl methyl sulfide	+	+
7	143	2-Methylquinoline	++	-
8	194	Geranylacetone	+	++

TESTING OF VOLATILE COMPOUNDS WITH WILD RED FOXES

     In order to test the effect of a mixture of these synthetic
compounds on wild red foxes, artificial scent trails were made in
the snow during the winter breeding season (Whitten et al., 1979).
These trails were placed in two areas on Mt. Desert Island, Maine
known to be frequented by red foxes. The first area was a pond of
about 10 hectares surrounded mainly by coniferous forest. The
second area was 13 Km. southeast in Acadia National Park and consisted of hiking trails, mainly through a hardwood forest between
a mountain and a brook on which there were several beaver dams.
Because of the geographical separation, it is unlikely that the
same foxes visited both areas during the study. Also because of
the absence of other fox species (Burt and Grossenheider, 1976)
and feral dogs in this area, the foxes' interest in the scent could
easily be determined by evidence of fox tracks and fox urine marks
with their characteristic "skunk-like" odor.

The eight odorous volatiles identified in red fox urine were dissolved in one liter of distilled water at the amounts shown in Table 2 for Test Solution. These represent an estimate of the concentration found in normal male fox urine before dilution with snow or loss from evaporation. In addition, both test and control solutions contained ethanol, used as a transfer solvent, and poly-ethylene glycol (PEG) as an odor fixative. Without the latter com-pound, the odor of the test solution was not detectable to the experimenters within a few hours after dispensing, whereas with the PEG the odor could be detected with careful sniffing 40 hours later. The PEG also acted as an effective antifreeze and permitted dis-pensing of the solutions at low temperatures.

Table 2. Composition of solutions used (Whitten et al., 1979).

Compound	Formula	Test[a] Solution	Control[b] Solution
4-Heptanone	$C_7H_{14}O$	10 mg	0
Δ^3-Isopentenyl methyl sulphide	$C_6H_{12}S$	50 mg	0
6-Methyl-Δ^5-hepten-2-one	$C_8H_{14}O$	2 mg	0
Benzaldehyde	C_7H_6O	1 mg	0
Acetophenone	C_8H_8O	50 mg	0
2-Phenylethyl methyl sulphide	$C_9H_{12}S$	2 mg	0
2-Methylquinoline	$C_{10}H_9N$	7 mg	0
Geranylacetone	$C_{13}H_{22}O$	7 mg	0
Polyethylene glycol (6000-7500 MW)	$H(OCH_2-CH_2)_nOH$	100 g	100 g
Ethyl alcohol	C_2H_6O	5 g	5 g
Water	H_2O	to 1 liter	to 1 liter

[a]The concentrations of the volatiles in the test solution are estimated values for freshly voided male urine and are approxi-mately three times those found by Jorgensen et al. (1978) thus allowing for loss from evaporation and dilution by snow in the samples they analyzed.

[b]Citronellal, 20 mg, was added to the control for two tests to give it a positive odor.

Prior to laying a trail marked with synthetic fox urine, experimental marking sites were identified by numbered labels tied to trees. These labels were placed at intervals of 20-40 meters, since intervals in this range are characteristic of the distance between urine marks reported for wild foxes (Montague, 1975; Rome, 1978). The numbered sites were also recorded on maps of the pond and hiking trail using a scale of 1:4000.

On late afternoons following significant (greater than 5 cm) snowfalls, two experimenters travelled single file wearing snow-shoes and shoveled a mound of snow (roughly a 20 x 20 x 20 cm cube) near each numbered site. These mounds were placed at least one meter from the snowshoe trail and reasonably remote from other potential scent posts such as tree trunks, dead stumps, rocks, or conifer seedlings. After the mound was made, 10-15 ml of test or control solution was placed on the top of the mound. To minimize cross contamination, one person always handled the control mixture and another the test mixture. To simplify record keeping and prevent errors in numbering under conditions of failing light, and to prevent contamination by lingering odors from previous trials, the control preparation was always placed on the odd-numbered sites and the test preparation on the even. Although this was a departure from strict randomization, it did not appear to bias the results since the foxes seemed to join and depart from the trails in a random manner. In the pond area, there were three trials using the test and control solutions with an average of 96 mounds per trial, and in the hiking trail area, there were four trials with an average of 58 mounds per trial. On the two to three mornings after setting out the scented mounds, the mounds and trails were examined for evidence of fox tracks and urine or fecal marks. The marked mounds were checked for the characteristic "skunk-like" odor and for indications of the sex and number of foxes involved.

Table 3 shows the results obtained with the test and control solutions. Here the data are recorded and analyzed according to the total number of mounds prepared and the total number of mounds that were either marked with urine or feces. These results show that the synthetic compounds induced an increase in marking activity in both study areas. The greater frequency of marking along the hiking trail in Area 2 may have been due either to a greater population of foxes in this area and/or to the confinement of the foxes to a clearly defined trail.

The presentation of the data in such a compact form as Table 3 leaves several unanswered questions: (1) Did male foxes mark more frequently than female foxes? (2) Were mounds investigated but not marked? (3) Did the foxes always mark a mound if they got

Table 3. The effect of synthetic volatile urinary compounds on marking of mounds by foxes in two areas (Whitten et al., 1979).

		Marked	Not Marked	x^2	P
Area 1	Test	4^a	139	4.06	< 0.05
	Control	0	143		
Area 2	Test	16^b	95	14.03	< 0.0001
	Control	1	110		
Combined	Test	20	234	17.93	< 0.00001
	Control	1	253		

[a]Included 1 fecal mark.

[b]Included 2 fecal marks.

near enough to smell it? (4) Was the response of the foxes to the synthetic urine mixture only due to the fact that it had a perceptible odor and the control mixture had none? (5) Which of the compounds in the test mixture were necessary to induce marking?

The first two questions can be answered fairly easily Most mound urinations appeared to involve leg raising and therefore, were presumably done by male foxes; however, there were at least three cases in which footprints characteristic of a squatting posture were observed. In one of these cases, the urination appeared to be part of a double marking by a male and female fox. Also, most mounds that were investigated were marked. In only two cases was there evidence of investigation of a mound without marking, once for a control mound and once for a test mound.

Tentative answers to the last three questions can be found in Table 4. In this table, the data from Table 3 are presented in a different form, including only the mounds that were passed within three meters, the approximate width of much of the hiking trail. In addition, subsequent trials, in which only the sulfur compounds were included in the test mixture, or in which the sulfur compounds were removed from the test mixture are presented. Here it is

evident that marking occurred in about 10% of the mounds passed in
Area 1 (4/45) and 30% of the test mounds passed in Area 2 (22/74).
Citronellal (at 20 mg/l), a fairly odorous substance, did not cause
marking (or avoidance) when it was placed in the control solution.
In two trials, a mixture with only the sulfur compounds caused
marking, and in one trial, the test mixture was marked 6 times,
while the test mixture without the sulfur compounds was not marked
at all. These data taken together seem to indicate that the unique
odorous sulfur compounds present in red fox urine induce overmarking
in wild red foxes. Further testing of these compounds against compounds found in other wild canids may eventually lead to the conclusion that these odorous sulfur compounds are indeed species
specific pheromones.

Table 4. Urine marking activity of red foxes passing within three
meters of artificially scented snow mounds.

Trial No.	Area	Test Mixture[a]	Vehicle alone	Vehicle plus citronellal	Vehicle plus "S" compounds	Mixture minus "S" compounds
1	1	1 (26)[b]	0 (26)			
2	1	1 (6)	0 (5)			
3	1	2[c](13)	0 (15)			
1	2	3[c](19)	0 (8)			
2	2	6[c](21)	1 (20)			
3	2	1 (2)	0 (2)			
3	2	4 (9)		0 (10)		
4	2	2 (10)		0 (10)		
5	2		0 (4)		1 (4)	
6	2		0 (?)[d]		1 (?)[d]	
7	2	6 (13)				0 (11)
Total	1	4 (45)	0 (46)			
Total	2	22 (74)	1 (34)	0 (20)	1 (4)	0 (11)

[a]The test mixture contained 8 volatile compounds, two of which contained sulfur ("S"). These compounds and the application vehicle
are listed in Table 2.

[b]The number of marked mounds is indicated, followed by the total
number of passed mounds in parentheses.

[c]Values include one scat.

[d]Footprints were not clear enough to tell how many mounds were
passed. Data are not included in totals.

CONCLUSIONS

The initial success of this approach to scent marking in foxes could be the starting point for many other investigations. It should be possible to analyze the chemical components in red fox urine throughout the year, so that compounds present in greater quantities during the breeding season can be identified and tested in the field. It may be that the two sulfur compounds are present in greatest quantity during the breeding season. In addition, it may be possible to find compounds present in females' urine that will be attractive to male foxes.

Another area of research should be identification of volatile chemical constituents in the urine of other species of foxes and other wild canids. These chemical compounds could be tested on red foxes, and if possible, vice versa, the red fox pheromones could be tested on other species of foxes and wild canids. With captive or tagged foxes, additional behavioral studies could be done to see whether animals with different social status show different marking patterns or different urinary chemistry.

Finally, the technique of laying artificial scent trails on the snow may be adaptable to other species living in cold climates and could lead to the identification of those substances which elicit marking in other species.

REFERENCES

Burt, W.H. and Grossenheider, R.P. 1976. A Field Guide to the Mammals, Peterson Field Guide, Series 5, (p. 72) Houghton Mifflin Co., Boston.

Fox, M.W. 1971. Behaviour of Wolves, Dogs, and Related Canids, pp. 75-76, 179, 186, Harper and Row, New York.

Heimburger, N. 1959. Das Markierungsverhalten einiger Caniden, Z. Tierpsychol. 16: 104.

Henry, J.D. 1977. The use of urine marking in the scavenging behavior of the red fox (Vulpes vulpes), Behaviour 61: 82.

Jorgenson, J.W., Novotny, M., Carmack, M., Copland, C.G., Wilson, S.R., Katona, S. and Whitten, W.K. 1978. Chemical scent constituents in the urine of the red fox (Vulpes vulpes L.) during the winter season, Science 199: 796.

Montague, F.H., Jr. 1975. The ecology and recreational value of red fox in Indiana, Thesis, Purdue University Diss. Abstr. 4836B.

Rome, A. 1978. Devising a method to study the responses of foxes to synthetic urinary compounds, Colby College and Jackson Laboratory Student Report.

Tembrock, G. 1957. Das Verhalten des Rotfuchses, Handbuch Zool.
    8(9), part 10 (15), pp. 3-4, 10-11, de Gruyter, Berlin.
Whitten, W.K., Wilson, M.C., Wilson, S.R., Jorgenson, J.W.,
    Novotny, M., and Carmack, M. 1979. Induction of marking
    behavior in wild red foxes (Vulpes vulpes L.) by synthetic
    urinary constituents, J. Chem. Ecol. In press.
Wilson, S.R., Carmack, M., Novotny, M., Jorgenson, J.W. and
    Whitten, W.K. 1978. Δ^3-Isopentenyl methyl sulfide, a new
    terpenoid, in the scent mark of the red fox (Vulpes vulpes),
    J. Organic Chemistry 43: 4675.

CHEMICAL SIGNALS IN ALARM BEHAVIOR OF DEER

Dietland Müller-Schwarze

College of Environmental Science & Forestry
State University of New York
Syracuse, NY 13210

Alarm pheromones in the widest sense are chemical signals emitted by animals that induce, maintain or enhance in conspecifics a state of alertness or alarm which in turn may result in defense, avoidance, or escape behavior.

Sea anemones (_Anthopleura_ sp.) and sea urchins (_Diadema antillarium_) move away from crushed conspecifics (Howe and Sheik, 1975; Snyder and Snyder, 1970). The earthworm _Lumbricus terrestris_, secretes in its mucus a pheromone that facilitates rapid dispersion of conspecifics and also repels predators (Ratner, 1971; Ressler et al., 1968). The marine mud snail _Nassarius obsoletus_, moves away from crushed conspecifics and buries itself (Atema and Burd, 1975).

A number of alarm pheromones have been described for insects, and many of them have been chemically identified, especially in the social hymenopterans (Blum, 1974, 1979; Wilson, 1975). In worker honey bees (_Apis mellifera_), isopentyl acetate from the sting apparatus releases alarm behavior (Boch et al., 1962). The alarm behavior of the weaver ant _Oecophylla longinoda_ consists of a series of steps: alert, alarm, attraction to the odor source, locomotor arrestment and vigorous biting. Alerting and alarm are released by hexanal which is secreted by the mandibular gland. Workers are attracted from at least 10 cm by 1-hexanol, and biting is released by 3-undecanone and 2-butyl-2-octenal, all from the mandibular gland. But alarm behavior is also released by formic acid and n-undecane from the abdomen (Bradshaw et al., 1975).

In addition to escalating levels of alarm behavior, alarm pheromones can serve as defense secretion vis-a-vis predators, as

the formic acid in ants of the genus Formica (Maschwitz, 1964).
Indeed, alarm behavior control may be a secondary function of defense secretions, since many defense secretions of solitary insects
serve as alarm odors in social insects (Blum, 1974).

Further behaviors connected with alarm pheromones are recruitment of conspecifics, as in ants of the genera Pogonomyrmex and
Camponotus (Hölldobler, 1971), marking or "tagging" intruders, as
in formicine ants (Bergström and Löfquist, 1971), or disarming
opponents during nest raids. The latter occurs when the Meliponine
bee Lestrimelitta limao raids colonies of stingless bees of the
genus Trigona or Melipona. The host workers' behavior becomes disrupted by citral, the alarm pheromone of the raiding Lestrimelitta
which is used as a "diabolical instrument of subjugation" (Blum,
1974).

The fright reaction of fish is well known (von Frisch, 1941;
Pfeiffer, 1967). Minnows and other fish flee, crowd together, retreat and hide in response to a substance from the "club cells" in
broken epidermis of group members. The response is found in the
Ostariophysi and Gonorhynchiformes (Pfeiffer, 1967), and varies
with the ecology of the fish species: some flee to the surface,
others to the bottom. A fright substance exists also in tadpoles
of toads of the genus Bufo and is confined to Bufonidae (Pfeiffer,
1974). None of these substances has been identified completely,
although Pfeiffer and Lemke (1973) found that the alarm substance
in the Cypriniformes (belonging to Ostariophysi) is probably a
pterin. Of 59 pteridine, purine, and pyrimidine derivatives
tested, isoxanthopterin and 6-acetonylisoxanthopterin elicited
fright reaction and produced bradycardia, in the giant danio (Danio
malabricus) and minnows (Phoxinus phoxinus) respectively (Pfeiffer,
1978).

In mammals, house mice (Mus musculus) avoid a stressed conspecific's "fright odor" ("Angstgeruch") which is contained in the
urine (Müller-Velten, 1966). In several studies, avoidance of
odors of stressed individuals by conspecifics has led investigators to classify the odor as an alarm pheromone. For instance,
Rottman and Snowdon (1972) "stressed" mice by injecting hypertonic
saline and found that other mice avoided their odor. In contrast
to Müller-Velten's study, urine was not found necessary for the
"alarm" response, but social experience was. Jones and Nowell
(1974) on the other hand, described an androgen-dependent "aversive
pheromone" that "discourages" prolonged investigation, at least
by subordinate males. Also, mice "preferred" the odor of non-
stressed to that of stressed mice, the "stress" being defeat in
fights between males, or electric shock (Carr et al., 1970). It
seems that the properties described in the latter two cases place

the odors more into social behavior and outside of what we ordinari-
ly consider alarm pheromones.

Laboratory rats retreat, freeze, and eliminate in response to
muscle homogenate and blood of conspecifics (Stevens and Gerzog-
Thomas, 1977), and discriminate between the odors of stressed and
non-stressed conspecifics (Valenta and Rigby, 1968; Stevens and
Koster, 1972). Sherman (1974) described for hamsters a locomotor
response to odors of stressed conspecifics.

Some species of mammals produce odors in the context of anti-
predator behavior. Pronghorns (Antilocapra americana) discharge an
odor from their ischiadic gland when fleeing (Buechner, 1950;
Müller-Schwarze and Müller-Schwarze, 1972). Also, shrews of the
genus Suncus discharge an odor when handled (Dryden and Conaway,
1967). Although effects of these odors on conspecifics are not
known, they are candidates for alarm pheromones.

In cervids, there is evidence that chemical stimuli operate
in alarm behavior. Caribou (Rangifer tarandus) have been observed
to escape after sniffing footprints of a conspecific that had been
alarmed previously at the same site (Lent, 1966). Roe deer
(Capreolus capreolus) bucks fled from footprints of other male roe
deer (Kurt, 1967), but it is not clear whether this constitutes an
alarm reaction in the usual sense or a response to dominant or
potentially dangerous conspecifics.

Black-tailed deer (Odocoileus hemionus columbianus) discharge
a garlic-like odor from their metatarsal gland when alarmed, cor-
nered, or handled (Müller-Schwarze, 1971). The metatarsal gland
is located on the outside of the hindleg, is present in both sexes,
and is 6 x 4 cm in area (Quay and Müller-Schwarze, 1970). The
effect of the odor on conspecifics under natural conditions is not
known. Captive black-tailed deer withdraw from food that is presented together with metatarsal secretion. When metatarsal secretion is present, about 40% of the approaches result in feeding,
while control food samples are eaten in 95 to 100% of the cases
(Müller-Schwarze, 1971). The closely related white-tailed deer
(Odocoileus virginianus) has a much smaller metatarsal gland that
lacks the strong odor. A cross-species experiment with several
different concentrations of metatarsal odors from both species
showed that feeding is not inhibited in white-tailed deer by concentrations that are active in black-tailed deer. Only concentrations one or two orders of magnitude higher inhibited feeding in
white-tailed deer (Müller-Schwarze and Volkman, 1977). This
indicates that chemical alarm signals from the metatarsal gland
are more highly evolved in the black-tailed deer (and in its sibling
subspecies, the mule deer, O. h. hemionus), than in white-tailed
deer.

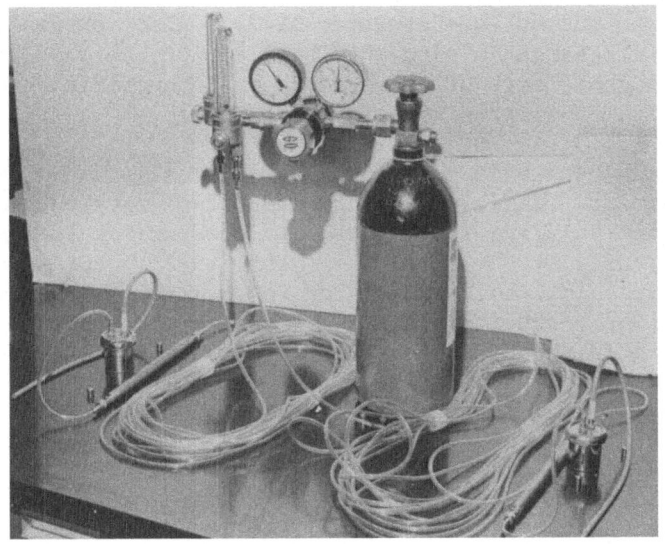

Figure 1. Odor delivery apparatus. Compressed air in tank serves
           as propellant. To the left of the gauges are two flow
           meters for two different air streams. Each stream runs
           through 15 m tubing. Charcoal filter is located between
           tubing and odor container. Odor container has two valves
           that are opened by air stream. Odor outlet hoses at the
           far right and left.

MATERIALS AND METHODS

The role of social chemical stimuli in alarm responses in deer was examined experimentally in captive black-tailed deer and by observation in free-living white-tailed deer.

Black-tailed deer. Eight male and 16 female captive black-tailed deer were arranged in six different groups which ranged in size from two males to one male and six females. Two groups consisted of two males and two females each, one of one male and two females and one of four females. They were exposed to metatarsal (M) and control odors while feeding near an odor outlet in their pen. The control odors used were clean air (A), male deer urine (U), beaver castor (C; 0.05 g in 1 ml ethyl alcohol), n-butyl-mercaptan (B; 2 μl in 1 ml ethyl alcohol), and gasoline (G). One cc of the odor solution was dripped on a cotton ball which was placed in an air-tight, cylindrical stainless steel container (4 cm high, 4 cm diameter). Two valves in the lid of the container permit a unidirectional flow of an air stream through the sample. The odor delivery apparatus (Fig. 1) consists of the odor container with 30 cm teflon tubing as the odor outlet; a charcoal filter to clean the entering air, 15 m of tygon tubing (3 mm inner diameter), and a tank with compressed air that serves as propellant for the odor. The tank has a pressure gauge with maximum scale of 15 psi, and a split line with two flowmeters permitting a choice of one or two air streams. For the experiment reported here, the apparatus was run in the one channel mode, and the airflow was 120 ml/min. The odor concentration at the outlet is estimated to be about 1/5 of saturated vapor. The air space around the outlet (1 m diameter, 1 m high), dilutes the odor further to about 10^{-3} to 10^{-6} of saturated vapor.

The odor is placed in the air-tight container, the tubing is connected and the odor outlet tube is fastened with wire to a post. To attract the deer, food such as hay, apple twigs or white pine fronds are placed around the post in an area of 1 m radius. On windy days the food is placed on the downwind side. The behavior of the deer is observed for two minutes with the valves closed ("pre-odor" responses as baseline), then the air is turned on, opening the valves, and releasing the odor for 2 min. After the valves are closed again, the deer are observed for another 2 min ("post-odor" responses). A day's test consists of a sequence of control odor, clean air, and metatarsal odor, each lasting for 6 min, as described above. The sequences for the different control odors were: M-A-C, A-M-U, A-B-M, and G-A-M. The intervals between the 6-min trials within each test varied between 5 and 50 min, depending on the time required to attract the deer back to the feeding/odor station (Fig. 2).

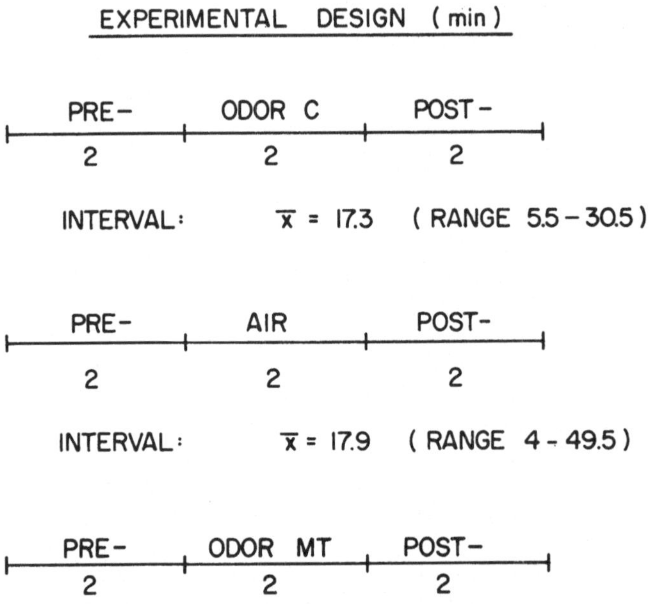

EXPERIMENTAL   DESIGN   (min)

PRE–	ODOR C	POST–
2	2	2

INTERVAL: $\bar{x}$ = 17.3   (RANGE 5.5 – 30.5)

PRE–	AIR	POST–
2	2	2

INTERVAL: $\bar{x}$ = 17.9   (RANGE 4 – 49.5)

PRE–	ODOR MT	POST–
2	2	2

Figure 2.   Experimental Design.

Two observers stand in a shed 10 m from the feeding station with the odor outlet. One operates the gauges and flow meters at the air tank, the other records the responses of the deer on pre-pared data sheets. The responses recorded are feeding, four intensity levels of the alert posture, sniffing toward or at the odor outlet, leaving the site, and returning. Since the number of individuals at the station changes during the 6 min of each trial, the number of animals is noted every 30 sec, and all responses are related to the number of "animal minutes".

The tests were performed during the months of October through December 1978, and March and April 1979.

White-tailed deer. Free-living white-tailed deer were observed for 59.6 hours between January 1 and 11, 1978 and for 57 hours be-tween December 27, 1978 and January 5, 1979 on a ranch near the Archbold Biological Station in central Florida. Of the total 116.6 hrs observation time, 85.3 hrs were on foot, and 31.3 hrs by car. Actual animal contact was made for a total of 31 hours (26.6% of observation time). A total of 474 deer contacts were made and the average group sizes were 2.61 in January 1978 and 2.75 in December 1978/January 1979. Group sizes and composition, cover type and social interactions, including social sniffing were recorded.

A caretaker fed the deer daily from a jeep along a 5-mile route, thus conditioning the deer to people and vehicles, so that observa-tion at close range was possible.

RESULTS

Black-tailed deer. First, a uniformity test was performed.
The frequencies of assuming the alert posture (AL) and leaving the
site (LV) did not change during the course of each daily test.
The pre-odor responses for the three consecutive 6 min trials (control odor, air, and MT secretion in the various permutations) did
not differ significantly (Table 1).

Table 1. Pre-odor responses to control odors, air and MT secretion.

Sequence	Odor	Animal Minutes (am)	AL frequency* Total	AL frequency* Per am	LV frequency** Total	LV frequency** Per am
1	Control	85	102	1.20	16	0.19
2	Air	66	84	1.27	16	0.24
3	MT	60	83	1.38	12	0.20

*Differences of AL frequencies for an average 70.3 am are not
 significant ($x^2 = 0.936$, df = 2).

**Differences of LV frequencies for an average 70.3 am are not
 significant ($x^2 = 3.644$, df = 2).

The baseline level of response did not differ between the
sexes. In males, the frequencies of AL and LV during clean air
trials were 1.19 and 0.20 per animal minute (am), respectively.
The values for females were 1.15 and 0.26, respectively.

Females were more sensitive to odors: in females, AL was
almost twice as frequent (1.73 AL/am) in the presence of MT than
in that of air (0.97 AL/am), while the males showed no significant
difference in their responses to MT (1.33 AL/am and 0.34 LV/am)
and air (1.19 AL/am and 0.20 LV/am). Therefore, further treatment
of the data will deal with the responses by the females exclusively.

The frequencies for LV in females were 0.16 LV/am in the presence of air only, and 0.45 LV/am in response to MT, an almost
threefold increase. The response levels in the presence of the
control odors (B, C, G and U) were intermediate. This is true for
both AL (Fig. 2) and LV (Fig. 3) and LV (Fig. 4).

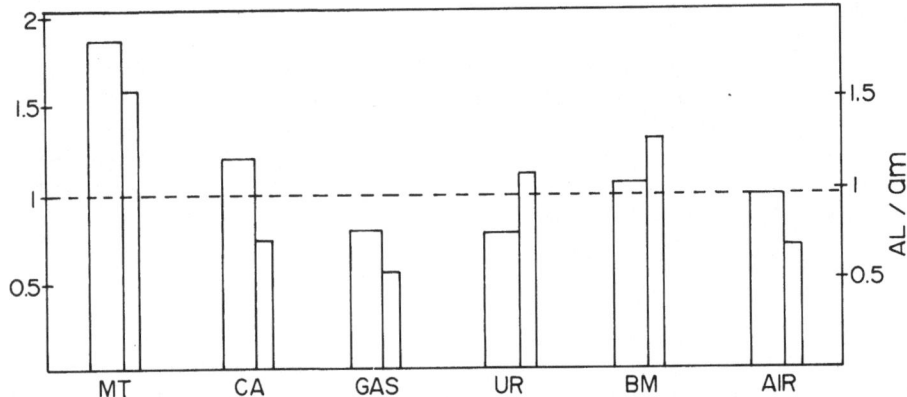

Figure 3. Frequencies of Alert posture (AL) per animal minute (am)
in response to MT secretion, air and four control odors,
beaver castor (CA), gasoline (GAS), male deer urine (UR),
and n-butyl-mercaptan (BM). Abscissa: The various
odors; right ordinate: AL per am; left ordinate:
response intensity in multiples of response to air.

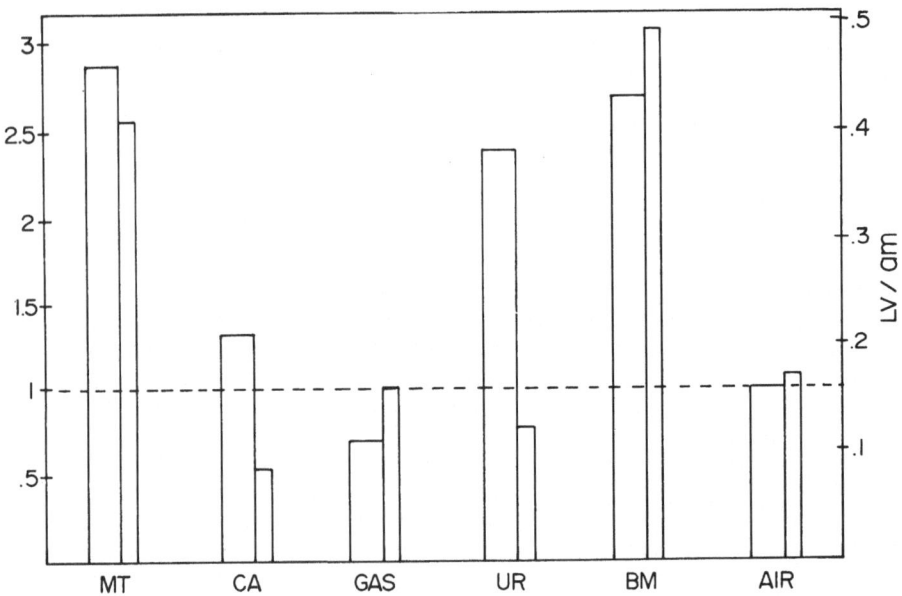

Figure 4. Frequencies of leaving the site (LV) per animal minute
(am) in response to MT secretion, air and four control
odors. As in Figure 3.

White-tailed deer. Since currently there is no evidence that
any particular secretion or excretion provides an alarm signal in
white-tailed deer, the alarm behavior of free-ranging white-tailed
deer was observed on a ranch in Central Florida.

Alarmed white-tailed deer raise and flag their tail (visual
signal) and snort (auditory signal), and also sniff one another.
Various body regions, such as nose, head, neck, back and tail, are
sniffed by conspecifics. White-tailed deer that are alert and about
to retreat into cover are 3-7 times more likely to sniff one another's
tail, than other body parts (Table 2). This seems to be due to a
decrease of sniffing other body parts in alarm situations rather
than increased tail sniffing.

Table 2. Frequencies of sniffing tail and other body parts in
          alert and non-alert white-tailed deer on a Florida ranch
          (sniffing and sniffed individuals combined).

Sniffed Body Part	Behavior Before Sniffing			Behavior After Sniffing		
	Alert	Not Alert	Total	Retreat	No Retreat	Total
Tail	17	17	34	18	15	33
Other Body Parts	6	72	78	11	59	70
Total	23	89	112	29	74	103
% Tail Sniffing	73.9	19.1	30.4	62.1	20.3	32.0
P	0.001			0.001		

The higher frequency of tail sniffing among alert deer does
not necessarily indicate the tail area as the source of an alarm
odor, since the deer are alert at the time of sniffing. However,
two possible functions are individual recognition of group members
and assessment of the state of excitation of group members.

DISCUSSION

The search for intraspecific chemical signals that modulate
behavior in alarm situations in mammals is still in its beginning.
It has to be based on a thorough knowledge of the alert, alarm,
avoidance, escape and defense responses of the species in question.

No active compound has been chemically identified; the release of odors in alarm situations is extremely difficult to record and to measure; and the effects of alarm odors on conspecifics (including specific responses as well as a wide range of possible responses) are virtually unknown. Alarm odors may potentially influence not only escape and avoidance behaviors, but also earlier stage of alarm such as interruption of ongoing behavior, becoming or staying alert, keeping contact with conspecifics, identification of group members, or assessing the level of arousal of other individuals. Some obvious functions of the signal may be to alert conspecifics to the presence of a predator, release or facilitate avoidance, escape or hide, or serve in group cohesion during alarm and flight by permitting species, group, or individual recognition. Thus, in alarm situations, social signals fall into three functional categories: warning, triggering avoidance, and recognition.

Thus far, the work with mammalian alarm odors has focused on the avoidance function. If, however, warning by chemical means takes place in a social group, modulation of the opposites of avoidance, such as mutual approach, and group cohesion, or staying hidden in a given area, may be major functions of alarm pheromones.

The stimuli used to stress experimental animals, such as injecting saline (Rottman and Snowdon, 1972) or blowing and shaking (Müller-Velten, 1966) seem rather remote from naturally occurring alarm stimuli. Also, avoidance responses to injured individuals or isolated tissues (Stevens and Gerzog-Thomas, 1977) lie outside the interactions of intact animals considered here, although they are analogous to the fright reactions in fish and amphibians.

The present experiment demonstrated an odor effect on the frequencies of the alert posture and leaving, while in our earlier bioassay the metatarsal secretion inhibited feeding (Müller-Schwarze, 1971). The latter effect can be viewed as the first step of alarm behavior, i.e., inhibition of ongoing or intended behavior. In any event, the responses observed thus far are subtle and odors alone did not release complete escape behavior as in fish, for instance.

Alarm behavior in deer is under multisensory control with alarm calls, tail-flagging, footstomping, locomotion noises, and alarm odors all acting to alert conspecifics and finally to induce them to flee. Hence, presentation of an odor alone will at best release only a partial response, a small step in the escalating sequence of alert and alarm responses.

The responses to metatarsal secretion in our experiment are remarkable since the odor is familiar to the deer and habituation to it would be expected, while control odors, such as beaver castor or gasoline were novel odors. Generally, mammals avoid novel stimuli (Bolles, 1970).

The concentrations of the control odors are arbitrary. However, many odors can be presented at concentrations that alert the experimental animals. The important point is that a secretion from the deer themselves facilitates alert responses.

The sex difference in the responsiveness to odors reflects a general trend; females are more cautious, more alert, and more excitable than males. For instance, in our earlier bioassay with metatarsal odor next to food, feeding was more inhibited in females than in males. Furthermore, in free-living white-tailed deer, 61% of the adult females flagged their tails (fawns: 86%), while only 40% of the mature bucks did so (Müller-Schwarze, in prep.).

A fuller appreciation of the function of alarm odors will only be possible after they have been studied under a variety of environmental conditions. For instance, during darkness alarm odors may possibly play a greater role than during the daylight hours. Also, it might be possible to generally characterize alarm stimuli by their physiological effects on heart and respiration rates, hair erection, and locomotion.

ACKNOWLEDGEMENTS

This work was supported by NSF Grant BNS 7618333. I thank Richard Altieri for technical assistance during the experiments with black-tailed deer.

REFERENCES

Atema, J. and G.D. Burd. 1975. A field study of chemotactic responses of the marine mud snail, Nassarius obsoletus. J. Chem. Ecol., 1, 243-251.

Bergström, G. and J. Löfquist. 1971. Camponotus ligniperda Latr. A model for the composite volatile secretions of Dufour's gland in formicine ants. In Chemical Releasers in Insects, Proc. 2nd Int. IUPAC Congr., Tel Aviv, Israel, 1971, Vol. 3, ed. A.S. Tahoi (Gordon and Breach, NY). pp. 195-223.

Blum, M.S. 1974. Pheromonal sociality in the Hymenoptera. Pheromones. Vol. 32, 222-249.

Blum, M.S. 1979. Hymenopterous pheromones. Optimizing the specificity and acuity of the signal. Chemical Ecology: Odour Communication in Animals. F.J. Ritter, ed. pp. 20.-211.

Boch, R., D.A. Shearer and B.C. Stone. 1962. Identification of iso-amyl acetate, as an active compound in the sting pheromone of the honey bee. Nature (London) 195, 1018-1020.

Bolles, R.C. 1970. Species specific defense reactions and avoidance learning. Psychol. Rev. 77: 32-48.

Bradshaw, J.W.S., R. Baker and P.E. Howse. 1975. "Multicomponent Alarm Pheromones of the Weaver Ant". Nature, 258, 230-231.

Buechner, H.K. 1950. Life history, ecology and range use of the pronghorn antelope in Trans-Pecos Texas. Amer. Mid. Nat. 43, 287-354.

Carr, W.J., R.D. Martorano, and L. Krames. 1970. Responses of mice to odors associated with stress. J. Comp. Physiol. Psychol. 71, 223-228.

Dryden, G.L. and C.H. Conaway. 1967. Origin and hormonal control of scent produced in Suncus murinus. J. Mammal. 48, 420-428.

Frisch, K. Von. 1941. Über einen Schreckstoff der Fischhaut und seine biologische Bedeutung. Z. Vergl. Physiol. 29, 46-145.

Hölldobler, B. 1971. Recruitment behavior in Camponotus socius (Hym. Formicidae). Z. Vergl. Physiol. 75, 123-142.

Howe, N.R. and Y.M. Sheikh. 1975. Anthropleurine: a sea anemone alarm pheromone. Science 189, 386-388.

Jones, R.Z. and N.W. Newell. 1974. The urinal aversive pheromone of mice: species, strain, and grouping effects. Animal behavior, 1974, 22, 187-191.

Kurt, F. 1967. Zur Rolle des Geruchs im Verhalten des Rehwildes. Verh. Schweiz. Naturforsch. Ges. Zürich 144, 140-142.

Lent, P. 1966. Calving and related social behavior in the barren-ground caribou. Z. Tierpsychol. 23, 702-756.

Maschwitz, U.W. 1964. Alarm substances and alarm behaviour in social Hymenoptera. Nature (London) 204, 324-327.

Müller-Schwarze, D. 1971. Pheromones in black-tailed deer, Odocoileus hemionus columbianus. Anim. Behav. 19, 141-152.

Müller-Schwarze, D. In prep. Communication in free-living white-tailed deer, Odocoileus virginianus seminolus.

Müller-Schwarze, D. and C. Müller-Schwarze. 1972. Social scents in hand-reared pronghorn (Antilocapra americana). Zool. Afric. 7, 251-271.

Müller-Schwarze, D. and N. Volkman. 1977. Responses of white-tailed and black-tailed deer (genus Odocoileus) to metatarsal gland secretions. Abstr. Anim. Beh. Soc. Meeting Penn. State Univ. University Park, Pennsylvania.

Müller-Velten, H. 1966. Über den Angstgeruch bei der Hausmaus (Mus musculus L.). Z. Vergl. Physiol. 52, 401-429.

Pfeiffer, W. 1967. Schreckreaktion und Schreckstoffzellen bei Ostariophysi and Gonorhynchiformes. Z. Vergl. Physiol. 56, 380-396.

Pfeiffer, W. 1974. Pheromones in fish and amphibia. Pheromones. M.C. Birch, ed., Amsterdam; North Holland Publ. Co., 269-296.

Pfeiffer, W. 1978. Heterocyclic compounds as releasers of the fright reaction in the giant danie, Danio malabaricus (Jerdon) (Cyprinidae, Ostariophysi, Pisces). J. Chem. Ecol. 4, 665–673.

Pfeiffer, W. and J. Lemke. 1973. Untersuchungen zur Isolierung and Identifizierung des Schreckstoffes aus der Haut der Elritze, Phoxinus phoxinus (L.) (Cyprinidae, Ostariophysi, Pisces). J. Comp. Physiol. 82, 407–410.

Quay, W.B. and D. Müller-Schwarze. 1970. Functional histology of integumentary glandular regions in black-tailed deer (Odocoileus hemionus columbianus). J. Mammal. 51, 675–694.

Ratner, S.C. 1971. Behavioral characteristics and functions of pheromones of earthworms. Psychol. Rec. 21, 363–371.

Ressler, R.H., R.B. Cialdini, M.L. Ghoca and S.M. Kleist. 1968. Alarm pheromone in the earthworm, Lumbricus terrestris. Science 161, 597–599.

Rottman, S.J. and C.T. Snowdon. 1972. Demonstration and analysis of an alarm pheromone in mice. J. Comp. Physiol. Psychol. 81, 483–490.

Sherman, R.A.W. 1974. Demonstration of the hamster alarm pheromone. Diss. Abstract Int. B. Science and Eng. 34 (8) 3656 (Feb. 74).

Snyder, N.F.R. 1967. An alarm reaction of aquatic gastropods to intraspecific extract. Cornell Univ. Agric. Exp. Station Memorandum 403.

Snyder, N. and Snyder, H. 1970. Alarm response of Diadema antillarum. Science 168, 276–278.

Stevens, D.A. and D.A. Gerzog-Thomas. 1977. Fright reactions in rats to conspecific tissue. Physiol. Behav. 18, 47–51.

Stevens, D.A. and Koster, E.P. 1972. Open-field responses of rats to odors from stressed and non-stressed predecessors. Behav. Biol. 7: 519–522.

Valenta, J.G. and M.C. Rigby. 1968. Discrimination of the odor of stressed rats. Science 161, 599–601.

Wilson, E.O. 1975. Sociobiology. Belknap/Harvard, Cambridge, Mass., 697 pp.

TERRITORIAL MARKING BEHAVIOR BY THE SOUTH AMERICAN VICUNA

William L. Franklin

Dept. of Wildlife Science
Utah State University
Logan, UT 86322

INTRODUCTION

Two wild members of the camel family occur on the aridlands of South America: the vicuna (_Vicugna vicugna_) and the guanaco (_Lama guanicoe_). The vicuna is found in the central Andean altiplano or puna, an equatorial short-grassland that is above treeline and below the snowline, despite its 3700m to 4800m elevation (Koford, 1957; Franklin, 1973 a, b). In contrast, the guanaco ranges from sea level to 4000m and is more widely distributed (Franklin, 1975).

Vicuna populations are socially organized into family groups, bachelor male groups, and solo males (Franklin, 1974 and 1978). Family groups are composed of an adult male, several females, and their offspring less than one-year old (called crias). Territorial behavior forms the foundation of spatial distribution and social organization for this alpine species. Family groups each occupy year-round feeding and sleeping territories actively defended by the resident male (Franklin, 1974). The feeding territory is located on a lower slope, flat or bottomland in good habitat, while the sleeping territory is on some nearby flattened ridge. The sleeping territory is an area of poor to fair habitat. Preferred forage plant communities are occupied by Permanent Territorial Family Groups (PTFGs), while marginal and secondary habitats are occupied by Marginal Territorial Family Groups (MTFGs) and Male Groups (MGs).

[1]Present Address: Dept. of Animal Ecology, 124 Science II, Iowa State University, Ames, IA 50011

Territorial borders are well defined and rigorously defended
by the occupant male several times daily. The 1 to 3m wide boundaries that separate territories, while not visible to an observer,
are clearly defined and learned by the resident family groups.
Two adjacent family groups often feed tranquilly side by side only
a few meters apart, each on its own side of their mutual territorial
boundary. If one group wanders across the border, it will be
promptly chased back into its own territory by the other group's
male. Some territorial boundaries conform to natural topographical
features such as streams and gullies or man-made structures such
as roads. But often the terrain lacks natural landmarks, except
for vicuna made dung piles. Feeding territories provide year-round
exclusive access to preferred and scarce forage. They also provide
predictable and socially stable sites for females to mate, give
birth and raise their young (Franklin, 1978). Thus, an area inhabited by vicuna family groups is socially divided up into a mosaic
of permanent year-round territories, whose borders are well defined
and defended.

Vicunas make obvious use of dung piles when defecating and
urinating (Koford, 1957). The overall goal of this paper is to
assess how dung piles are used by vicuna and their relationship to
territorial behavior. Specific questions include:

1) What individuals use dung piles?

2) What standardized behaviors and postures are done on dung
piles?

3) Do behaviors on dung piles differ between sex and age
classes?

4) Under what socioecological contexts are dung piles used?

5) What is the spatial distribution of dung piles within the
territory?

Acknowledgement is given to sources of funding and collabora-
tors who made this research possible: The Conservation Foundation,
World Wildlife Fund, Direccion General Forestal y de Caza of Peru,
and Utah State University.

METHODOLOGY

Vicunas were studied at the Pampa Galeras National Vicuna
Researve in southern Peru (14° 40'S, 74° 25'W) from 1968 to 1971.
Pampa Galeras is located on the western edge of the Andean altiplano at 4100m above sea level. This treeless alpine grassland is
characterized by broad valleys separated by flattened ridges.

Animal density in the Cupitay Valley study area averaged 20 vicuna/ km^2. Intensive observations were made on local family group 211 from August 1970 to January 1971 totalling 142 hours. Detailed data was taken on all defecations and urinations and the male's behavior during territorial defense. Because of the openness of the terrain, vicunas could be continuously observed from sunrise, when animals arose in their sleeping territory, to sunset, when they returned to the sleeping territory after having spent the day in their feeding territory.

RESULTS

Evidence suggests that defecation-urination behavior plays a key role in the rigid territorial system and spatial distribution of the vicuna. The following results examine this behavior in a free-ranging wild population.

The Vicuna: Defecation-Urination Behavior and Territorial Marking

The vicuna is one of the few hoofed animals that always uses traditional dung piles, a behavior that has profound environmental and social consequences for the species. When animals defecated or urinated, they used a dung pile. It was most unusual to observe a vicuna not eliminating on a dung pile.

Animals in all vicuna social groups eliminated on dung piles. Members of family groups defecated and urinated on dung piles in their sleeping territory, feeding territory, between the two terri-tories and in unoccupied zones of neutrality between their feeding territory and dry season watering sites. Even crias used dung piles from the time they first started eliminating. Male Groups used dung piles in neutral zones or even within a family group if they happened to be passing through one.

Description of Vicuna Defecation-Urination Behavior

All vicunas showed basically the same preliminary behaviors of smelling a dung pile, forefeet kneading, turning and positioning before defecating or urinating. Pilters (1956) observed the same behaviors in captive vicunas. I distinguished two different types of defecation-urination behaviors in free-ranging wild vicunas: 1) Defecation-Urination (DUs) were done by all animals primarily as routine elimination, and 2) Defecation-Urination Displays (DUDs) were special forms of DUs done by territorial adult males during the defense of territories.

Vicuna Defecation-Urination. Vicuna DU behavior was highly
stereotyped. The first step in DU was smelling, occurring over 90
percent of the time (Table 1). The adult male smelled slightly more
regularly in DU than did others (P < .20). Next came kneading of
the dung pile with the forefeet while standing in place; this
occurred about a quarter of the time. Crias kneaded the most (39
percent, n = 104), adult females intermediate (27 percent , n =
169), and the adult male the least (18 percent, n = 56). Kneading
may have been a means of spreading individual or group scent
throughout the territory. Following the kneading came turning.
Adult females and crias turned about half the time, primarily with
a one-quarter turn. The adult male turned slightly more frequently
than other group members (P < .20).

Table 1. Preliminary behaviors connected with Defecation-Urination
          (DU) and Defecation-Urination Displays (DUD) in family
          group 211.

	Percent Occurrence		
	DU		DUD
	Adult Females and Crias	Adult Male (No obvious Territorial Defense)	Adult Male (During Territorial Defense)
N =	392	64	84
Smelled	92	97	96
Kneaded	28	20	4
Turn 1/4	41	39	42
1/2	7 }49	17 }58	36 }83
3/4	1	2	5
Head Shake	2	6	4
Flehmen	1	13	6

A common DU sequence for adult females and crias was to walk
up to and top immediately over a dung pile, lower the head to smell
the site, make a turn, position the body with the hindquarters
lowered and defecate or urinate. This smell-turn-position-defecate-
urinate sequence was observed 27 percent of the time for adult females and crias (n = 407), and 42 percent for the adult male (n =
64). The closely related smell-position-DU sequence without turning was equally common for females and crias (28 percent), but less
for the male (14 percent).

Once in position to eliminate, vicuna defecated 60 percent of
the time, urinated only 5 percent, defecated and urinated 19 percent
and positioned without eliminating 16 percent (n = 561). In the
feeding territory, females positioned without eliminating one
percent of the time (n = 150) and crias four percent (n = 95).
There was a marked increase in the frequency of animals going
through the preliminary DU behavioral sequence without defecating
or urinating when preparing to leave the feeding territory: 33
percent of the time by females (n = 24) and 64 percent by crias (n =
14). Once the group was outside the territory and moving through
neutral zones, animals went through the preliminary DU sequence
even more frequently without eliminating: 83 percent of the time
(n = 24) for females and 100 percent (n = 16) for crias.

Vicuna Defecation-Urination Display. The form of DUD
was similar to the DU (Table 1). It always occurred, however, during the defense of a territory and the behavior itself differed
from DU by the male in that he kneaded less (P < .01), turned more
often, turned more than a quarter turn (P < .01), and went through
the preliminary DU behaviors more often without eliminating (39
percent vs. 15 percent, n = 79 and 64; P < .01). Male 211 stopped
at a dung pile during 58 percent (n = 196) for all intergroup
aggressive encounters for a DUD: 47 percent were the smell-turn-
position-DU sequence and 27 percent the smell-turn-position without eliminating.

Male 211 'DUDed' primarily (92 percent, n = 102) when defending
his territory against trespassers (i.e., initiator of the aggressive interaction). The other eight percent occurred when the male
had trespassed onto a neighbor's territory; he was (as the recipient of the aggressive interaction) chased out and retreated into
his own territory where he DUDed. Most DUDs occurred at the terminal point of an encounter (47 percent) or during the male's
return toward his group (29 percent). Thus, DUDs tended to be performed towards the outer portion of the territory, but not
necessarily on the border. Pilters (1956) reported that captive
male vicunas conspicuously defecated along their enclosing fence,
behind which other camelids were penned.

The male was laterally oriented to his opponent 56 percent of
the time (n = 80) when he stopped for a DUD during territorial defense. But for male 211, body position and balance appeared to be
more important than orientation since he almost always (90 percent,
n = 39) faced uphill with his hindquarters downhill. During DUDs,
the male defecated 60 percent of the time (n = 87) and positioned
without eliminating 39 percent, twice as frequent as in DUs.

When two neighboring males were disputing a common boundary,
they often approached within a few meters, stopped on dung piles
within their own territory and performed a DUD. Sometimes they
trotted from one dung pile to the next with their heads lowered to
the ground, abruptly stopping over a dung pile, smelling, turning,
and positioning without eliminating. During intensive or continuous territorial defense, the DUD position was often maintained for
prolonged periods atop a dung pile. For example, on 5 March 1970,
when group 214 was being harassed and disrupted by Solo Males and
Male Groups of vicuna, the adjacent newly established territorial
male 228 stood near the common border. Changing dung piles frequently, male 228 positioned in a DUD over the same dung pile for
as long as 28 minutes (DU positioning without eliminating lasted
no more than five seconds). He appeared to be posting himself to
keep marauding males from his own area.

The frequency of DUDs was in relation to the potential threat
or challenge that a social group offered to the defending territorial male. For example, 65 percent (n = 81) of encounters by
male 211 with groups sharing a common border with him
elicited a DUD, 60 percent (n = 15) with unfamiliar marginal territorial family groups moving through his area, but only 33 percent
(n = 15) with familiar non-neighboring family groups that were
being forced away.

Defecation-Urination Displays appeared to reinforce the male's
possession of his territory through visual and scent marking behavi-
or. The function of this behavior is hypothesized to be a display
to the male's opponents and to assist the male in orientation to
his own territorial boundaries. The display also prolongs territorial defense, perhaps affording the recipient additional time for
retreat and reducing the defender's chances for a longer chase or
even direct physical confrontation.

Context and Rate of Vicuna Defecation-Urination

The rate of Defecation-Urinations was greatly influenced by the
socioecological context and site, and appeared to be an indicator of
individual anxiety in stressful situations. Changes in the group's
location, activity, and movement to or across the group's territorial borders caused a marked increase in the rate of DUs. Nearly

all group members commonly used the <u>same</u> dung pile in succession when leaving the sleeping territory or feeding territory, or when moving between the feeding territory and a neutral zone.

The daily cycle of activities and movements started each day when the group awoke in the morning after passing the night in their sleeping territory. Almost immediately after rising, each member walked to a nearby dung pile within a few meters and defecated. Although not observed, I suspected animals also frequented the dung pile during the night. On one occasion group 211 passed the night slightly outside its sleeping territory and upon rising walked 50m back into its sleeping territory before eliminating. When the group was within its sleeping territory during those 15 to 60 minutes after rising in the morning, the DU rate was 2.5 times greater than when they were in the feeding territory (Table 2).

Table 2.   Vicuna Defecation-Urination rate during daylight hours for all animals in group 211 and change in rate relative to feeding territory (FT) by age class and activity. August 1970 to January 1971.

Socio-Ecological Context	DU/Animal/Hour	Rate Relative to FT	
		Adults	Crias
Sleeping Territory	.033	2	4
Leaving Sleeping Territory	.315	22	28
Between Sleeping Territory and Feeding Territory	.017	1	2
Feeding Territory	.013	1	1
Leaving Feeding Territory	.646	46	54
Between Feeding Territory and Zone of Neutrality	.143	10	12
Zone of Neutrality	.423	28	40

As the group started to leave their sleeping territory, members filed up to the same dung pile on the periphery of the territory and DUed at a rate 10 times higher than when in sleeping territory itself and 24 times higher than when in the feeding territory. While moving between their sleeping territory and feeding territory, an undefended corridor and part of their total home range, the DU rate was low and averaged only slightly more than the feeding territory rate.

Once within their feeding territory the mean DU rate was at its lowest and nearly equivalent for both adults and crias. The low DU rate possibly reflected the group's security within familiar and safe terrain. If the group moved to the territory's periphery, such as a watering spring or mutual boundary, the animals commonly defecated on their way. When they did so, it appeared to me that the animals were orienting themselves or checking their location. For example, on 31 October 1970, when groups 211 and 208 came together at their mutual border, most members of both groups stopped and defecated on a dung pile within their own territory just before coming to the border. The two groups then fed 5 to 10m apart, separated by their common boundary. In the latter half of the dry season, groups on ridges and slopes commonly had to make brief movements out of their territories onto the flats and bottomlands to drink from springs and small streams. When group 211 was preparing to leave its feeding territory and cross the flat to water 2km away, DUs were at their highest rate and 50 times more than when in the feeding territory.

Members frequently stopped at dung piles while moving through the narrow neutral corridor towards the neutral zone on the lower flat. The DUs were short, hurried and abbreviated forms of the usual complete DUs. Movement from the feeding territory to a neutral zone was direct and the animals walked fast. If the group strayed outside the neutral corridor and into adjacent territories, they were subject to attack from the territorial owners. I had the impression that the brief and truncated DUs were "checks" or confirmations of location. Since they regularly used the same route between their feeding territory and watering site during this time of year, it was quite possible animals were checking for previously used dung piles with their own or group odor, to insure they were in the safe neutral corridor.

While in the neutral zone animals DUed at the second highest rate compared to other social contexts; crias especially made frequent use of dung piles, most likely confirming their location and as a consequence of being in unfamiliar terrain.

Juvenile DU Behavior

Crias consistently showed a higher DU rate than adults in a variety of different social situations, especially when preparing to leave and when outside their territory (Table 2). Laboratory animals also have higher defecation rates under stressful situations (Bindra and Thompson, 1953). The crias' general level of stress was apparently higher as they prepared for and faced unfamiliar situations out of their home range.

There were two additional situations where a high rate of
Defecation-Urinations among crias, both of which appeared stressful, also occurred. The first situation was when crias temporarily
left their family group to associate with crias in a neighboring
group. They frequently DUed as they left their area or arrived at
the neighbor's territory. The second stressful situation was
during the final weeks before the female cria was forcefully ex-
pelled from the family group, a time when the adult male and females were frequently aggressive toward the cria. Thus, the final
remaining female cria in group 211 showed a sharp increase in the
frequency of defecation-urination in her final weeks before her
expulsion from the group by the adult male (Figure 1).

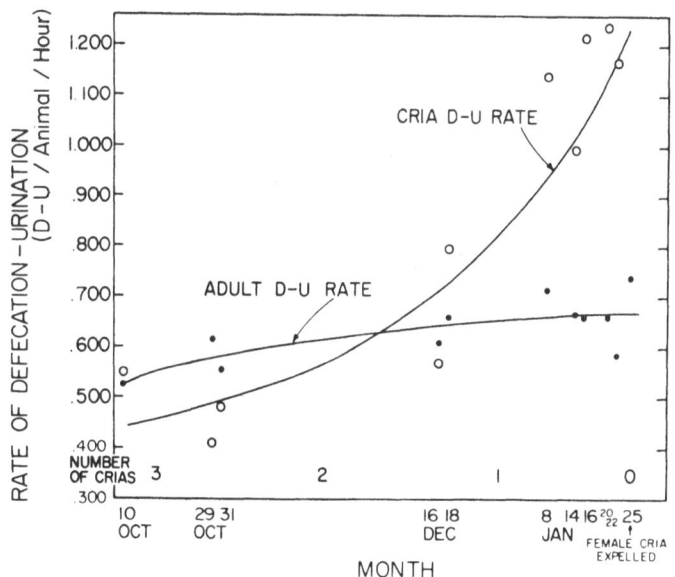

Figure 1.   Comparison between vicuna adult and cria defecation-
            urination rate leading up to expulsion of the female
            cria from group 211. Adult n = 64, cria n = 78.

Distribution, Density and Weight of Vicuna Dung Piles

    All dung piles within 128 ha of the study area were mapped,
counted and weighed before the fall summer rains started in January
1971. Excrement had accumulated since the previous year's rainy
season, so essentially one annual accumulation of fecal material
was measured.

    Distribution of dung piles on the slopes and flats coincided
with the downhill lines of Excrement Influenced Vegetation (Franklin, 1974 and in prep) and even tended to be in undulating rows on
the level bottomland(Figure 2). Few dung piles were found within
the coarse bunch grass community (Ichu = Festuca dolichophylla) on
the upper slopes. Dung piles were found throughout the territory,
and there was no indication that they were more common near
boundaries, although boundaries sometimes followed lines of dung
piles.

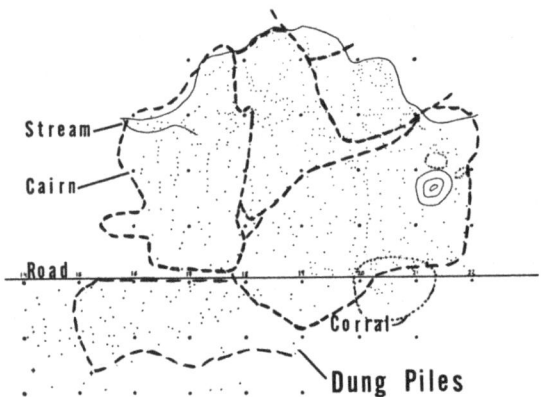

Figure 2. Distribution of vicuna dung piles relative to
           boundaries (dashed lines) of feeding territories.

On the communal sleeping ground where family groups 208, 209, 211, 212, 219, and 227 had their sleeping territories, the density of dung piles was slightly higher (but not significantly) and their mean weights substantially greater (P < .005) than was found on the slope, flat and bottomland where feeding territories were located (Table 3). Total fecal material per unit area was over six times greater within sleeping territories than within feeding territories. This is what one would expect since vicuna were spending about as much time on sleeping territories as on feeding territories and were presumably moving about much less on the sleeping territory. Yet, the mean total amount of accumulated fecal material per average territory was greater in feeding territories than sleeping territories (Table 3), since the average feeding territory was seven times as large as the average sleeping territory.

Table 3. Density and dry matter weight of vicuna dung piles within a 150 ha portion of the study area. Weights represent one year's accumulation of fecal organic material.

Measurement	Sleeping Territories	Feeding Territories
Hectares Sampled	11.2	138.8
Dung Piles Present	75	588
Dung Piles/Ha	6.7	4.3
S.E.	1.48	0.43
Mean Dung Pile Weight in Kg	28.5	6.8
Range	2-80	2-56
S.E.	0.97	0.35
Mean Kg of Fecal Material/Ha	191	29
Mean Kg of Fecal Material Per		
Average Territory	497	534

DISCUSSION

     Whether deposited in a scattered or localized manner, defeca-
tion and urination is perhaps the most common form of scent marking in mammals (see Altmann, 1969; Eisenberg and Kleiman, 1972;
and Leuthold, 1977). For those ungulates that localize defecation-
urination so that there is an accumulation of material on a dung
pile (heap, midden), one (usually the male) or both sexes of the
species may do so. In Leuthold's (1977) review of African ungulate ethology and behavioral ecology, 25 to 30 percent of the
species localize their defecations-urinations. For native western
hemisphere ungulates, localization of feces and urine is most
unusual; besides the South American camelids, only the pronghorn
antelope (<u>Antilocapra americana</u>) has a tendency to do so (Müller-
Schwarze et al., 1973).

     For the vicuna, localized defecation-urination behavior has
both social and environmental significance, the latter of which
is covered elsewhere (Franklin, 1974 and In Press). Vicunas of
both sexes, of all ages, in all social groups eliminated only on
traditional dung piles. Although the function of territorial
defecation has long been assumed to be for territorial defense
(warning outside animals away), there is a lack of evidence to
support this view (Johnson, 1973). Dung piles within vicuna
territories did not prevent outside groups from trespassing onto
and using a site while an owner was temporarily absent. In
Johnson's (1973) review of mammalian scent marking, he reported
growing evidence for a number of functions within the same species
and different functions between species. Johnson (1973, p. 522)
gave the six most frequently proposed functions of scent marks,
the third of which closely supports my observations as one of the
important functions of this behavior for vicuna: "a system of
labelling the habitat for an animal's own use in orientation".

     It was vital that vicuna group members remain within their
own territory or in familiar neutral areas to avoid attack from
other territorial males. Vicuna communal dung piles were large
and conspicuous throughout areas of usable habitat. Existing
dung piles were used within newly established territories; unused
sites were uncommon, and new ones were rare. Family group members
appeared to use dung piles for their own orientation as an aid to
remaining within safe and familiar areas (Franklin, 1971 and 1974).
When a group left a safe area and trespassed onto another territory, it was the adult female members who were attacked and
chased by the resident male. The rate of dung pile use (smelling,
turning, positioning, etc.) greatly increased when animals
approached their territorial boundaries, left the territory or
moved through neutral areas: situations in which animals needed
frequently to confirm their location. The greater stress of

crias leaving, or being outside familiar terrain was reflected by even higher rates of defecation-urination in these situations.

My observations on vicuna male Defecation-Urination Displays agree with Rall's (1971, p. 449) statement that mammals appear to "mark frequently in a situation where they are both intolerant of and dominant to other members of the same species". Key characteristics of the DUD behavior that occurred during the defense of a territory by the vicuna male were 1) 92% of DUDs observed were by the individual that initiated and dominated the aggressive encounter, 2) about half of the time the male was laterally oriented to his opponent during DUD, 3) there was less kneading behavior with the forefeet, he turned more, and went through the preliminary DU behaviors without eliminating compared to a regular defecation-urination when not defending his territory, 4) the male tended to perform the DUDs towards the outer portion of his territory, but not necessarily on the border, and 5) DUDs were done by the male more frequently if the trespassing social group offered a potential threat or challenge to the male, i.e., familiar family groups sharing a common border, unfamiliar family groups not sharing a common border, vs familiar family groups not sharing a common border. The regular occurrence of exaggerated defecation-urination behaviors in conjunction with the male vicuna's territorial defense and aggressiveness to rivals suggested it was a form of threat display. The absence of any elimination 40 percent of the time supports this hypothesis.

Thus, use of dung piles by vicunas was an important social behavior. Elimination on dung piles by both sexes of all age classes was especially noteworthy. The male's regular use of dung piles during the defense of his territory appeared to reinforce his possession of his territory through visual and scent marking behavior. Yet, as so often assumed, dung piles as territorial markers did not "keep outsiders out", but were suspected to be functioning more for "keeping insiders in" the territory through individual spatial orientation.

LITERATURE CITED

Altmann, D. 1969. Harnen und Koten bei Säugetieren. Neue Brehm-Bücherei No. 404. Wittenberg-Lutherstadt: Ziemsen.

Bindra, D. and W.R. Thompson. 1953. An evaluation of defecation and urination as measures of fearfulness. J. Comp. Physiol. Psychol. 461: 43-45.

Eisenberg, J.F. and D.G. Kleiman. 1972. Olfactory communication in mammals. Ann. Rev. Ecol. and Syst. 3: 1-32.

Franklin, W.L. 1971. Vicuna survey, Peru. Pp. 145-148. In P. Jackson (Ed.). World Wildlife Yearbook 1970-71. World Wildlife Fund, Morges. 300 pp.

Franklin, W.L.   1973a. High, wild world of the vicuna.   Natl.
     Geogr. 143: 76-91.
Franklin, W.L.   1973b. To save the vicuna.   Americas 25: 2-10.
Franklin, W.L.   1974.   The social behavior of the vicuna.   Pp. 477-
     487.   In V. Geist and F. Walther (Eds.) The behaviour of
     ungulates and its relation to management.   I.U.C.N., Morges.
     940 pp.
Franklin, W.L.   1975.   Guanacos in Peru.   Oryx 13: 191-202.
Franklin, W.L.   1978.   The socioecology of the vicuna.   Ph.D.
     Thesis, Utah State University, Logan, Utah.   167 pp.
Johnson, R.P.   1973.   Scent marking in mammals.   Anim. Behav. 21:
     521-535.
Koford, C.B.   1957.   The vicuna and puna.   Ecol. Monogr. 27: 153-
     219.
Leuthold, W.   1977.   African ungulates: a comparative review of
     their ethology and behavioral ecology.   Springer-Verlag, New
     York.   307 pp.
Müller-Schwarze, D., C. Müller-Schwarze, and W.L. Franklin.   1973.
     Factors influencing scent-marking in the pronghorn (Antilo-
     capra americana).   Verhandlungen der Deutschen Zoologischen
     Gesellschaft 66: 146-150.
Pilters, H.   1956.   Das Verhalten der Tylopoden.   Handbuch der
     Zoolgie 8(10) No. 27, 1-24.
Ralls, K.   1971.   Mammalian scent marking.   Science 171: 443-449.

INDUCTION OF SETTLING AND METAMORPHOSIS OF PLANKTONIC MOLLUSCAN (HALIOTIS) LARVAE. III: SIGNALING BY METABOLITES OF INTACT ALGAE IS DEPENDENT ON CONTACT

Daniel E. Morse[1], Mia Tegner[2], Helen Duncan[1], Neal Hooker[1], George Trevelyan[1], and Andrew Cameron[4]

[1]Marine Science Institute and Department of Biological Sciences, University of California, Santa Barbara, CA 93106

[2]Ocean Research Division, Scripps Institution of Oceanography, La Jolla, CA 92093

[3]Center for Coastal Marine Studies, University of California, Santa Cruz, CA 95064

INTRODUCTION

Classical models of chemical signaling mechanisms which govern various aspects of the behavior, reproduction and development of animals in aqueous environments have been based largely upon well-characterized systems of chemoreception, chemotaxis and chemosensory behavior in response to concentration gradients of freely diffusible chemical inducers in aqueous or gaseous media (see, for example, Grant and Mackie, 1974; Shorey, 1976). In contrast, we recently have identified a mechanism of chemical signaling and induction governing the behavioral and developmental metamorphosis of certain planktonic molluscan larvae which is mediated, at least in part, by contact-dependent recognition of covalently complexed (or sequestered) non-diffusing algal metabolites.

We have shown previously that competent planktonic (drifting and swimming) larvae of the red abalone (Haliotis rufescens) are induced to settle from the water and begin genetically programmed metamorphosis to the benthic (bottom-dwelling) juvenile gastropod form by specific crustose red algae, and by a group of simple and homologous chemical inducers contained within these algae (Morse et al., 1979a, 1979b; Morse, 1978a, 1978b, 1978c, 1979). We present evidence here to demonstrate that induction of the settling response ("recruitment") of the planktonic larvae is mediated by a contact-dependent recognition process, rather than by chemotaxis. These results thus are congruent with our biochemical findings (Morse et al., 1979a), indicating that the most potent chemical inducers yet identified within the active algae (γ-aminobutyric acid and/or its analogs, including the algal pigment, phycoerythrobilin) are present almost exclusively in covalently complexed or sequestered form.

MATERIALS AND METHODS

Swimming, planktonic larvae of H. rufescens (ca. 0.28 mm in length) are produced in the laboratory by fertilization of gametes obtained by the peroxide induction of spawning in gravid adults, as described previously (Morse et al., 1977, 1978). Larvae produced by this technique, when cultivated at 15°C in gently flowing, filtered, fresh seawater, are uniformly healthy and become fully competent for induction of settling, rapid metamorphosis and benthic development 7-8 days following fertilization (Morse et al., 1979a, 1979b; Morse, 1979).

Settling is quantitated in aliquots of 200-500 competent larvae tested in clean glass dishes (10 cm diameter, 250 ml filtered seawater, 15° ± 1°C), following addition of potential inducers as indicated in the text (Morse et al., 1979a). Settling is scored as the percentage of animals showing firm plantigrade attachment to the indicated substrate, as determined by microscopic examination.

RESULTS

Observations in the Native Habitat

In SCUBA-reconaissance surveys off the coast of San Diego and Santa Barbara, CA, it was observed that small rocks encrusted with patches of coralline red algae frequently served as nursery grounds for juveniles of H. rufescens and other Haliotis species at depths ranging from 2-20 m. A sample of 24 such rocks (8-15 cm in longest dimension, silt-free, and encrusted with patches of coralline red algae), collected from a depth of 15 m off Point Loma, San Diego, contained 33 juvenile H. rufescens (1-22 mm in longest dimension), adhering primarily to the algae-encrusted undersides of the rocks. This sample represents a density of approximately 224 juvenile H. rufescens per m^2 of bottom surface area of the nursery rocks collected (Tegner, in preparation). Similar densities were found in comparable samples collected off the coast of Santa Barbara.

Other observations from Point Loma, San Diego, confirmed the finding that areas of rock cobble encrusted with red algae serve as nursery grounds for juvenile H. rufescens, and that one such habitat continued to recruit new juveniles over a period of one year (June 1977 - June 1978). As reported previously (Tegner and Dayton, 1977), many of the juvenile abalones were found under the spine canopies of adult urchins (Stronglyocentrotus franciscanus and S. purpuratus). As many as 10 juvenile abalones were observed under the spines of one S. franciscanus (Tegner, in preparation). The crustose red algal habitat also was observed to serve as a nursery for juvenile S. purpuratus.

Experimental Demonstration of Biological Specificity of the Settling (Recruitment) and Metamorphosis Responses

The potential significance of the observed association of juvenile Haliotis species with various crustose red algae, and with other algal and animal species found in the same habitat, was investigated experimentally (Table 1). Aliquots of competent larvae of H. rufescens exposed to these algae, a variety of micro-epiphytes and small invertebrate animals (from the recruiting habitat) exhibit rapid and substrate-specific settlement primarily on species of crustose red algae including Lithothamnium spp., Lithophyllum spp., and Hildenbrandia spp. (Table 1, Figures 1 and 2) (see also Morse et al., 1979a, 1979b; Morse, 1979). Lithothamnium and Lithophyllum are genera of calcified crustose red algae, whereas Hildenbrandia represents a non-calcified genus. Bossiella, an erect and branching genus of calcified red algae, shows significantly less recruitment than the crustose species even when presented to the larvae in the horizontal plane on the bottom of the test vessel.

The preferred foods of adult H. rufescens, including the giant kelp Macrocystis pyrifera (a foliose brown alga), Egregia laevigata (foliose brown) and Ulva sp. (foliose green) all fail to induce settlement when tested either alone or in combination with attached and feeding juvenile abalones (H. rufescens) or urchins (Stronglyo-centrotus franciscanus) with which the newly settled abalones often are associated in the field. None of the other epiphytes or animal cohabitants tested displayed any activity in inducing settling. These data extend our earlier findings (Morse et al., 1979a, Morse, 1978, 1979) indicating a high degree of substrate-specificity in recruitment of planktonic Haliotis larvae to the crustose red algae.

Table 1. Specificity of induction[a].

Inducer	Settled (%)
None	0
Lithothamnium spp.	93
Lithophyllum spp.	96
Hildenbrandia dawsonii	87
Bossiella californica	8
Rock, ceramic tile and glass:	
Clean	≤ 1
Colonized with diatoms, bacteria, and other microflora	≤ 1
Diatoms (on glass):	
Cyclotella nana	≤ 1
Nitzschia closterium menutissima	≤ 1
Cocconeis sp.	≤ 1
Tetraselmis sp.	≤ 1
Adult food algae ($\pm$ feeding abalones and urchins):	
Macrocystis pyrifera	≤ 1
Egregia laevigata	≤ 1
Ulva sp.	≤ 1
Animal cohabitants:	
Haliotis rufescens (N = 3, 1 cm)	0
Strongylocentrotus franciscanus (N = 1, 5 cm)	0
Spirorbis sp. (N $\geq$ 20, ca. 1 mm)	0
Bryozoan sp. (N $\gtrsim$ 100, ca. 0.5 mm)	0

[a]Table 1. Biological specificity of the induction of settling of planktonic H. rufescens larvae. Competent larvae (400 $\pm$ 100 per sample) were tested for responsiveness to potential inducers of settling as described in Materials and Methods; plantigrade attachment was quantitated 24 hr after presentation of potential inducers. Algae and bacteria each were tested at 0.2–1.0 g (wet weight)/sample; Lithothamnium, Lithophyllum, Hildenbrandia, Bossiella, Spirorbis and Bryozoa were presented as adherent to rock surfaces. Unfractionated bacteria, diatoms and microalgae are those cultured from local seawater; specific diatoms are those isolated from the natural habitat of juvenile H. rufescens. Mortality of larvae 1% in all samples.

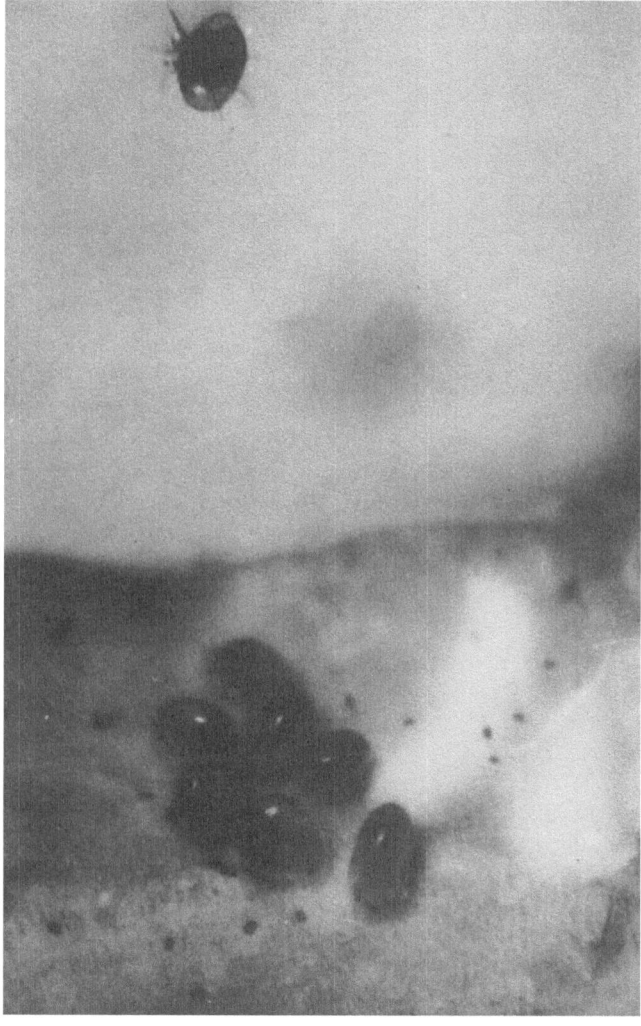

Figure 1. A swimming, planktonic larvae of H. rufescens (above)
comes in for a landing among newly settled siblings,
minutes after the introduction of a crustose coralline
red alga. The settled larvae have begun active feeding
on the red algal surface. Larvae are ca. 0.28 mm in
diameter. (Photograph by L. Friesen.)

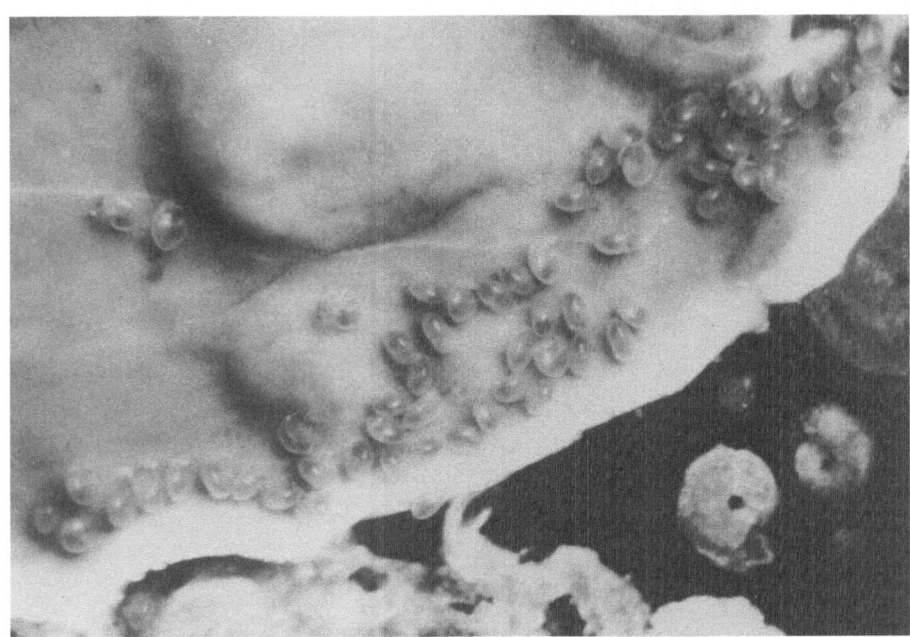

Figure 2.   Larvae of H. rufescens have settled densely and commenced
            feeding minutes after introduction of the crustose corra-
            line red alga, Lithophyllum sp. Note that settling occurs
            specifically on the red algal surface; adjacent, non-algal
            surfaces are uncolonized. H. refescens larvae are ca. 0.28
            mm in diameter; the two larger coiled shells in the lower
            right are those of the polychaete, Spirobis sp. (Photograph
            by L. Friesen.)

Following settling of H. rufescens larvae on the recruiting
red algae, firm plantigrade attachment and characteristically snail-
like gliding locomotion and active feeding quickly ensue (Figures 1
and 2). Morphogenetic changes occur rapidly and fairly synchronously
in the settled population over the next 48 hours, with loss of the
cilia and velar lobes (larval swimming appendages), yolk-granule
translocation associated with maturation of the digestive gland,
progressive development of opacity in the shell gland, and development and rapid growth of the flaring peristomal shell characteristic
of the juvenile and adult (Figure 3). Active feeding with radular
teeth supports rapid growth and further maturation on the red algal
substrate, with beating of the developing heart detectable within
3-4 days after settlement. The rapid course and success of this
feeding behavior and sequence of developmental changes indicates
that the recruiting algae and/or associated epiphytes provide adequate nutrition and other inducers which may be required for early
development and rapid growth. In contrast, uninduced sibling larvae
incubated under otherwise identical conditions fail to show synchronous, rapid or significant settling, development, or growth; most
eventually succumb to microbial overgrowth while developmentally
arrested or retarded (Morse et al., 1979b; Morse, 1978, 1979). Algal
inducers thus are **required** for normal behavioral and developmental
metamorphosis.

Induction of Settling by Extracted Algal Metabolites

The addition of small portions of cell-free extract prepared
from the recruiting crustose coralline red algae (consisting primarily of Lithothamnium spp. and Lithophyllum spp.) is sufficient
to induce settling of competent H. rufescens larvae on clean glass
(Table 2; see also Morse et al., 1979a). An initial, temporary
paralytic response was observed with all concentrations of the
extract tested here; this was followed, over a period of several
hours, by a resumption of swimming or benthic locomotion. Final
settling (plantigrade attachment and benthic locomotion) showed a
dose - and time - dependent response to algal extract up to a
maximum of 100% at 10-15 µg algal protein per ml; additions above
this concentration were less effective, and apparently partially
toxic. Addition of dilute γ-aminobutyric acid (GABA, 10^{-6}M), a
potent inducer of settling previously identified in extracts of
the inducing algae (Morse et al., 1979a), induced significant
settling in aliquots of these larvae over the period tested. There
was no settling in the absence of added inducers. The inclusion
of antibacterial penicillin and streptomycin in this experiment may
have been instrumental in preventing the mortality previously
observed in larvae exposed to high concentrations of crude algal
homogenate (cf. Morse et al., 1979a).

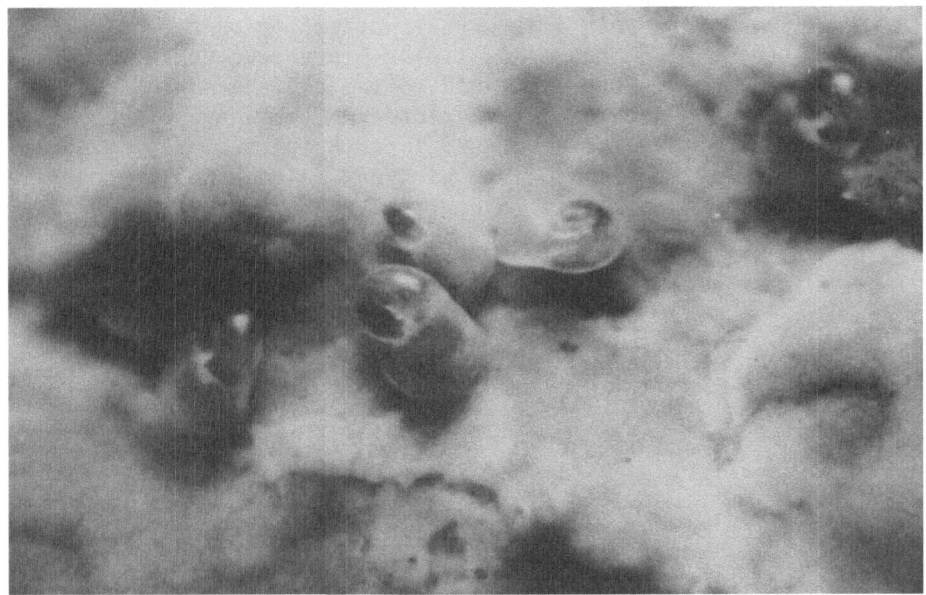

Figure 3. Metamorphosed H. rufescens continue feeding, growth and
          development 2 weeks after settling on Lithophyllum sp.
          Development of the flared shell characteristic of the
          juvenile abalone is conspicuously apparent as an outgrowth
          from the original, larval shell. The eyes are visible
          through the shell as these animals graze the algal surface.
          Longest dimension of these post-larvae is ca. 0.60 mm.

Table 2. Induction of settling by extracted algal metabolites[a].

Inducer (µg/ml)	Settled (%)	
	14 hr	38 hr
None	1 ± 1	1 ± 1
Algal Extract, (1)	8 ± 4	7 ± 2
Algal Extract, (5)	59 ± 2	82 ± 8
Algal Extract, (10)	86 ± 2	100 ± 0
Algal Extract, (15)	21 ± 4	100 ± 0
Algal Extract, (20)	19 ± 2	76 ± 1
GABA, 10^{-6}M	59 ± 17	26 --

[a]Table 2. Induction of settling by extracted algal metabolites. Settling of competent H. rufescens larvae was determined as described in Table 1 and Materials and Methods, with the following exceptions: Aliquots of larvae (500 ± 100) were tested, in duplicate, in vials (2.2 cm diam.) containing 10 ml filtered seawater with penicillin G and dihydrostreptomycin (150 ppm each). The algal extract was prepared by scraping crustose coralline red algae (consisting principally of Lithothomnium spp. and Lithphyllum spp.) from rocks, and homogenizing the algal tissue in approximately 2 volumes chilled filtered seawater, using a ground-glass homogenizer at 0°C. Particulate material was removed from the extract by centrifugation (15,000 x g, 10 min at 2°C); protein concentration in the extract was then determined by a modification of the method of Lowry, et al. (1951). Settling was measured 14 and 38 hr after presentation of the inducers as indicated; there was no mortality of larvae in any of the experimental samples. Each value represents the mean ($\pm$ deviation) obtained in duplicate determinations.

Dependence of Recruitment Upon Algal Metabolites

We have identified a family of simple chemical inducers of
settling and metamorphosis from crude extracts of the inducing red
algae (Morse et al., 1979a; Morse, 1978a, 1978b, 1979). The action
of the most potent of these - all structurally related to the neurotransmitter, γ-aminobutyric acid - appears sufficient to account
for the inducing activities of the recruiting algae (Morse et al.,
1979a).

However, to determine whether other sensory modalities of the
larvae may be involved in their discrimination of, and substrate-
specific recruitment to, the inducing algae, analyses were performed
as a function of the presence or absence of light and of algal
modification (Table 3). As seen in these data, the initial rate,
final yield and specificity of settling on the inducing algae are
not affected significantly by the absence or presence of light.
This indicates that photic discrimination (via visual or other
systems) is not significantly involved in the recognition of the
recruiting algae by the competent larvae.

Denaturation of the algae (by boiling; drying; freeze-drying;
or osmotic shock) followed by gentle washing, depletes the algae
of active inducing substances or textures (Table 3). Efficacies of
intact and denatured algae were compared in the absence of light,
because the oxidation of algal pigments (especially pronounced after
boiling) converts the red phycoerythrobilin moiety to a green structural isomer. The exclusion of light, and of photic discrimination
in the inducer-recognition process, excludes the possibility that it
is this color change per se which affects recruitment.

In a separate experiment, it was observed that exogenously provided GABA (10^{-3}M) was capable of inducing settling in 100% of the
larvae tested in the presence or absence of rocks bearing each form
of denatured alga investigated. This latter result demonstrates
that the reduction in settling observed in the presence of the
denatured algae alone (Table 3) is not the result of inhibitors of
settling which might have been produced by denaturation.

Induction of Settling Normally Requires Direct Contact with
Recruiting Algae

Several observations indicate that the induction of settling
requires direct contact with the naturally recruiting algae. First,
we observe no chemotaxis or oriented swimming of the larvae to
these algae. The behavior of the competent   larvae constitutes a

Table 3. Dependence upon algal metabolites[a].

Inducer	Settled (%)			
	Dark		Light	
	2 hr	36 hr	2 hr	36 hr
Diatoms	0 ± 0	$\leq 1 \pm 0$	0 ± 0	0 ± 0
Crustose Red Algae	21 ± 10	81 ± 10	26 ± 7	95 ± 3
Crustose Red Algae - Dried	2 ± 2	-	2 ± 1	-
Crustose Red Algae - Boiled	7 ± 0	-	7 ± 3	-
Crustose Red Algae - Osmotic Shock	1 ± 0	-	5 ± 3	-
Crustose Red Algae - Lyophilized	0 ± 0	-	1 ± 0	-

[a]Table 3. Dependence of recruitment upon algal metabolites. Settling of competent H. rufescens larvae was determined as described in Table 1 and Materials and Methods, with the following exceptions: Aliquots of larvae tested in parallel included 200 ± 50 larvae; crustose red algae (including Lithothamnium, Lithophyllum and Hildenbrandia spp.) and diatoms were presented to larvae in each sample on a single rock substrate of standardized size and shape. As indicated, algae were inactivated by drying (18°C, in air, 4 days); boiling (in 2 changes of deionized water, 15 min each); osmotic shock (exposure to deionized freshwater at 15°C for 3 days); or lyophilization. Samples were incubated either in the total absence of light or exposed to ambient lighting (indirect daylight and overhead fluorescent). Larvae settled and firmly attached were quantitated after transfer of each rock substrate from its test vessel at the time indicated; this transfer admitted no light to those samples in which settling was tested in the dark. Results are the averages of duplicate determinations, with the exception of the 36-hr samples, which were quantitated in quadruplicate.

pattern of random but efficient search for suitable (inducing) algal substrates; this behavior consists of alternate, negatively geotropic upward swimming, followed by cessation of swimming, passive sinking to the bottom, and resumption of upward swimming shortly afterward if inducer is not contacted. We find no evidence for directed perturbation of this swimming and sinking behavioral pattern in the presence or close vicinity of inducing algae, suggesting that the algae release little or no freely diffusible inducer into the water. Larvae which come in direct contact with the recruiting algal substrate first exhibit localized, exploratory swimming and repeated collisional contact with the substrate; sampling by exploratory feeding is possible.

The obligate requirement for direct algal contact and the absence of significant release of soluble and freely diffusible inducers from the recruiting algae are demonstrated directly by the data in Table 4. Thus, prolonged soaking of inducing algae in seawater fails to deplete the algae of their recruiting activity, and fails to extract detectable inducers of settling into the seawater supernatant. Similarly, no active inducer is released from the intact algae which is capable of freely diffusing through a semipermeable dialysis membrane (although GABA placed in such a membrane does diffuse out and induce settling of larvae outside the dialysis bag). Controls shown in Table 4 indicate that neither dialysis membrane alone nor membrane enclosing crustose red algae on rock inhibit the settling of larvae induced by exogenously added GABA. [Such controls do indicate, however, that supernatants from the algae may contain such inhibitors, which would appear to be macromolecular. The identity and significance of any such inhibitor remain to be determined, as there is no apparent interference with settling induced either by the intact algae (Table 1, 4 and 5) or by extracts of these algae (Table 2).]

Data in Table 5 extend our findings indicating that there are no significant levels of freely soluble, diffusible inducers of settling released by the recruiting algae. In these data we see that the proximity of the crustose red algae does not lead to the induction of "heterologous settling" on adjacent non-algal surfaces; in contrast, the addition of exogenous soluble inducer (GABA) does lead to heterologous or non-substrate-specific settling. Thus, the intact algae do not release inducing levels of GABA or similar compounds into the water.

The absence of heterologous settling on adjacent non-algal surfaces is seen directly in Figure 2, illustrating the high fidelity of substrate specificity demonstrated in the recruitment of Haliotis larvae.

Table 4. No detectable release of diffusible inducers[a].

Addition	Settling (%)	
	0	GABA
None	11	100
Crustose Red Algae	98	100
Crustose Red Algae - Washed	99	100
Algal Supernatant	9	47
Dialysis Tubing	8	99
Crustose Red Algae in Dialysis Tubing	7	100

[a]Table 4. Absence of detectable, diffusible inducers released from the recruiting algae is demonstrated in assays as described in Table 1 and Materials and Methods, with the following modifications: Crustose red algae (Lithothamnium and Lithophyllum spp.) adherent to rock were used either directly, after soaking in 250 ml seawater (at 15°C, for 24 hr), or sealed in a semipermeable membrane dialysis tubing. The supernatant from the soaked algae was used directly. Samples were evaluated in paired duplicates, in the presence and absence of added GABA (10^{-3} M). The percentage of larvae settled upon all surfaces was scored after 24 hr.

Table 5. Algae do not induce heterologous settling[a].

Addition	Relative Distribution (%)		
	Coralline Red	Diatoms	Glass
0	$\geq$99	0	0
GABA	32	41	27

[a]Table 5. The recruiting algae do not normally induce heterologous settling on adjacent, non-algal surfaces. Settling was quantitated as described in Table 3, except that only two dishes were used: each contained adjacent rocks coated with crustose red algae and with diatoms; GABA (10^{-3} M) was added to one of the dishes. The relative distribution of larvae settled on comparable areas of crustose red algae, diatoms or glass substrate was quantitated after 24 hr. Absolute numbers of larvae settled (total) in the two dishes were comparable.

DISCUSSION

Substrate-specific Recruitment of Planktonic Larvae by Algal Metabolites

It has been appreciated for some time that the control of settling and metamorphosis of planktonically dispersed larvae of benthic marine species is in many cases dependent upon substrate-specific recruitment, although the precise chemical and/or other sensory basis for this recognition and control has in most cases only partially been defined (Crisp, 1974; Chia and Rice, 1978). Numerous benthic invertebrates have been found to be recruited to various crustose red algae, including among the recruited species: the polychaete, Spirorbis rupestris (Gee, 1965); the chitons, Tonicella lineatta (Barnes and Gonor, 1973), and Mopalia muscosa (Morse et al., 1979a); and the starfish, Stichaster australis (Barker, 1977). Crofts (1929) had observed that juveniles of the European abalone H. tuberculata were cryptically associated with camouflaging crustose coralline red algae, although the biological significance of this association (beyond that of protection conferred by the camouflaging coloration) was then not recognized or explored.

The results reported here and shown in Table 1, and those reported previously (Morse et al., 1979a, 1979b; Morse, 1978a, 1978b, 1978c, 1979), confirm the significance of the association of juvenile H. rufescens with the crustose red algae observed in the natural coastal environment. These results identify several genera of crustose red algae as components of the complex natural community which are sufficient in themselves to provide the required inducers of settling (recruitment) and metamorphosis of the planktonically dispersed larval abalones. The presence of other algae or (feeding) juvenile abalones or urchins (S. franciscanus, with which juvenile abalones are associated in their natural habitat), was not required for, nor did it enhance, recruitment. The data in Table 1 extend those previously reported (Morse et al., 1979a), in showing that efficient induction of settling of larval H. rufescens is not confined to the coralline genera of crustose red algae, but is demonstrated here by the non-coralline genus, Hildenbrandia, as well.

The data in Tables 2 and 3 demonstrate that the activity of non-denatured metabolites of the recruiting algae is sufficient to induce behavioral metamorphosis, and thus account for the recruitment of competent H. rufescens larvae by the intact algae. Chemical recognition alone appears sufficient for rapid and specific induction of settling by the intact algae or its inducing metabolites;

there is no requirement for visual cueing or any other light-dependent
process (Table 3). Simple and defined algal inducers, including
GABA and its homologs (Morse et al., 1979a; Morse, 1979), are capable
themselves of inducing the complete behavioral (Tables 2 and 3) and
developmental (Morse et al., 1979a, 1979b) metamorphosis associated
with this recruitment and the normal transition of planktonic H.
rufescens larvae to the benthic juveniles. These findings represent
the first in which the chemical basis sufficient for the substrate-
specific recruitment and metamorphosis of any planktonically dispersed molluscan larvae has been defined.

Contact-dependent Induction of Behavioral and Developmental Metamorphosis by the Intact Algae

What is further significant (and perhaps unanticipated) in
these findings, with respect to the elucidation of chemical signaling
mechanisms controlling reproduction, planktonic dispersion, recruitment and larval metamorphosis, is the demonstration that this
signaling process is mediated not primarily by gradient-dependent
chemotaxis, but by contact-dependent recognition (possibly even
dependent upon exploratory ingestion). (A possible combination of
extremely short-range chemotaxis with obligate contact-dependent
induction of settling is not excluded by our observations, however.)

Contact-dependence of recognition and recruitment is established
by: (a) the absence of any detectable chemotaxis or oriented swimming of competent larvae toward the inducing algae; (b) the absence
of any released, soluble or diffusible inducer detectable in the
dialysate or supernatant of the intact, recruiting algae (Table 4);
and (c) the absence of any "heterologous settling" on adjacent,
non-algal surfaces (as induced by the presence of exogenously
provided, freely diffusible inducer; cf. Table 5 and Figure 2).

The recruiting algae thus appear to "trap" Haliotis larvae by
a contact-dependent recognition and induction process; the larvae
are not attracted to the algae from a distance, but must encounter
the inducing substrate in their random search and drift. With their
repeated, alternate upward swimming and sinking behavior, the competent veliger larvae effectively and repeatedly "sample" the bottom,
as they are carried in the water current, until suitable (inducing)
substrates are found. This mechanism appears to ensure a highly
specific, fail-safe protection against induction of settling and
irreversible benthic commitment (through loss of the swimming cilia,
an early event in metamorphosis) in sub-optimal (i.e., non-algal)
microhabitats. Such lethal "errors" in settling ("heterologous
settling") might be expected to be frequent, if settling of the

weakly swimming larvae were triggered by the release of diffusible inducers from the algae into water subject to any significant turbulence or flow.

Mutual Advantages and Selection

Substrate-specific recruitment, or "trapping", of Haliotis larvae on the crustose red algae appears to confer selective advantages to both the plant and animal partners in what appears to be a mutualistic relationship between inter-dependent species. The larvae are provided with strictly required, stereo-chemically specific inducers of behavioral and developmental metamorphosis, suitable microhabitats, nutrition adequate to support early growth and development, and a source of protective coloration ultimately derived from algal pigments. Prolonged grazing by post-larval or juvenile H. rufescens confers little or no detectable damage to the supporting red algae, suggesting that principal sources of nutrients for the abalone may include the carbohydrate-rich mucus exudate regularly sloughed from the epithelia of certain of the crustose red algae (cf. Giraud and Cabioch, 1976), and/or the diatoms (Navicula spp. and others), bacteria and other epiphytes which we have observed on these algae with scanning electron microscopy. The inducing algae appear to benefit reciprocally from this relationship, possibly depending upon the shallow and non-destructive grazing by the recruited herbivores, which continually free the algal surface from fouling epiphytes which otherwise can overgrow and kill the algae (Adey, 1973). The selective and coevolutionary perfection of this relationship, cemented by the algal synthesis and deployment of animal neurotransmitter (GABA) homologs and conjugates capable of inducing the settlement and metamorphosis of planktonically dispersed larval herbivores, thus can be understood in terms of the enhanced differential reproductive advantage to both the inducing algal host and the recruited animals (Morse et al., 1979a).

Several species of herbivorous animals (including limpets, chitons and urchins, as well as several species of abalones) are found in close association with the crustose red algae. Survival of these algae may well depend upon continuous grazing by members of this entire group of animals. Of these herbivores, in addition to results reported here with H. rufescens, we have found that planktonic larvae of the chiton, Mopalia muscosa, and of the green abalone (H. fulgens) also are induced to settle and undergo metamorphosis by crustose red algae and by GABA (Morse et al., 1979a, and in preparation).

Ecological Implications

These results appear to explain the commonly observed association of juvenile H. rufescens (and other species of Haliotis) with
certain crustose red algae in California coastal waters. Nevertheless, many factors, both biological and physical in nature, undoubtedly are involved in the successful settlement, metamorphosis
and survival of young abalones in the benthic habitat; such complex
interactions probably affect the isolated interactions characterized
here. The distribution of small predators (including annelids and
nematodes) cryptically associated with the recruiting algal habitat
is very important in determining the final success of juvenile H.
rufescens survival, after induction of settling and metamorphosis
on the algal surfaces (Morse et al., 1979b). Sea urchins are not
by themselves required for the recruitment of competent H. rufescens
larvae (Table 1), but the frequent occurrence of juvenile abalones
under the spine canopies of the two local species of Strongylocentro-
tus suggests that this association confers certain advantages on the
abalones. It has been shown (Tegner and Dayton, 1977) that juvenile
urchins living under the spine canopies of adult urchins receive
protection from predators and a share of the adult's macro-algal food.
In addition, the adult urchins apparently offer a third advantage to
abalones, namely the provision of substrate suitable for settlement
and metamorphosis. Grazing by urchins maintains patches of encrusting coralline algae free from overgrowth by other organisms (reviewed by Lawrence, 1975; Tegner, in preparation). Thus the urchins
indirectly provide substrate for abalone settlement, and offer
shelter and food during the critical period when the post-larvae
begin to grow. Similarly, the occurrence of juvenile abalones on
cobble-sized ("nursery") rocks may be explained not only by the
presence of the inducing crustose red algae, but by habitat dimen-
sions affording the post-larvae protection from predation (Tegner,
in preparation).

Correlation with Biochemical Findings

We have shown that extracts of the recruiting algae which retain activity in inducing Haliotis larval settling (Table 2) contain
several potent inducers of settlement and metamorphosis, the activities of which are enhanced upon proteolytic release from algal proteins (Morse et al., 1979a; Morse, 1979). We have identified a
variety of simple chemical inducers from the active algal extracts,
most potent and abundant of which are the animal neurotransmitter
substance GABA, and certain of its precursors, metabolites and
homologs, including the red photosynthetic accessory pigment
phycoerythrobilin (Morse et al., 1979a, 1979b; Morse, 1978a-c, 1979).
GABA and phycoerythrobilin are the most potent (50% maximal activity
in settling and developmental metamorphosis induced at $\leq 10^{-6}$ M in

free solution) and most abundant of the inducers yet found in the
active algal extracts (both ca. 10^{-2} M total intracellular concentration, if free and uniformly distributed). Both GABA and
phycoerthrobilin are present virtually exclusively as covalent
conjugates of algal proteins, and not released in diffusible form
from the intact inducing or recruiting algae (Morse et al., 1979a).

The biological observations reported here - which demonstrate
that recruitment is contact-dependent, and not mediated by diffusible
inducers - thus are congruent with our biochemical findings which
indicate that the induction of behavioral and developmental metamorphosis by the active algae can be accounted for by their content
of covalently complexed (and/or internally sequestered) potent
inducing molecules, which are not released from the algae in diffusible form. The contact-dependent induction of settling ("trapping") and metamorphosis of Haliotis larvae by the algae, with its
inherent fail-safe protection against induction in suboptimal, non-
algal habitats, thus can be explained by the synthesis and deployment of non-diffusible macromolecular conjugates of GABA and its
homologs in the recruiting algal tissues.

Although the foregoing biochemical explanation is sufficient
to explain the recruiting and metamorphosis-inducing activities of
the active algae, the possible roles of other inducing substances
or discriminants (e.g., texture) in these and other algae are not,
of course, excluded, and remain to be further elucidated.

The mechanisms by which the covalently complexed algal inducers are contacted, incorporated, processed or otherwise recognized by the planktonic molluscan larvae, and the means by which
these signaling events are transduced to activate behavioral and
developmental metamorphosis in these animals, are the objects of
our current investigations. Our preliminary analysis of the cephalic
or apical sensory organ of larval H. rufescens suggests that this
structure may be especially adapted for contact-dependent recognition of inducing signal from the recruiting algae, possibly by
some chemosensory mechanism analogous to taste or odor perception
(Morse et al., in preparation).

In Conclusion

These findings extend to the aquatic environment, and the control of marine species reproduction, planktonic dispersion, and
developmental metamorphosis, insights into the mechanisms of biochemical coevolution, signaling and genetic interactions between
plant and animal mutualists as are now being elucidated for
terrestrial species (see, for example, Harborne, 1978). Indeed,
it had been the high mortality of Haliotis larvae cultivated under
artificial conditions which first led us to recognize the existence

of a requirement for obligate and normally available inducer(s) of
the genetically programmed behavioral and developmental metamorphosis in these species (Morse et al., 1979a, 1979b; Morse, 1978c, 1979),
and prompted our search for these inducers from the anticipated
specific algal associates of the juveniles.

ACKNOWLEDGMENTS

This work was supported by grants from the U.S. Department
of Commerce NOAA Sea Grant Program #04-6-158-44021-R/NP-1A and
#04-7-158-44121-R/NP-1A, and assistance from the Marine Science
Institute of the University of California at Santa Barbara.

REFERENCES

Adey, W.H. 1973. Temperature control of reproduction and pro-
    ductivity in a subarctic coralline alga, Phycologia 12 111.
Barker, M.F. 1977. Observations on the settlement of the brachio-
    laria larvae of Strichaster australis (Verrill) and Coscinas-
    terias calamaria (Gray) (Echinodermata: asteroidea) in the
    laboratory and on the shore. J. Exp. Mar. Biol. Ecol. 30 95.
Barnes, J.R. and J.J. Gonor. 1973. The larval settling response
    of the lined chiton Tonicella lineata. Marine Biol. 20 259.
Chia, F.-S. and Rice, M. eds. 1978. "Settling and Metamorphosis of
    Marine Invertebrate Larvae". Elsevier Press, North Holland.
Crisp, D.J. 1974. Factors influencing the settling of marine
    invertebrate larvae, in "Chemoreception in Marine Organisms"
    (P.T. Grant and A.M. Mackie, eds.), pp. 177-266, Academic
    Press, New York.
Crofts, D.R. 1929. "Haliotis", University Press of Liverpool,
    Hodder and Stroughton, Ltd., Liverpool.
Gee, J.M. 1965. Chemical stimulation of settlement in larvae of
    Spirorbis rupestris (Serpulidae). Animal Behavior 13 181.
Giraud, G. and J. Cabioch. 1976. Etude ultrastructurale de l'
    activité des cellules superficielles du thalle des Corallinacées
    (Rhodophycées). Phycologia 15 405.
Grant, P.T. and A.M. Mackie, eds. 1974. "Chemoreception in Marine
    Organisms". Academic Press, New York.
Harborne, J.B., ed. 1978. "Biochemical Aspects of Plant and Animal
    Coevolution". Academic Press, New York.
Lawrence, J.M. 1975. On the relationships between marine plants
    and sea urchins. Oceanography and Marine Biol. Ann. Rev.
    13 213.
Lowry, O.H., N.J. Rosebrough, A.L. Farr, and R.J. Randall. 1951.
    Protein measurement with the Folin phenolreagent. J. Biol.
    Chem. 193 265.

Morse, D.E. 1978a. Biochemical control of recruitment, settling and development in planktonic molluscan larvae, with potential applicability to species distributions, evolution, cultivation and reseeding. Proceedings International Symposium Marine Biogeography and Evolution, University of Auckland (July 1978), in press.

Morse, D.E. 1978b. Biochemical control of reproduction, recruitment and metamorphosis in Haliotis. Proceedings Symposium Western Society of Naturalists, Tacoma (December 1978). p. 17.

Morse, D.E. 1978c. Biochemical control applied to critical stages in culturing abalone, II: Control of settling, metamorphosis and early development. Annual Report University of California Sea Grant Program (1977-1978), University of California, San Diego. p. 71.

Morse, D.E. 1979. Biochemical and genetic control of critical physiological processes in molluscan life-cycles: Basic mechanisms, water-quality requirements and sensitivities to pollutants. Annual Report University of California Sea Grant Program (1978-1979), University of California, San Diego, p. 239-252.

Morse, D.E., H. Duncan, N. Hooker, and A. Morse. 1977. Hydrogen peroxide induces spawning in molluscs, with activation of prostaglandin endoperoxide synthetase. Science 196 298.

Morse, D.E., N. Hooker, and A. Morse. 1978. Chemical control of reproduction in bivalve and gastropod molluscs, III: An inexpensive technique for mariculture of many species. Proc. World Maricult. Soc. 9 543.

Morse, D.E., N. Hooker, H. Duncan, and L. Jensen. 1979a. γ - Aminobutyric acid, a neurotransmitter, induces planktonic abalone larvae to settle and begin metamorphosis. Science 204 407.

Morse, D.E., N. Hooker, L. Jensen, and H. Duncan. 1979b. Induction of larval abalone settling and metamorphosis by γ-aminobutyric acid and its congeners from crustose red algae, II: Applications to cultivation, seed-production and bioassays; principal causes of mortality and interference. Proc. World Maricult. Soc. 10 in press.

Shorey, H.H. 1976. "Animal Communication by Pheromones". Academic Press, New York.

Tegner, M.J. and P.K. Dayton. 1977. Sea urchin recruitment patterns and implications of commercial fishing. Science 196 324.

RELATIONSHIPS BETWEEN AGGRESSION, SCENT MARKING AND GONADAL

STATE IN A PRIMATE, THE TAMARIN <u>SAGUINUS</u> <u>FUSCICOLLIS</u>

Gisela Epple

Monell Chemical Senses Center and Department
of Biology, University of Pennsylvania
Philadelphia, PA 19104

INTRODUCTION

In many mammals chemical signals are important in the control
of aggressive behavior and social dominance (c.f. Ralls, 1971;
Thiessen, 1976). Aggressive behavior and the ability to establish
and maintain a high social rank, on the other hand, are influenced
by an individual's endocrine condition and, in turn, affect its
endocrinology. In male rodents gonadal hormones are of crucial
importance in the development and maintenance of intraspecific
aggression and in chemical communication (Bronson and Desjardins,
1971; Hart, 1974; Leshner, 1975; Quadagno et al., 1977; Thiessen,
1976). In primates our understanding of the hormonal control of
aggression and social dominance, and their relation to chemical
communication, is poorer than in rodents. However, reports on
changes in aggressive behavior and scent marking which coincide
with gonadal activation during the breeding season (Harrington,
1975; Jolly, 1967; Mazur, 1976; Schilling, 1974), on the influence
of androgens on the fetal nervous system (c.f. Eaton et al., 1973;
Rose et al., 1974) and on the effects of gonadectomy and hormone
therapy (c.f. Mazur, 1976) show that the gonads are of some importance in controlling primate aggression and chemical communication.

This paper reviews some of our studies on the relationship
between the gonads, aggressive behavior and chemical communication
in the saddle back tamarin. This South American primate, a member
of the marmoset family, inhabits the primary and secondary forests
of the upper Amazonian region (Hershkovitz, 1977). Laboratory
studies (Box, 1978; Epple, 1972, 1975, 1978a; Kleiman, 1977; Vogt,

1978 and others) and field work on a number of callitrichid species
(Dawson, 1978; Izawa, 1978; Neyman, 1978; Thorington, 1968) have
suggested that most marmosets live in small groups consisting of
an adult breeding female, a male with whom she has a monogamous
pair bond, and several sets of her offspring. Wild groups may
also contain a few individuals who are not the offspring of the
breeding female and whose kinship relations to the nuclear family
are unknown.

Callitrichids are aggressive primates. At least under some
ecological conditions they occupy distinct territories, and contact
aggression, a number of visual and vocal displays and scent marking
are involved in territorial defense (Dawson, 1978; Neyman, 1978;
Thorington, 1968). In the laboratory they show strong overt
aggression against all adult conspecifics not belonging to the
same social group, particularly if they are members of the same
sex (Epple, 1970; 1978a, b).

Male and female saddle back tamarins mark by rubbing with
large scent glands located in the circumgenital-suprapubic area
and above the sternum. The sternal scent pad consists mainly of
large apocrine glands, the suprapubic-circumgenital pad of apocrine
glands and sebaceous follicles (Perkins, 1966). The circumgenital-
suprapubic scent gland involves the skin of the external genitalia
(scrotum and labia) and extends as a thick, glandular cushion into
the suprapubic area. It is darkly pigmented, largely hairless
and usually moist with oily secretions (Figs. 1, 2). The sternal
gland is much less obvious since it is covered by the dense fur of
the chest.

The scent glands begin to develop as macroscopic differentia-
tions in late juveniles at approximately 7-10 months of age and
grow until they reach their fully adult size at 2 years of age
or later. Scent marking behavior is occasionally seen in infants
and juveniles but reaches its adult level around the time of
puberty, i.e., at 12-18 months of age. Circumgenital marking is
much more frequent than sternal marking and therefore was the
only response evaluated in all studies reported here. Scent marking activity is extremely variable, being easily evoked by a wide
spectrum of arousing stimuli, as long as these do not elicit fear
responses. Social status is the most important factor in determining the long term marking activity of adult males and females
(Epple, 1970, and unpublished). Marking occurs in a variety of
behavioral contexts and appears to be involved in a number of
sexual and social interactions such as courtship and pair bonding
as well as intragroup aggression (Epple, 1974 a, b). The scent
identifies the species and subspecies, the individual, its gender
and gonadal condition and perhaps also its social status (Epple,
1974 a, b, 1978c; Epple et al., 1979; Epple, 1979).

MATERIAL AND METHODS

Effects of Gonadectomy in Adulthood

Most of the data reviewed here are derived from a study of the effects of gonadectomy in adulthood on aggressive behavior and scent marking. A detailed description of all methods was published previously (Epple, 1978b) and shall only briefly be summarized here. Fourteen permanently cohabiting, adult male-female pairs served as subjects. All partners had lived together for at least six months prior to being used in the experiments. The aggressive behavior and scent marking were studied in all subjects prior to gonadectomy. Following these presurgical studies the males of seven pairs and the females of two pairs were surgically gonadecto-mized while their partners remained intact. The males of the re-maining five pairs underwent sham castration, replicating all surgical procedures except removal of the testes.

Each pair lived in a 3 x 3 m testing room, equipped with a home cage containing natural branches and a sleeping box. The tamarins were not confined to their cage most of the time but had the full use of the room, which contained no other monkeys. Each pair remained in the test room throughout the period of presurgical testing, but was placed into a 61x122x122 cm cage in a colony room after gonadectomy and control surgery. Seven months after surgery, each pair was placed into a test room of its own and re-mained there for all postsurgical testing. The behavior of each pair was tested in an identical manner before and after surgery under three conditions.

A. Trial free situation. The scent marking activity of each undisturbed pair was recorded in a series of twenty 20-min observa-tion sessions. During each session the observer focused on the behavior of one of the two subjects for 10 minutes, counterbalanc-ing the order of observation across tests. The test period was divided into 80 intervals of 15 sec each. For each interval the focus animal received a score of one if it scent marked, regard-less of the actual frequency of performance (Hansen frequencies). The animals were confined to their home cage during these observa-tions. In a separate series of control tests we established that the scent marking activity per 10 min of pairs confined to their home cages did not differ from that of pairs which had the full use of the test room, and therefore the scores of confined animals are used here.

B. Social encounters with intruders. Every pair was given a series of 20 social encounters with 20 different adult, intact conspecifics, both before and after surgery. Ten adult males and ten adult females, belonging to other groups, were each intro-duced into the subjects' room for ten minutes once before and

once after surgery. To protect the tamarins from severe injuries
the intruder was wheeled into the test room in a 61x91x122 cm
wire mesh cage and remained confined to the cage during the entire
encounter. Subjects and intruder could interact through the 1/4
6 mm wire mesh, which allowed a limited amount of contact. The
subjects were never tested more than once a day and usually only 3
encounters per week were given.

Each 10 min session was divided into 40 intervals of 15 sec
each, indicated by an audible signal. Two observers were present,
each narrating the behavior of one of the subjects on tape, which
also recorded the audible timer signal. The records were trans-
cribed onto data sheets and a score of 1 per interval was given if
the subjects scent marked and displayed patterns of overt aggression, regardless of actual frequency of occurrence. For final
analysis of the raw aggression data an accumulative score for
injurious aggression was computed by summing the scores per interval of attacking the intruder, fighting with it and chasing it.
Average scores for each individual before and after castration
were computed and all data were analyzed by nonparametric statistics following Siegel (1956).

C. Five minutes after the intruder had been removed from the
test room, the scent marking activity of the pair was recorded for
another ten minutes, using Hansen frequencies per 15 sec intervals.

In addition to assessing the scent marking behavior of the
subjects before and 7 months after surgery, long term records on
marking by castrates were also kept (Epple and Cerny, in press).
The marking activity of the castrates was followed for many months
from the time of gonadectomy up to a maximum of 4 1/2 years under
conditions different from those outlined above. Except for the
time the pairs spent in single test rooms, the animals lived in
cages (usually 61x122z122 cm) in a colony room. All marking
tests described below were given while the subjects were confined
to one half of their 2 compartment home cage while their mates
remained in the adjacent compartment. Marmoset monkeys usually
scent mark any new object placed into their cages, particularly if
it carries the scent of conspecifics. Therefore, the marking
activity of all subjects was assessed in series of 5 min tests,
during which the animal received a 61x5x0.6 cm bar of
aluminum which had been marked several times by an adult, intact
male immediately before the test. The 5 min test was divided into
60 intervals of 5 sec each. For each interval the subject received a score of 1 if it scent marked the stimulus bar or any
other place in the cage. The castrates were not tested continuously during the years covered by this study. Several weeks of
testing, usually featuring four tests per week, alternated with
several weeks of rest.

Three of the seven castrates were tested as intacts (a total
of 64 tests were given before castration), and all seven subjects
were tested between two weeks and 12 months after castration (total
of 442 tests). Testing of 2 of the males was extended until 21
and 25 months after surgery. During these months all seven males
cohabited with their original females. In the course of an experiment on the social behavior of castrates in complex groups, four
of the subjects were removed from their females 13, 13, 14 and 25
months after surgery and each was placed into an experimental trio
consisting of an intact male, a castrated male and a female, all
nonrelated adults. The four castrates were given a total of 184
marking tests while living in the groups. At 26, 38 and 51 months
after surgery three of the group living castrates were again placed
with a permanent female partner alone, either by removing the
intact male from the group or by removing the castrate and pairing
him with another female. These males received a total of 140 marking tests after being replaced into a pair situation. The changes
in the social setting provided an opportunity to assess the relative
importance of gonadal hormones and social conditions in influencing
scent marking behavior.

The marking activity of the sham castrated males was tested
before and after surgery in the same manner as that of the castrates. The control males remained with their original females.
Each male was given 20-27 tests in a 6 week block. Four sham
castrates received a total of 94 tests before surgery and all 5
sham castrates received a total of 99 tests after surgery.

Effects of Prepubertal Castration

In addition to studies on the effects on gonadectomy in adulthood, preliminary results of a study on the effects of prepubertal
castration on male behavior are also discussed here. All males
used in this study lived in their families until 5 1/2 months old.
At this age, five juvenile males were castrated, four juvenile
males sham castrated. After a recovery period of two weeks which
the males spent in their families, each male was paired with an
adult, nonrelated female. These pairs lived together until the
males were ten months old. Each pair was then placed into a test
room and between the ages of ten and twelve months the subject
underwent the same post surgical testing routine described for
the adult pairs. Twenty trial-free observation sessions (A) were
performed and each pair was given seven aggressive encounters with
different adult strange males and seven encounters with different
adult strange females (B). Five minutes after the intruder had
been removed from the test room the marking activity of the pair
was recorded for ten minutes (C).

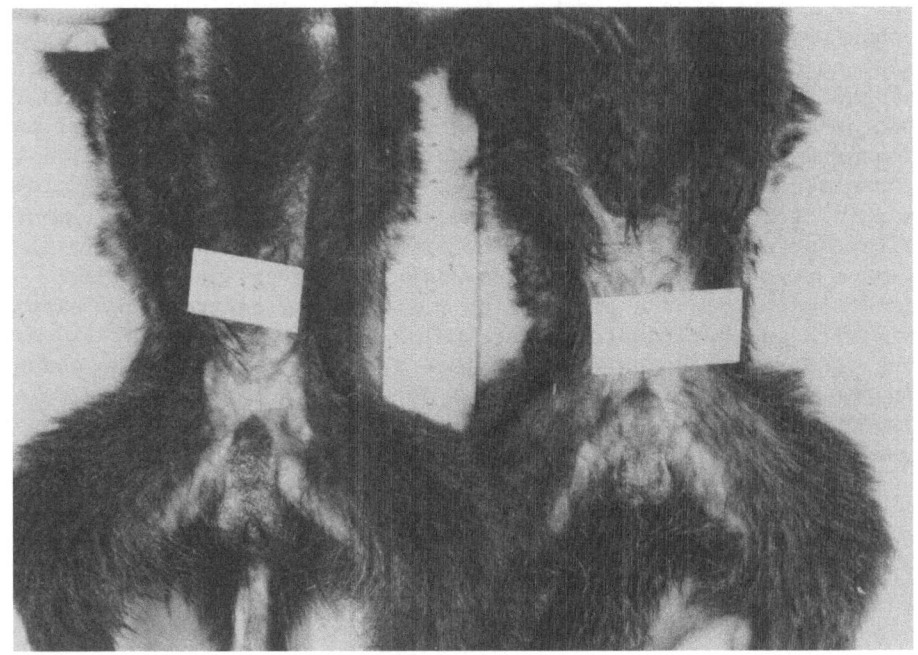

Figure 1.   Circumgenital-suprapubic scent glands of an adult
            castrate (left) four years post castration and of a
            sham castrate (right).

        After this experiment was completed, the partners cohabited
in a colony room until the males were two years old. At that time
all aggressive encounter tests (B) were repeated in a test room
under the same conditions as before. However, one of the controls had died meanwhile, and therefore only three sham castrates
were tested when two years old.

        The scent marking activity of the subjects was also assessed
following the method used during our long term study of adult
castrates. Each one year old castrate and sham castrate received
twenty 5 min marking tests, offering an aluminum bar which had
been marked by an adult, intact male as described above.

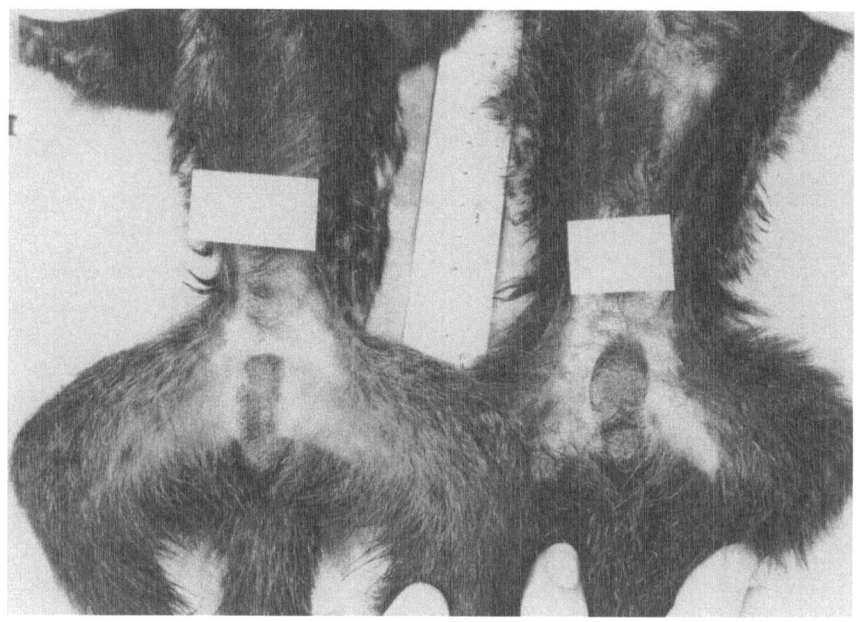

Figure 2.   Circumgenital-suprapubic scent glands of an adult
            female four years post ovariectomy (right) and an
            intact female (left).

RESULTS

The Scent Glands

      The effects of gonadectomy on the morphological appearance of
the circumgenital scent glands are illustrated in Figs. 1-3.
Gonadectomy of adult males resulted in atrophy of the scrotum which
forms part of the gland, thereby decreasing its surface area.
However, the appearance of the suprapubic part of the glands of
adult males was little changed by gonadectomy (Fig. 1). Fig. 2
shows that the morphology of female scent glands does not seem to
be effected by ovariectomy in adulthood.

      Castration prior to puberty severely retarded the development
of the male scent gland. By one year of age, castrates showed an

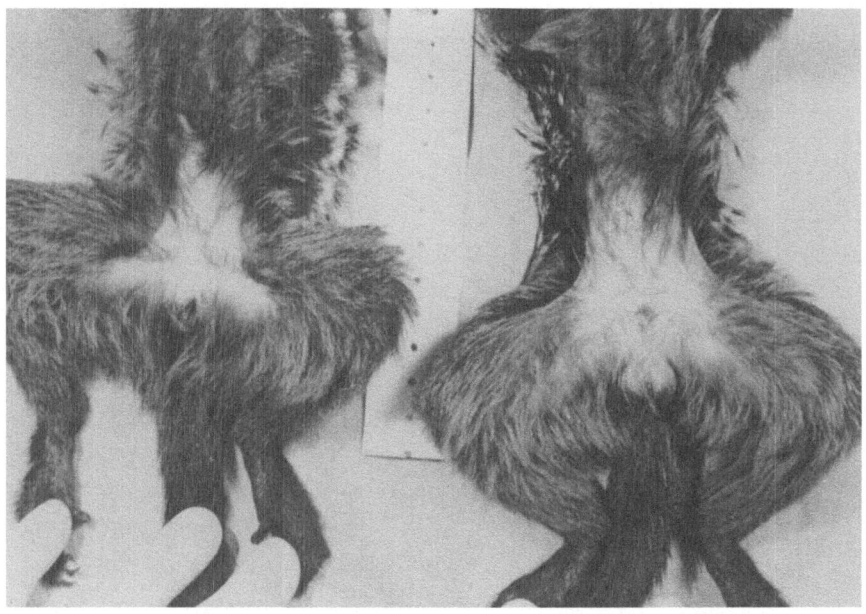

Figure 3.   Circumgenital-suprapubic scent gland of one year old
             twin brothers castrated (right) and sham castrated
             (left) at 5 1/2 months of age. Note that at one year of
             age the gland of the intact male has not yet reached its
             adult size.

approximately 15-20 mm^2 spot of unpigmented, glandular skin above
the symphysis pubis, while their sham castrated twin brothers
showed normally developed, darkly pigmented scent glands (Fig. 3).
By two years of age the scent glands of one of the castrates had
grown slightly larger than at one year while that of the others had
remained largely unchanged. All two year old control males had
developed glands of adult size.

     These observations suggest that gonadectomy in adulthood has
little effect on male and female scent glands while the castration
prior to puberty inhibits the development of these glands in males.
This hypothesis awaits histological confirmation which is presently
in progress.

Aggressive Behavior

In adult pairs before and after surgery as well as in pairs consisting of adult females and prepubertally castrated males no injurious aggression was observed under trial free conditions. The introduction of an intruder resulted in the display of overt aggression, almost all of which was directed at the intruder. Gonadectomy and sham castration of adult, socially experienced subjects had no statistically significant effect on their aggressive behavior. However, the postsurgical aggression scores of the adult males tended to be higher than their presurgical scores, particularly in the castrates (Fig. 4). The pre- and postsurgical aggression scores of the intact females did not differ. The two ovariectomized females showed a mean aggression score of 3.3 before and of 8.4 seven months after surgery. However, no statistical analysis was done on only two subjects.

Pairs containing prepubertally castrated and sham castrated one year old males showed considerably less aggression than adult male-female pairs (Fig. 4). The sham castrated juvenile males tended to have higher aggression scores than the castrates. However, the average scores of the subject groups were not statistically different. When all aggressive encounters were replicated at the time the males were two years old, each castrate and sham castrate received a higher aggression score than at one year of age. The average scores of two year old castrates and sham castrates did not differ significantly from each other nor did they differ significantly from the scores obtained by the 14 adult males during presurgical testing.

Scent Marking Behavior

In adult, gonadally intact pairs females scent marked significantly more frequently than males. This sexual dimorphism in marking behavior was displayed during trial free observations as well as during and after aggressive encounters with strange conspecifics (Fig. 5). Surgery did not cause any statistically significant changes in marking behavior. Under all conditions the postsurgical marking scores of the castrated males tended to be higher than their presurgical scores. Although this increase was not statistically significant, it resulted in the disappearance of the sexual dimorphism of marking, except during encounters with strangers when females still marked significantly more frequently than castrates and sham castrates.

The marking scores of the sham castrates increased postsurgically only during aggressive encounters and those of the females remained the same The scent marking activity of adult

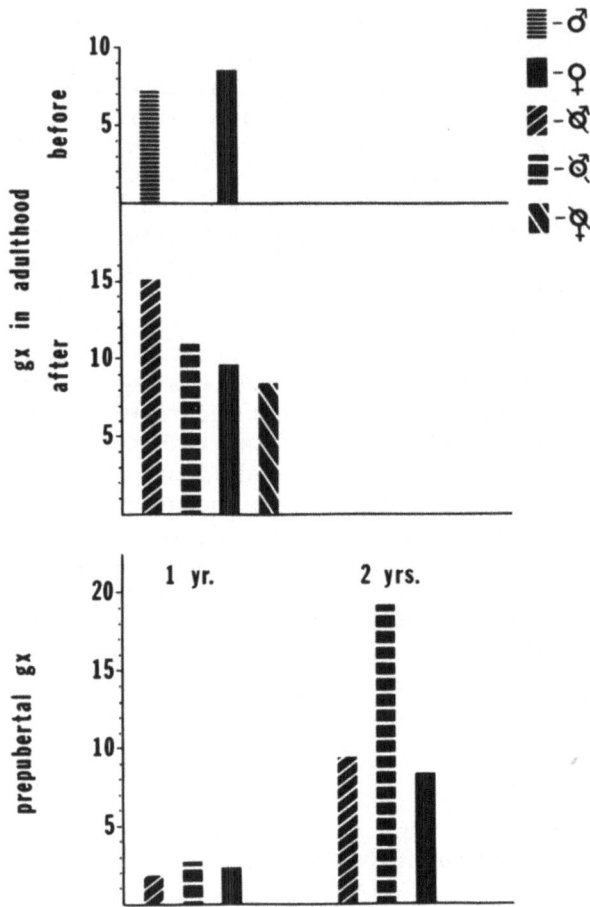

Figure 4. Mean accumulative aggression scores of subjects undergoing surgery (gx) in adulthood and prior to puberty. The scores of adults before and after surgery and of prepubertal castrates and controls tested at one and two years of age are given. Presurgical scores of adults combine experimental and control subject. Castrates – ♂, sham castrates – ♂, ovariectomized females – ♀.

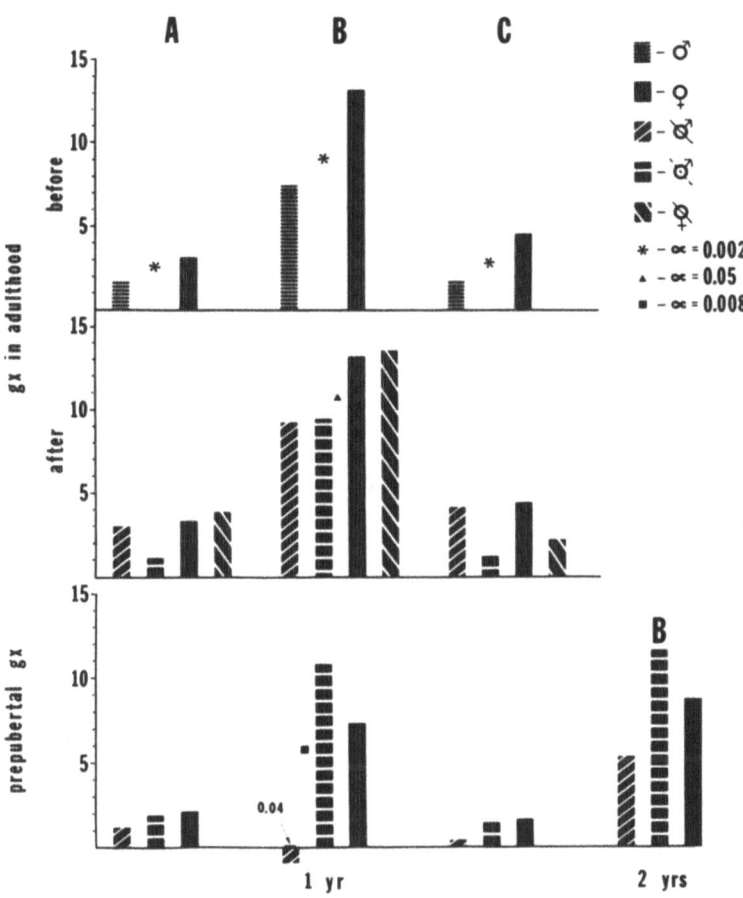

Figure 5.   Mean marking scores of subjects undergoing surgery (gx) in adulthood and prior to puberty. The scores of adults before and after surgery and of one year old prepubertal castrates and controls under trial free conditions (A) during (B) and after (C) aggressive encounters and those of two year old castrates and controls during aggressive encounters (B) are given. Presurgical scores of adults combine experimental and control subjects. Castrates – ⚥, sham castrates – ⚥, ovariectomized females – ⚥; Mann Whitney U test.

pairs was relatively low under trial free conditions, both before
and after surgery (Fig. 5). Before and after surgery adult pairs
increased their scores significantly in the presence of an intruder
(all subjects prior to surgery: $\alpha = 0.011$; after surgery: castrates
$\alpha = 0.046$; sham castrates $\alpha = 0.062$; females $\alpha = 0.011$, Walsh test;
Fig. 5).

Long term studies of the marking behavior of the adult castrates
on prescented plates, as described above, also showed no changes in
marking activity attributable to gonadectomy. The overall average
marking scores of castrates and sham castrates before and after
surgery did not differ (Fig. 6). The overall score of the adult
castrates (Fig. 6, o) includes data obtained in the course of the
4 1/2 year study during which the males lived in 3 different social
settings. When the marking scores obtained by the castrates under
each social condition are considered separately (Fig. 6, ptp) it
becomes clear that marking activity was significantly different
under the 3 social conditions (Kruskal-Wallis One Way Analysis of
Variance; $p < 0.05$). While the marking activity of the castrates
during cohabitation with their original females (Fig. 6, p) was in
the range of their overall scores, each of the 4 males studied
when living with an intact male and an intact female decreased his
scent marking behavior (Fig. 6, t). The decline in the marking
activity of the castrates during life in trios was completely
reversible when their social situation changed. After placing 2
of them with a single female partner again, their marking activity
increased strongly. Another castrate killed the intact male living
in his group during a dominance fight. Thereafter he remained with
the adult female and her offspring. His scent marking activity
also recovered significantly after the death of the intact male
but did not increase to its original height. The average score of
these 3 males who had been castrated 2, 4 and 4 1/2 years ago was
now above the overall average for all castrates (Fig. 6, p). The
marking activity of prepubertally castrated and sham castrated one
year old males on prescented plates is also illustrated in Fig. 6.
All of these subjects lived with one adult female during the time
of testing. As the figure shows, the marking activity of castrates
and sham castrates did not differ under these conditions and
tended to be slightly but not significantly higher than the overall
scores of the adult males.

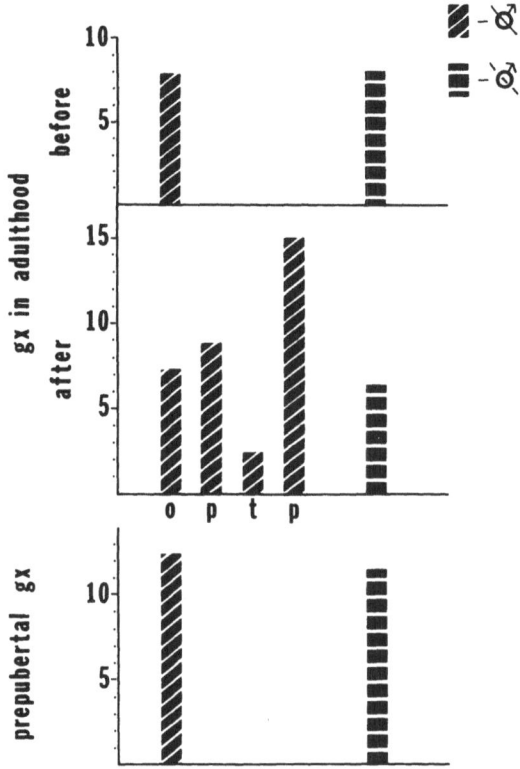

Figure 6. Mean marking scores on prescented plates obtained by
males undergoing surgery (gx) in adulthood and prior
to puberty. The scores of adults before and after
surgery and of prepubertal castrates and controls at
one year of age are given. The overall (o) postsurgical scores of adult castrates are compared with
their scores obtained when living with their original
partners (p), in trios (t) and when replaced into a
pair setting (p). Castrates – ♀ , sham castrates –
♂ .

DISCUSSION

Intact Subjects

Saddle back tamarins show high levels of injurious aggression against conspecifics who do not belong to their own group. Under the conditions of this study overt aggression prevented any friendly social contact with the strangers. As discussed in detail by Epple (1978 a, b) the adult, intact pairs showed a tendency toward a sexual dimorphism of aggressive behavior, which was slightly more frequent in females. Aggression strongly stimulated scent marking activity in adults of both sexes but under all conditions in which the adult, intact pairs were studied females marked more frequently than males.

Somewhat different results were obtained during a previous study on three large groups of common marmosets, Callithrix jacchus (Epple, 1970). In these groups, scent marking activity during trial free conditions was higher in dominant males and females than in subdominant animals and the relative marking activity of high ranking animals depended on the presence of nonrelated adult group members of the same sex. In two groups containing two adult, nonrelated females the α-female was the most active marker while in one group containing only one adult female but two adult, nonrelated males, the α-male was the most active marker (Epple, 1970). In saddle back tamarin groups containing several nonrelated adults, marking during trial free condition is also higher in dominant than in subdominant animals (Epple, unpublished). However, in contrast to the common marmoset, even in trios of 2 adult, intact males and one adult intact female, the female scent marked significantly more frequently than both males (Epple, unpublished). In both species, aggression stimulated scent marking above trial free levels. In common marmosets, encounters with a strange male only stimulated marking in the α-male of each group while encounters with strange females stimulated it in the α-female. Subdominant group members did not generally show an increase in marking (Epple, 1970). In the tamarins, the scent marking scores of both, intact males and females were higher during encounters than under trial free conditions. However, both sexes showed a tendency to mark more frequently in the presence of a stranger of the same sex than in the presence of an oppositely sexed stranger, but under each condition females marked more than males (Epple, 1978 a, b).

The results of both studies suggest that one of the functions of scent marking in saddle back tamarins and common marmosets is the communication of aggressive motivation. Quality and/or quantity of the odors produced during aggressive marking may communicate the motivation of the marker to group members and

strange conspecifics. Since <u>Saguinus</u> <u>fuscicollis</u> communicates
species, subspecies, gender, hormonal condition, individual
identity and perhaps even rank by scent (Epple, 1974 a, b, 1978
c, 1979; Epple et al., 1979) it is easy to envision how odors
might serve as long distance threat signals. High marking
activity, regardless of its motivation, automatically results in
a large amount of this particular individual's scent in the environment. The scent identifies the individual, its sex, maybe
even its social status and advertises its presence throughout the
living space, even if the animal itself is not present. The
monkeys sniff the marks of conspecifics very frequently. Therefore, any group mate or stranger must be aware of the amount and
identity of an active marker's scent throughout most of the area
in which the group lives. A submissive group member or an intruder
who has experienced aggression by one or all of the group members
might be quite strongly affected by exposure to the superior's
odor. In this way the scent marks may function as substitutes
for injurious aggression, maintaining the hierarchy within groups,
enforcing the pair bond and group integrity and perhaps functioning
in territorial defense.

Effects of Gonadectomy

Gonadectomy in adulthood had little effect on the appearence
of male and female scent glands and did not significantly influence
the display of injurious aggression and the level of scent marking
in either sex. Moreover, 4 1/2 years after gonadectomy, male scent
marking activity on prescented perches was high. On the other
hand,the social condition under which the castrated males were
living appeared to be more important in determining the level at
which they scent marked than their gonadal state.

Castration prior to puberty severely and permanently retarded
the development of male scent glands but appeared to have little
effect on aggressive behavior and scent marking. Although one
year old castrates and sham castrates were less aggressive than
adult males, their level of aggression increased as they matured.
The marking scores of one year old sham castrates during all
conditions of observation were in the range of those of adults.
One year old castrates on the other hand showed almost no marking
during aggressive encounters but marked at approximately the same
level as the control males under trial free conditions and on prescented perches. By two years of age marking was at adult level
in all males although castrates tended to mark less frequently
than sham castrates. It appears possible that this tendency
reflects a real, though slight inhibition of marking by prepubertal castration, an effect which would have stood out more
clearly if a larger number of subjects had been used.

These results suggest that the full development of the male gland depends on the presence of the gonads and therefore may be controlled by gonadal androgens. Prepubertal castrates develop a trace of a scent gland which is possibly stimulated by extra-gonadal steroids, perhaps of adrenal origin. Once the gland is fully developed in an intact adult, it might become either independent of steroid hormones or, in the absence of gonads, be more or less fully supported by extragonadal steroids. In contrast to the scent glands, marking behavior and aggression appear to be largely independent of gonadal hormones. They may be controlled by the central nervous system alone or depend on extragonadal steroids.

Our findings differ from those of a number of studies on rodents, where the gonads are often necessary for the development and for the maintenance of aggressive behavior and scent marking in adulthood (Bronson and Desjardins, 1971; Hart, 1974; Leshner, 1975; Thiessen, 1976). In several primate species it has also been shown that gonadal steroids affect or reflect social status and/or aggressiveness (c.f. Cochran and Perachio, 1977; Dixson and Herbert, 1977; Mazur, 1976; Rose et al., 1974). In accordance with our findings, on the other hand, several investigators failed to detect an effect of gonadectomy on aggressive behavior and social rank in a number of primates (Green et al., 1972; Dixson and Herbert, 1977; Mirsky, 1955; Wilson and Vessey, 1968). The hormonal control of primate scent marking has been studied in only a few species. These studies have shown, however, that scent marking in male tree shrews (Tupaia belangeri) and the activity of the male scent glands in two prosimians, Galago crassicaudatus and Lemur catta, are also androgen   dependent (Andriamiandra and Rumpler, 1968; Dixson, 1976; von Holst and Buergel-Goodwin, 1975).

ACKNOWLEDGEMENTS

This study was supported by grants No. GB 12660 and No. BNS 7606838 from the National Science Foundation and by Research Career Development Award No. K04HD 70575.

REFERENCES

Andriamiandra, A. and Rumpler, Y. 1968. Rôle de la testostérone sur le déterminisme des glandes brachiales et antébrachiales chez le Lemur catta. Compt. red. sean. soc. Biol. 162: 1651.

Box, H. 1978. Social interactions in family groups of captive marmosets (Callithrix jacchus). In The Biology and Conservation of the Callitrichidae (D. Kleiman, Ed.). pp. 239-249, Smithsonian Press, Washington, D.C.

Bronson, F.H. and Desjardins, C. 1971. Steroid hormones and aggressive behavior in mammals. In The Physiology of Aggression and Defeat (B.E. Eleftheriou and J.P. Scott, Eds.) pp. 43-63, Plenum Press, New York.

Cochran, C.A. and Perachio, A.A. 1977. Dihydrotestosterone proprionate effects on dominance and sexual behaviors in gonadectomized male and female rhesus monkeys. Horm. Behav. 8: 175.

Dawson, F. 1978. Troop size, composition and instability in the Panamanian tamarin (S. oedipus geoffroyi). In The Biology and Conservation of the Callitrichidae (D.G. Kleiman, Ed.) pp. 23-37, Smithsonian Press, Washington, D.C.

Dixson, A.F. 1976. Effects of testosterone on the sternal cutaneous glands and genitalia of the male greater galago (Galago crassicaudatus crassicaudatus). Folia primat. 26: 207.

Dixson, A.F. and Herbert, J. 1977. Testosterone, aggressive behavior and dominance rank in captive adult male talapoin monkeys (Miopithecus talapoin). Physiol. Behav. 18: 539.

Eaton, G.G., Goy, R.W. and Phoenix, C.H. 1973. Effects of testosterone treatment in adulthood on sexual behaviour of female pseudohermaphrodite rhesus monkeys. Nature 242: 119.

Epple, G. 1970. Quantitative studies on scent marking in the marmoset (Callithrix jacchus). Folia primat. 13: 48.

Epple, G. 1972. Social behavior of laboratory groups of Saguinus fuscicollis. In Saving the Lion Marmoset. Proc. WAPT Golden Lion Marmoset Conference (D.D. Bridgwater, Ed.), pp. 50-58, WAPT. Oglebay Park, Wheeling, West Virginia.

Epple, G. 1974a. Primate pheromones. In Pheromones (M.C. Birch, Ed.). pp. 366-385, North Holland Publ. Co., Amsterdam.

Epple, G. 1974b. Olfactory communication in South American primates. Ann. N.Y. Acad. Sci. 237: 261.

Epple, G. 1975. The behavior of marmoset monkeys (Callithricidae). In Primate Behavior, Vol. 4 (L.A. Rosenblum, Ed.). pp. 195-239, Academic Press, New York.

Epple, G. 1978a. Notes on the establishment and maintenance of the pair bond in Saguinus fuscicollis. In The Biology and Conservation of the Callitrichidae (D.G. Kleiman, Ed.), pp. 231-237, Smithsonian Press, Washington, D.C.

Epple, G. 1978b. Lack of effects of castration on scent marking, displays and aggression in a South American primate (Saguinus fuscicollis). Hormones and Behav. 11: 139.

Epple, G. 1978c. Studies on the nature of chemical signals in scent marks and urine of Saguinus fuscicollis (Callitrichidae, Primates). J. Chem. Ecol. 4: 383.

Epple, G. 1979. Gonadal control of male scent in the tamarin, Seguinus fuscicollis. Chemical Senses and Flavour 4:15.

Epple, G. and Cerny, V.A. In press. Gonadal and social influences on scent marking in a primate, Saguinus fuscicollis. Folia primat.

Epple, G., Golob, N.F. and Smith, A.B. III. 1979. Odor communica-
    tion in the tamarin Saguinus fuscicollis (Callitrichidae):
    Behavioral and chemical studies. In Chemical Ecology: Odour
    Communication in Animals (F.J. Ritter, Ed.). pp. 117-130,
    Elsevier/North Holland Biomed. Press, Amsterdam.
Green, R., Whalen, R.E., Rutley, B. and Battie, C. 1972. Dominance
    hierarchy in squirrel monkeys (Saimiri sciureus). Folia
    primat. 18: 185.
Harrington, J.E. 1975. Field observations of social behavior of
    Lemur fulvus fulvus E. Geoffrey 1812. In Lemur Biology (I.
    Tattersall and R.W. Sussman, Eds.), pp. 259-279, Plenum Press,
    New York.
Hart, B. 1974. Gonadal androgens and sociosexual behavior of
    male mammals. Psychol. Bull. 81: 383.
Hershkovitz, P. 1977. Living New World Monkeys (Platyrrhini).
    Vol. I, Univ. of Chicago Press, Chicago.
Holst, von D. and Buergel-Goodwin, U. 1975. The influence of sex
    hormones on chinning by male Tupaia belangeri. J. Com. Physiol.
    103: 123.
Izawa, K. 1978. A field study of the ecology and behavior of the
    black-mantle tamarin (Saguinus nigricollis). Primates 19: 241.
Jolly, A. 1967. Breeding synchrony in wild Lemur catta In Social
    Communication Among Primates. (S. Altmann, Ed.), pp. 3-14,
    Univ. of Chicago Press.
Kleiman, D. 1977. Monogamy in mammals. Quart. Rev. Biol. 52: 39.
Leshner, A.J. 1975. A model of hormones and agonistic behavior.
    Physiol. Behav. 15: 225.
Mazur, A. 1976. Effects of testosterone on status in primate
    groups. Folia primat. 26: 214.
Mirsky, A.F. 1955. The influence of sex hormones on social
    behavior of monkeys. J. Comp. Physiol. Psychol. 48: 327.
Neyman, P.F. 1978. Aspects of the ecology and social organization
    of freeranging cotton top tamarins (Saguinus oedipus) and the
    conservation status of the species In Biology and Conservation
    of the Callithrichidae. (D.G. Kleiman, Ed.), pp. 39-71,
    Smithsonian Press, Washington, D.C.
Perkins, E.M. 1966. The skin of the black-collared tamarin
    (Tamarinus nigricollis). Amer. J. Phys. Anthrop. 25: 41.
Quadagno, D.M., Briscoe, R. and Quadagno, J.S. 1977. Effect of
    perinatal gonadal hormones on selected nonsexual behavior
    patterns: a critical assessment of the nonhuman and human
    literature. Psychol. Bull. 84: 62.
Ralls, K. 1971. Mammalian scent marking. Science 171: 443.
Rose, R.M., Bernstein, I.S., Gordon, T.P. and Catlin, S.F. 1974.
    Androgens and aggression: a review and recent findings in
    primates In Primate Aggression, Territoriality and Xenophobia.
    (R.L. Halloway, Ed.), pp. 275-304 , Acad. Press, New York.

Schilling, A. 1974. A study of marking behaviour in _Lemur catta._ In Prosimian Biology (R.D. Martin, G.A. Doyle and A.C. Walker, Eds.), pp. 347–362, Duckworth, Gloucester-Crescent.

Siegel, S. 1956. Nonparametric Statistics for the Behavioral Sciences, McGraw-Hill, New York.

Thiessen, D.D. 1976. The Evolution and Chemistry of Aggression, Charles C. Thomas, Springfield, Ill.

Thorington, R.W., Jr. 1968. Observations of the tamarin _Saguinus midas._ Folia primat. 9: 95.

Vogt, J.L. 1978. The social behavior of a marmoset (_Saguinus fuscicollis_) group: III Special analysis of social structure. Folia primat. 29: 250.

Wilson, A.P. and Vessey, S.H. 1968. Behavior of free-ranging castrated rhesus monkeys. Folia primat. 9: 1.

OLFACTORY ASPECTS OF RUTTING BEHAVIOR IN THE BACTRIAN CAMEL

(CAMELUS BACTRIANUS FERUS)

Chris Wemmer and James Murtaugh

Conservation and Research Center
National Zoological Park
Front Royal, VA 22630

The Bactrian camel was once widespread across Central Asia, but its remnant wild population of less than 1000 animals is now restricted to a small segment of the Gobi Desert (Bannikov, 1976a, b). The biology of these wild camels remains a mystery, but it is surprising that so little has been published about the much larger domesticated population that for more than 2500 years has served nomads as draught animals and provenders of meat and milk (Bulliet, 1975).

Domesticated Bactrian camels show a number of changes associated with a long history of captive husbandry. There has been some relaxation for example in the seasonality of ovulation and the subsequent birth period (Novoa, 1970). Sexually mature males, however, exhibit a pronounced winter rutting season. Rutting in male ungulates is a transient annual phenomenon associated with polygynous mating systems, and has been defined as "a form of hypersexual activity of limited duration...directly controlled by the level of testosterone" (Lincoln et al., 1970:475). Distinctive attributes of the rut include loss of appetite, the appearance of secondary sex traits and sexually dimorphic behavior associated with heightened sexual interest and rivalry between males. The purpose of this study is to present information on the behavior patterns associated with olfactory signaling by males during the rutting season and the influence of social context on the elicitation of these patterns.

METHODS AND MATERIALS

Table I lists particulars about the subjects and the observation time devoted to them. The animals are maintained in two hilly pastures measuring 16.8 and 23.6 hectares (41.6-58.4 acres). A large barn having four wings allowed the manipulation of herd composition as well as the isolation of certain individuals under conditions of reduced visual, olfactory, and auditory contact.

Table I. Subjects of study and observation time.

Camel	Birthdate	Winter Study Periods...	1977-78	1978-79
Beauregard	27/03/78	1978-79		---
Number One	09/04/75	1977-78, 78-79	20 hr 9 min	62 hr
Alfred	00/00/72	1977-78, 78-79	25 hr 37 min	55 hr
Clyde	00/00/72	1977,78, 78-79	27 hr 16 min	51 hr
Tex	1965	1976, 77, 77-78,78-79	24 hr 24 min	61 hr
Humphrey*	1969	1976-77, 77-78	22 hr 31 min	---
Jimmy	23/03/61	1976-77, 77-78	21 hr 46 min	---

Observations were made by six and four member teams in the winters of 1977-78 and 1978-79. Stopwatch times and observations were recorded with pen and ink or dictaphone. Because of the camel's generally deliberate pace of life, notation of behavior is not particularly challenging, and observer reliability criteria were reached after four days. Observers recorded the actions of each individual male for 20 minute periods in the 1977-78 season. In the 1978-79 season 30 minute observation sessions were devoted to each of four focal animals on a daily basis between 4 January and 4 March. If possible each male was observed in three distinct social contexts (1) in isolation from other animals, (2) while encountering another male, and (3) within a female herd. Table II summarizes the duration of each context for each male. Figure 1 is a diagram of the barn showing the areas used in the study. Isolated males were unable to physically interact with other animals, but had limited visual input from other camels through small discolored windows. Harem masters for the most part restricted their activity to the vicinity of the females who spent most of their time near the barn, but occasional interaction between harem masters took place across a small section of adjacent chain-link fence. Encounters between males were staged by opening the door between adjacent wings of the barn and escorting one camel into the other's enclosure. During the 1978-79 season 11 variables were measured in frequency or duration and analyzed by computer.

Table II.   Herd composition during the study period.

Camel	Duration of Social Grouping During Study Period

1977-78

Alfred	herdmaster (w/3 females): 5-I-78--16-III-78
Clyde	herdmaster (w/5 females and one 2 1/2 year old male, No. 1):6-I-78--16-III-78
Tex	bachelor; (w/Humphrey and Jimmy): 9-I-78-7-II-78; (alone): 7-II-78-16-III-78
Humphrey	bachelor (w/Tex and Jimmy): 9-I-78-II-78; alone: 7-II-78--16-III-78
Jimmy	bachelor (w/Humphrey and/or Tex): 9-I-78--7-II-78; alone: 28-II-78--16-III-78

1978-79

Alfred	herdmaster (w/3 females): 29-I-79--2-III-79; bachelor (alone): 26-XII-78--29-I-79
Clyde	herdmaster (w/4 females and 3 calves, including Beauregard): 12-XII-78--22-I-79; bachelor (alone): 22-I-79--2-III-79
Tex	herdmaster (w/4 females): 11-XII-78--2-III-78
No. 1	bachelor (alone): 26-XII-78--26-I-79; 29-I-78--2-III-78; herdmaster (w/4 females and 3 calves, including Beauregard): 26-I-78--29-I-78

The sonogram was made on a Kay Sonograph 7029A, and a General
Radio Type 1523 Level Recorder was used for measuring the intervals
between the successive onsets of tail flaps. The following abbreviations are used in the text and tables: SBF = sniff body frequency;
FF = Flehmen frequency; NUF = normal urination frequency; NDF =
normal defecation frequency; NYF = normal yawning frequency; TYF =
threat yawning frequency; HSPID = hindleg straddle posture, inclusive duration (the time devoted to the hindleg straddle posture
including when it was concurrent with other behaviors); HSPED =
hindleg straddle posture, exclusive duration (the amount of time the
hindleg straddle posture occurred in the absence of other concurrent
events); HSPTFUDD = hindleg straddle posture, tail flicking, urination, and defecation duration (the amount of time all these acts were
simultaneous); PROHF = poll rub on hump frequency; PROOD = poll rub
on object duration.

Descriptions of Olfactory Signals

Urine, feces, and occipital or poll gland secretion are
apparent chemical substrates of several visual and olfactory patterns. In addition, the saliva of male camels may have significance
in visual and olfactory signaling. Specific behavior patterns
utilizing these scent sources are described below. While body
rubbing against objects and rolling may leave incidental olfactory
information these patterns do not appear to be specialized forms
of olfactory signaling.

Poll Gland-Hump Rubbing (Figure 1a). Standing, walking or
reclining males touch or rub the occipital gland against the base
of the anterior hump by leaning the head backwards. The duration
is usually less than one or two seconds. The pattern is infrequently
seen in females.

Object-Marking with the Poll Gland. The poll gland may be
rubbed against objects with a vertical or rotational motion while
standing, reclining on the knees of the forelegs, or while reclining on the ventrum or side. Walls and fences are the usual
targets of this form of marking, but occasional attempts may be
made to mark a brush pile. Marked surfaces are easily identified
by large dark stains and bits of hair.

The Tail-Urine Flapping and Defecation Display (TUFD display)
(Figure 1b). Normal urination and defecation in male and female
camels takes place with the hindlegs spread (hindleg straddle
position) and the tail held away from the body. Males in peak
rutting condition frequently assume a more exaggerated hindleg
straddle stance for prolonged periods and indulge in bouts of

Figure 1a.   Rubbing the anterior surface of the hump
with the poll gland.

Figure 1b.   The TUFD display in a male Bactrian camel.
Notice the white area saturated with urine
on the posterior hump.

rhythmic tail-flapping, urination and defecation. Figure 1b illustrates several characteristics of the TUFD display which are discussed below.

Tail-flapping. Sporadic bouts of tail flapping may occur
during normal activity and immediately before or after normal
urination and defecation. Typically the tail is moved through a
vertical arc, slapped against the back of the rear hump and then
returned to a partially extended or normally relaxed posture. In
display, however, the tail is held momentarily between the hindlegs and often anointed with a small amount of urine from the
posteriorly directed penis. Although the duration of tail flapping
usually exceeds that of urination the result of this behavior is
saturation with urine of the tail, hindlegs, rump and posterior
surface of the rear hump. Urine is also thrown forward beyond the
camel's body. Tail flapping is audible (Figure 2) and has a consistency of rhythm that justifies including it with a limited number of mammalian examples of "typical intensity". Urination and
defecation do not modify the rhythm, but the performance of other
patterns such as hump rubbing, threat-yawning, or adjustment of
body position interrupt the tail-flapping cadence. The mean interval between tail flap onsets was 3.056 ± 0.146 seconds (N = 90,
r = 1.47 - 8.87 seconds, Tex). Variations in tail-flapping
frequency during the display seen in Figure 3 result more from
interruptions in tail flapping than from changes in rhythm.

Urination. Males that are TUFD displaying appear to ration
their urine, usually dribbling only small amounts as the tail moves
in and out between the hindlegs. During long display bouts urination may cease for a variable period then resume at a slow drip.

Defecation. The pumping motion of the tail during the TUFD
display prolongs the time required for defecation because pellets
are not voided when the tail is down. However, the total number of
fecal pellets does not consistently differ between normal and display defecations. Thirteen display defecations by Tex averaged
27.5 pellets (s = 6.46), while 15 normal defecations by the other
three males averaged 30 pellets (s = 13.89). Males in strong rut
seem to defecate with greater than normal frequency during the
TUFD display. As many as four separate defecation bouts have
taken place during one display. Normal defecation rates ranged
from 0.2 bouts/hr (Tex) to 2.1 bouts/hr (No. 1).

Integration of the Elements. The TUFD display shows several
levels of complexity (Figure 3). The hindleg straddle stance (HSP),
its simplest form, has a predominantly visual impact. However,
the straddled stance is most commonly concurrent with tail flapping
which enhances the display's effect at least visually, and auditoral-
ly. Maximal complexity is achieved when all four elements occur

Figure 2. Sonogram of the sound of the tail flapping the base of the hump (left: narrow-, right: wide band filter; scale = 80-8000 kHz).

C. WEMMER AND J. MURTAUGH

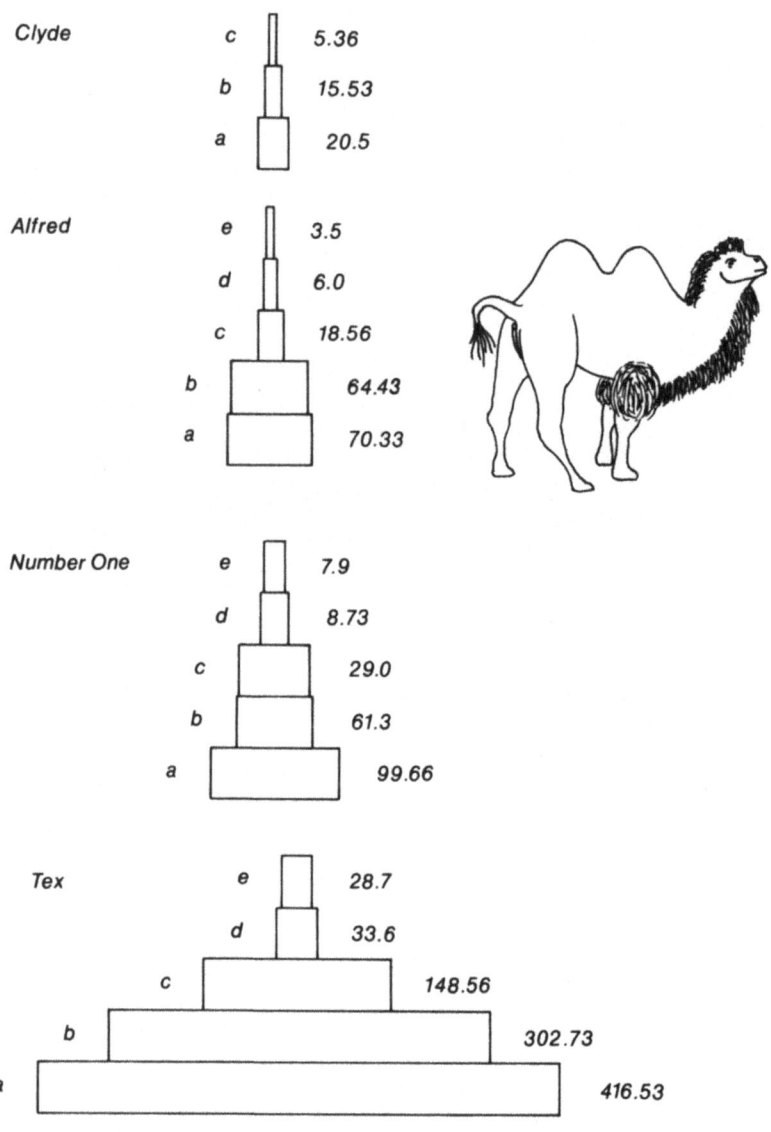

Clyde

c   5.36
b   15.53
a   20.5

Alfred

e   3.5
d   6.0
c   18.56
b   64.43
a   70.33

Number One

e   7.9
d   8.73
c   29.0
b   61.3
a   99.66

Tex

e   28.7
d   33.6
c   148.56
b   302.73
a   416.53

Figure 3. A time analysis of the TUFD display in four male Bactrian camels. The base level (a) of each pyramid represents the total amount of time each animal spent in the hindleg straddle stance; overlying levels represent the portion of time that the hindleg straddle stance was concurrent with (b) tail flapping, (c) tail flapping and urination; (d) tail flapping and defecation, and (e) tail flapping, urination, and defecation. Figures beside each horizon represent time in minutes.

together, but this takes place only about 5% of the time spent in the display stance. Several other visual, olfactory, and auditory patterns may occur during the display, as can be seen in Figure 4. Interaction with females occurs between displays.

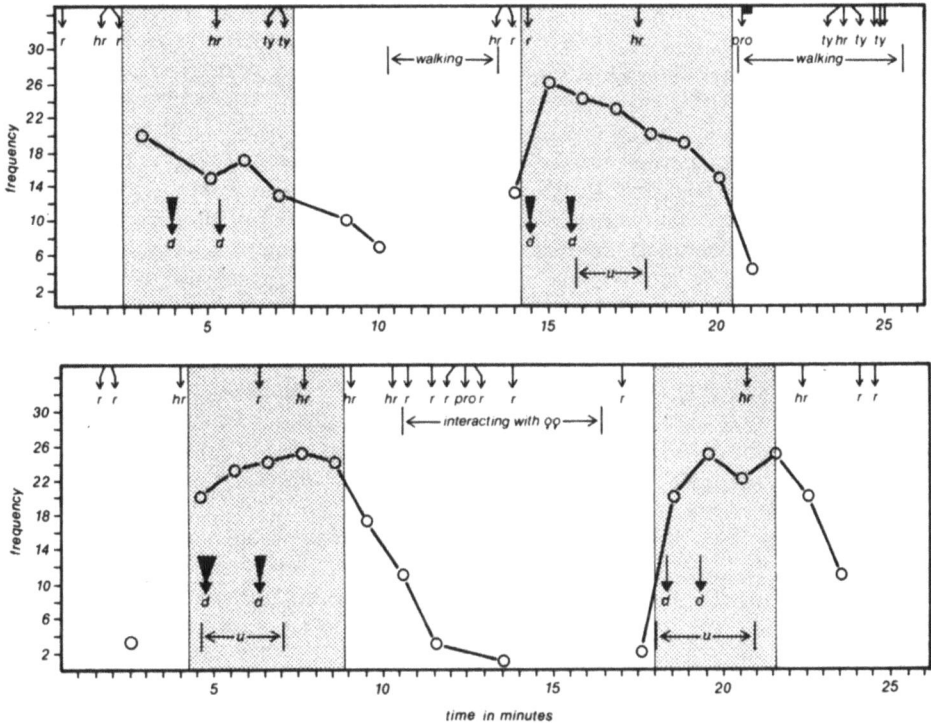

Figure 4.   Variations in a male Bactrian camel's TUFD display and associated behaviors. Both graphs are of Tex as he was with the female herd on 28 February (upper) and 10 February 1979 (lower). Strippled areas designate the duration of the hindleg straddle stance. Connected circles plot the number of tail-flaps each minute. Abbreviations: u = the duration of urination; d = defecation; app = approach; dep = depart; r = rumble; hr = hump rub with poll gland; proo= poll gland rubbing on object; ty = threat yawn.

Saliva. The saliva of the Bactrian camel is normally viscous
and somewhat clear, but excited animals of both sexes work it into
a white fluffy meringue consistency by repeated tooth gnashing.
This foamy saliva is seen most commonly and most copiously in males
during the rut. Head shaking distributes it on the face and neck.
The saliva may possibly have olfactory signal properties.

Secretion of the Occipital Gland. Secretion of the occipital
gland in Bactrian camels is a seasonal phenomenon characteristic
of mature males, Pocock (1923) reported the gland in both sexes,
but recent study shows it to be absent in female dromedaries (Singh
and Baradwaj, 1978), and there is no evidence of the structure in
female Bactrians. The gland is a paired macroscopically discrete
structure overlying the skull's occipital ridges. On the basis of
our limited histological evidence, the Bactrian poll gland appears
similar to that of the dromedary camel. Glandular lobules containing tubulo-alveolar glands are embedded in a fibrous stroma. The
glands give rise to secretory ducts which open into the necks of
hair follicles (Figure 5a). The secretion has a water consistency
and dark brown color.

The occipital gland shares a number of features in common with
certain secondary sex traits of other ungulates. Onset of secretion and secretory flow seems to be age related. Older camels
tend to begin secreting earlier than younger ones. In 1978-79 at
the age of 14 years Tex exhibited initial glandular flow on 30
October, Clyde and Alfred (nearly seven years) followed on 11
December and 21 January while No. 1 (nearly four years) showed
initial flow on 24 January. The initial secretion of the gland
is characterized by moistening of only the hair overlying the gland.
In two or three and a half weeks the glandular flow is sufficient
to trickle down the neck and saturate the long hair of the nape.
Neither nape hair nor the hump hair to which scent is transferred
appear to be structurally modified to hold the secretion, but the
electron photomicrographs of Figure 5b, c shows the coagulated
exudate on a single hair shaft from the nape.

The Circumstances of Olfactory Signaling

Differences Between Males. The behavioral differences between
individual males should be considered before examining the other
results. Camels are long-lived ungulates. Sexual maturity is
reached by the age of four and adult body size by five, but sexual
maturation continues probably until at least eight to ten years.
We compared the behavior of the four males when in the company of
females during the 1978-79 season. For each of the 11 behavioral
measures a t-test was performed on the mean values for each of the
six possible male-male combinations using the pooled sample variance
of the four males (Table III).

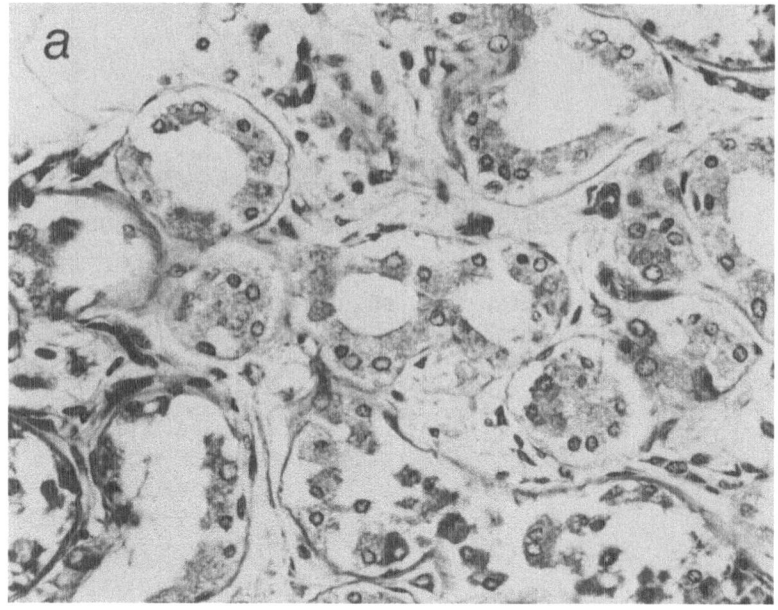

Figure 5.   a. Photomicrograph of the poll gland from a one year
               (non-secreting) male Bactrian camel (blurred areas
               are the result of antolysis). Dark staining rings
               of cells are tubulo-alveolar glands which drain
               into ducts leading to hair follicles.

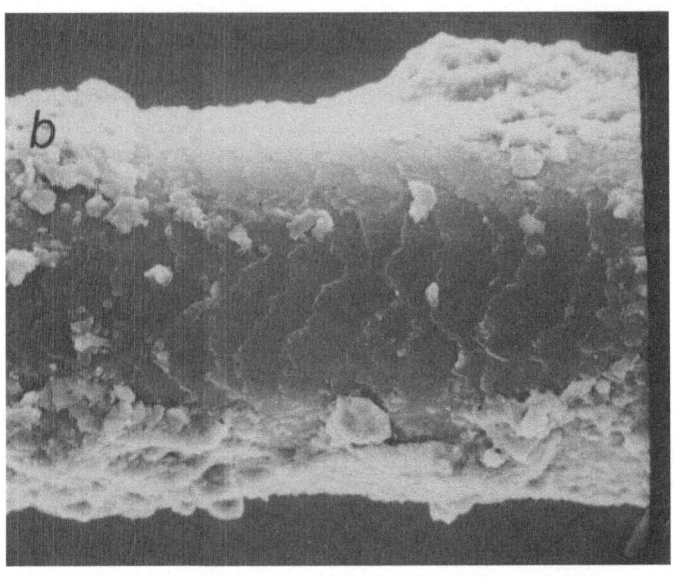

Figure 5b. A shaft of hair from the nape mane showing unspecialized
plate-like scales from the cuticle.

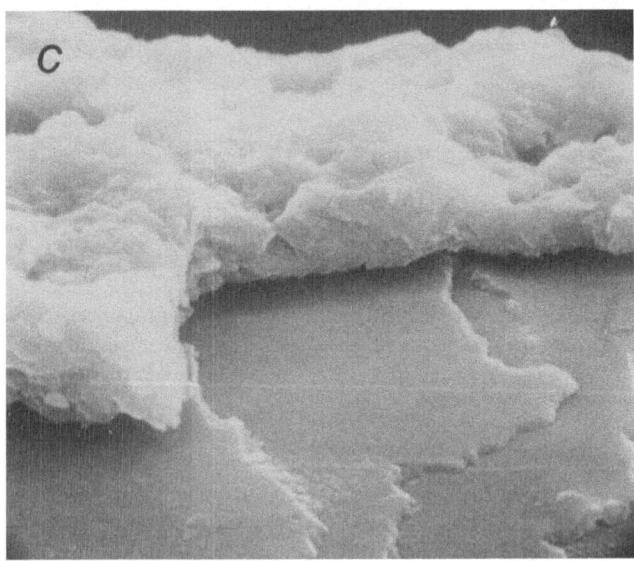

Figure 5c. A magnified view of a nape hair showing encrusting
coagulum from the poll gland.

Table III. T-values arising from a comparison between male pairs
             of 11 behavior measures.

Male Comparisons

Behavior	Tex vs Clyde	Tex vs No. 1	Tex vs Alfred	Clyde vs No. 1	Clyde vs Alfred	No. 1 vs Alfred
SBF[1]	1.259	5.332***	2.239*	3.679***	0.421	3.795***
FF	0.112	4.268*	1.917	3.574***	1.272	2.961***
NUF	2.479*	4.949***	0.524	2.489*	1.985*	4.476***
NDF	1.896	5.407***	1.210	3.290**	0.931	4.494***
NYF	1.584	1.965	2.375*	0.560	0.210	0.454
TYF	0.796	2.872**	1.672	3.029**	1.970	1.760
HSPID	1.848	0.252	2.019*	1.530	0.642	1.185
HSPED	0.112	0.268	0.868	0.150	0.892	0.580
HSPTFUDD	1.915	1.384	1.939	0.174	0.421	0.118
PROHF	2.430*	4.183***	0.320	5.912***	3.349**	3.852***
PROOD	1.127	1.658	0.302	2.232**	2.285*	1.412

[1]See Methods section for abbreviations.

*P < .05.

**P < .01.

***P < .001.

Clyde and Number One differed significantly in seven out of the 11 variables, while Tex and Alfred differed significantly from Number One in six and five variables respectively. The three older animals (Tex, Alfred and Clyde) differed from one another the least (two or three variables being significantly different). Hump rubbing and normal urination frequency showed great variation between individuals, differing significantly in five out of the six combinations. In summary, male behavior in the presence of females was not homogeneous. The older camels were more similar to one another than to the youngest male who differed the most.

Experiments with Social Setting. Tex was placed in a variety of social situations during the 1977-78 season (Table II). The sequence and duration of these social circumstances were dictated by considerations of animal management rather than experimental design, and behavioral data were routinely gathered. Several differences in the observed and expected frequency and duration of two patterns are evident in Table IV. Hump rubbing for example occurred four times more often than expected when Tex was with females, and nearly half as often when alone. The other two "with male" categories differed little from expected. The observed occurrence of hump rubbing in the four categories deviated significantly from expected (X^2 = 38.69, df = 3, P < .005). It was clear that even our imperfect method of isolating Tex from females had a dampening effect on his level of rutting behavior, including hump rubbing and the TUFD display. However, HSPID levels remained high for one day after Tex was separated from the females and confined alone in the barn, and this one day accounts for the 415 seconds recorded in the category "alone".

To extend these investigations males were subjected to different social conditions as much as possible in the winter of 1978-79. The pooled data of the four males are compared in Table V in three different circumstances.

Table IV. Observed and expected (in parentheses) frequencies and durations (in seconds) of two olfactory signals in four social situations (subject = Tex).

Social Situations and Time

Behaviors	W/females (90 min)	W/males (812 min)	In view of males (322 min)	Alone (210 min)	Totals
PROHF	14 (3.3)	28 (30.0)	7 (11.9)	4 (7.8)	53
HSPID	715 (149.41)	325 (1349.25)	928 (534.98)	415 (348.87)	2383 sec

Table V. Comparisons of seven measures of olfactory behavior in three social situations using the T-test (2-tailed) (data pooled for 4 males).

Behavior Measure[1]	I W/females (N=109,113)	II W/male (N=7)	III Alone (N=97,100)	T-values and Significance Level[2]		
				I x II	I x III	II x III
FF	2.088 ±0.335	0.143 ±.143	0.230 ±0.075	†5.34***	†5.41***	0.30 ns
HSPID	88.920 ±19.049	110.714 ±62.039	16.740 ±6.461	0.28 ns	†3.59***	-3.25**
HSPED	17.017 ±6.090	0	11.140 ±4.855	†2.79**	0.74 ns	†2.29*
HSPTFD	60.115 ±12.760	79.857 ±36.282	14.040 ±5.653	-0.38 ns	3.16***	-2.38**
HSPTFUD	32.902 ±6.614	35.428 ±25.859	6.410 ±2.170	0.09 ns	†3.81***	†-1.12 ns
PROHF	2.752 ±0.280	1.428 ±0.841	1.280 ±0.278	1.16 ns	3.72***	-0.14 ns
PROOD	13.256 ±2.155	3.857 ±3.857	6.896 ±1.580	1.09 ns	2.33*	0.51 ns

[1]See "Methods and Materials" for abbreviations; 2/† T-value based on separate variance estimate;

*P < .05; **P < .01; ***P < .001; ns P > .05.

        <u>With Females vs. With Another Male</u>. Most behavioral measures
were moderately high and showed little difference between these two
conditions, even though much of the time during male-male encounters
was spent fighting. The number of Flehmens (FF), and the total
amount of time devoted exclusively to the hindleg straddle position
(HSPED) were significantly greater when with females than when encountering another male.

        <u>With Females vs. Alone</u>. Males kept alone displayed olfactory
patterns ranging from one half to nearly a tenth of the magnitude
exhibited in the presence of females. All but one measure (HSPED)
were significantly different. The hindleg straddle posture in the
absence of other TUFD display components (HSPED) was adopted sub-
equally when males were alone or in the company of females.

        <u>With Male vs. Alone</u>. Three measures showed significant differ-
ences between these two conditions: the time spent in the TUFD display including all its variants (HSPID) was significantly greater
when with another male than when alone. The duration of tail-
flapping (HSPTFD) showed a similar trend, but the hindleg straddle
posture in the absence of other components (HSPED) was more common
in lone camels than in males encountering another male.

DISCUSSION

        Rutting behavior in Bactrian camels is characterized by three
patterns having probable functions in scent communication: namely
the TUFD display, hump-rubbing and object marking with the poll
gland. The TUFD display consumes the most time of the three patterns and contains design features for signaling over long and short
distance. The body posture is distinctive at long range, and the
exaggerated straddling of the hindlegs identifies the signaling
animal as a male. Both hump-rubbing and tail flapping anoint
the highest parts of the body with scent, which according to
Bossert and Wilson's (1963) model would effectively increase the
active space of scent emanating from point sources on the poll
gland, anterior and posterior humps. The rhythmicity of the
camel's tail flapping is possibly a means of atomizing urine at a
constant rate. Saturated hair on the posterior hump serves in
effect as a urine reservoir which keeps the tail wet when the
bladder is empty. While a boundary of ice may encircle the
saturated region during cold weather, constant agitation by tail
flapping prevents the area from freezing. The sound of the tail
smacking the hump may also broadcast the displaying male's activity to animals who are nearby but upwind. In addition, the TUFD
display's apparent coincidence with hump rubbing, rumbling
vocalizations, and threat yawning may function in breaking the
monotony of an otherwise static display.

This study shows that while TUFD display is not specific to one situation significantly more time is spent performing it in the presence of females than with males or when alone. This is different from the dromedary where tail flapping is unseen in courting males (Gauthier-Pilters, 1959). Gauthier-Pilters identified the dromedaries' hindleg straddle stance as a threat display, and reported tail-flapping as a component whose rate and strength of delivery is proportional to the camel's level of excitement. TUFD displaying in this study also took place when Bactrian male's encountered each other, but these episodes always quickly escalated into fights, possibly as a result of the confining circumstances. As in the dromedary, rival Bactrian camels exhibit the most extreme spreading of the hindlegs, but contrary to the dromedary the rate of tail flapping is relatively constant. Anointing the tail with products of the uro-genital system is uncommon in ungulates, but Krishnan (1972) describes an interesting analogue in the cow Asian elephant: "Visually flagrant signs of her being in season are not manifested..., but the cow bends her tail between her hind legs sharply so as to slap the abdomen with its tip, and then draws the brush at the tip firmly up, rubbing it against the vagina: the tail is then held aloft in the air and waved about in a scent flag". The hippopotomas also uses the tail as a feces beater during the dunging display which was postulated by Olivier and Laurie (1974) to function in dominance assertion.

Coblentz (1976) hypothesized that dominant male ungulates could achieve two reproductive advantages by anointing themselves with their own urine. By advertising a high plane of physical condition a male could deter challenge by potential rivals, or estrus could be pheromonally synchronized in females being tended. Tests of the first hypothesis are lacking, but evidence from domestic species support the latter hypothesis.

More controlled experiments are needed to clarify the significance of the poll gland. Rubbing of both the hump and objects occurred more often or for longer periods when males were with females, but not significantly so. Poll gland rubbing of objects, however, was concentrated in the small area where males could interact across the fence, and mutual fence rubbing by adjacent males was not uncommon. Furthermore fighting males occasionally snap at the dorsal neck mane which is saturated with poll gland secretion. So despite contradictory results the poll gland seems to play a role in male rivalry. It also exhibits many similarities to the temporal gland of the Asiatic elephant. Present only in males, its rate of flow increases with age until they are reproductively prime, and its annual secretory phase is characterized by aggressive behavior, increased libido and scent marking (Eisenbergh, McKay, and Jainudeen, 1971; Scheuermann and Jainudeen, 1972).

ACKNOWLEDGEMENTS

This study would not have been successful without the dedication of the following camel watchers: Ann Hedrick, Makiko Buma, Colm Dobbyn, Carolyn Emerick, Rondle Sampson, Elizabeth Glassco, Douglas Lebrun and Larry Collins. Peggy Murtaugh assisted in tabulating data. We thank the following Smithsonian employees: Dr. Charles Roberts for statistical advice; Dr. Richard Montali for histological services and commentary; Karl Kranz and Mary J. Mann for preparation and electron microphotography of hair specimens.

REFERENCES

Bannikov, A.G. 1976a. Wild camels, Wildlife 18(9): 398-403.

Bannikov, A.G. 1976b. Last refuge of the wild camel, International Wildlife 6(1): 31.

Bossert, W. and Wilson, E.O. 1963. An analysis of olfactory communication among animals, J. Theoret. Biol. 5: 443-469.

Bulliet, R.W. 1975. The camel and the wheel, Harvard University Press.

Coblentz, B. 1976. Functions of scent urination in ungulates with special reference to feral goats, Amer. Nat. 110: 549-557.

Eisenberg, J.F., McKay, G.M. and Jainudeen, M.R. 1971. Reproductive behavior of the Asiatic Elephant (Elephas maximus maximus L.), Behaviour 34: 193-225.

Gauthier-Pilters, H. 1959. Einige Beobachtungen zum Droh-, Angriffs- und Kampfverhalten des Dromedarhengstes, sowie über Geburt und Verhaltensenwicklung des Jungtieres, in der nordwestlichen Sahara, Z. f. Tierpsychol. 16(5): 593-604.

Krishnan, M. 1972. An ecological survey of the larger mammals of peninsular India, J. Bombay Nat. Hist. Soc. 69(2): 297-351.

Lincoln, G.A., Youngson, R.W. and Short, R.V. 1970. The social and sexual behaviour of the red deer stag, J. Reprod. Fert. Suppl. 11: 71-103.

Novoa, C. 1970. Reproduction in Camelidae, J. Reprod. Fert. 22: 3-20.

Olivier, R.C.D. and Laurie, W.A. 1974. Habitat utilization by hippopotamus in the Mara River, E. Afr. Wildl. J. 12: 249-271.

Pocock, R.I. 1923. The external characters of the pigmy hippopotamus (Choeropsis liberiensis) and of the Suidae and Camelidae, Proc. Zool. Soc. London 1923: 531-549.

Scheurmann, E. and Jainudeen, M.R. 1972. "Musth" beim Asiatischen Elephanten (Elephas maximus), Zool. Garten 42: 131-142.

Singh, U.B. and Barawaj, M.B. 1978. Anatomical, histological and histochemical observations and changes in the poll glands of the camel (Camelus dromedarius). Part 3. Acta Anatomica 102, 74-83.

CHEMOSENSORY SEARCHING BY RATTLESNAKES DURING PREDATORY EPISODES

David Chiszar and Kent M. Scudder [1]
University of Colorado
Boulder, CO   80309

ABSTRACT

Striking rodent prey is a prerequisite for chemosensory searching (prey trailing using the tongue – Jacobson's organ system) in rattlesnakes. High tongue flick rates occur after striking prey but not after seeing, smelling or detecting thermal cues arising from a rodent. At least two consequences of striking appear to be involved in the activation of chemosensory searching: (1) proprioception arising from the biomechanical aspects of the strike, and (2) chemical input to the vomeronasal organs concomitant with the strike. These two effects seem to be additive.

Striking also induces a preparedness to ingest prey. Furthermore, envenomation appears to alter chemical aspects of the prey (especially oral and/or nasal tissues) in a way that is detectable by rattlesnakes. In fact, snakes given a choice between envenomated prey and prey killed by other means reliably selected the former. Since this selection depends upon chemical modifications in the prey's oral and/or nasal tissues, it is suggested that this may also contribute to the well-known tendency for rattlesnakes to swallow their prey head first. Furthermore, since rattlesnakes prefer to attend to chemical cues arising from envenomated rodents, we may also be able to use this fact to explain why snakes often fail to strike a second live mouse. Namely, striking is released by visual and/or thermal cues, and if the snake is preferentially attending to other cues (i.e., those coming from an envenomated mouse), then the strike releasers will simply not be processed at the level of the innate releasing mechanism.

Previous work in our laboratory showed that adult garter snakes explored a novel environment more vigorously if it contained food odors than if it did not (Chiszar, Scudder and Knight, 1976; Scudder, 1976a; see also Burghardt, 1967, 1969, 1970; Kubie and Halpern, 1975). Our measure of exploratory behavior was tongue flick rate (TFR), although heart rate has been used in more recent experiments with essentially the same results (Chiszar, Lipetz, Scudder and Pasanello, under review). However, adult rattlesnakes of three taxa showed statistically equal TFRs in an analogous study: in the experimental condition snakes were placed into a novel environment containing mouse odors, while in the control condition snakes were placed into an identical environment containing no mouse odors. TFRs for all snakes were recorded for 20 minutes, and ANOVA revealed no significant differences as a function of chemical cues for Crotalus v. viridis, Sistrurus catenatus tergeminus and S. c. edwardsi (Chiszar, Scudder and Knight, 1976). Also, using Burghardt's cotton swab technique for presentation of chemical stimuli, we found that newborn rattlesnakes of three taxa showed no preference for prey vs. non-prey chemicals. Several different prey chemicals were offered at several different temperatures, and we recorded TFR and strike latency. No baby rattlesnakes struck any cotton swabs, nor did the animals exhibit differential TFRs contingent upon the chemical contents of the swabs (Chiszar and Radcliffe, 1977). It is significant that these neonates accepted live food immediately after the swab tests. Hence, they were hungry and ready to strike and swallow mouse pups. We concluded that predatory strikes and chemosensory searching by rattlesnakes were not released by prey-derived chemical cues (at least under the conditions of the present experiments).

Early work with European vipers indicated that prey trailing movements, including tongue flicking, occurred with greater frequency and precision if the snakes had struck rodent prey before being tested than if they had not (Baumann, 1927, 1928; Wiedemann, 1932; Naulleau, 1964, 1966; see also Dullemeijer, 1961, for a comparable experiment with rattlesnakes). These papers clearly suggest that striking is a prerequisite for chemosensory searching or prey trailing. If this is true, then the finding that rattlesnakes did not exhibit higher TFRs in novel environments containing prey odors might be explained by the fact that none of the snakes had struck prey before being tested. Perhaps the same experiment would have a very different outcome if snakes were allowed to strike a mouse before being transported to the test situation.

Striking and the Induction of Chemosensory Searching

Our first step in determining if striking is a prerequisite for chemosensory searching in rattlesnakes was to observe a variety of snakes during normal feeding episodes. We simply counted the TFRs occurring prior to and after striking and found average rates of 0.0 and 30.7 per min, respectively (Chiszar and Radcliffe, 1976). TFR began to rise within two or three minutes after striking, and was sustained at a high asymptotic rate until the commencement of swallowing. These data strongly suggest that striking releases chemosensory searching which facilitates the location of dead or dying prey.

However, it is important to recognize that all snakes in this study not only struck prey but they also saw, smelled and detected thermal cues arising from prey prior to striking. Hence, it is possible that these latter stimuli released chemosensory searching. In fact, most previous studies dealing with chemosensory searching (in Viperidae and Crotalidae) have confounded the effect of striking with effects of prey detection. Until these potential effects are separated, we will not be able to say with certainty that TFR increases are due to striking or to detection of visual, thermal or olfactory stimuli associated with prey.

Our first attempt to separate these effects (Chiszar, Radcliffe and Scudder, 1977; Scudder, 1976b) used a simple two-condition experimental design. In one condition (Control) a live mouse was suspended into the snake's cage via forceps and was held just out of striking range for three seconds. The mouse was then removed. In the experimental condition, a mouse was suspended into the snake's cage for three seconds and then brought into striking range. The snake was then allowed to strike the mouse, after which the mouse was removed. If prey detection is the significant factor for the activation of the chemosensory searching, then both conditions should produce increased TFRs. On the other hand, if striking is the important event, then only the experimental condition should generate a high TFR.

Subjects in this experiment were four C. enyo, two C. lepidus and two C. polystictus. Each snake was observed in each of the conditions on alternate weeks, just prior to regularly scheduled feeding sessions.

A summary of the results is shown in Table 1 where it can be seen that the experimental condition produced a far higher TFR during the ten minute post-mouse-presentation period. In fact, snakes generally coiled up and returned to a resting state in the control condition whereas they became strongly aroused and engaged in continuous searching behavior in the experimental condition. This pattern of results was the same in each of the taxa.

We have subsequently replicated this finding in another set of C. enyo (Chiszar, Radcliffe and Smith, 1978) and in C. viridis,

Table 1. Mean TFRs for eight rattlesnakes during 10 minutes
          after striking a mouse (condition S) and during 10
          minutes after seeing, smelling and detecting thermal
          cues arising from a mouse (condition NS).

| | Minutes | | | | | | | | | |
	1	2	3	4	5	6	7	8	9	10
Condition S	20	28	35	40	47	41	41	44	44	41
Condition NS	21	14	13	11	8	9	6	7	6	6

C. durissus, S. miliarius and S. catenatus (Scudder, in preparation; Scudder, Short and Chiszar, 1978). In most of these experiments at least six specimens of each taxon were used. Every
individual exhibited a strong difference between TFRs in the strike
vs. no-strike conditions. Accordingly, we conclude that (1) striking releases chemosensory searching, and (2) the effect is general
across both genera of rattlesnakes. [2]

     This finding helps to explain two commonly observed phenomena.
First, rattlesnake feeding episodes typically contain several responses which usually occur in a fixed sequence: Prey detection,
striking, chemosensory searching (trailing), location of the dead
prey, location of the head, and head-first swallowing. Whenever
responses occur in a regular sequence, questions naturally arise
concerning integrative mechanisms (e.g., why don't responses occur
out of the usual order?). Part of the answer in the case of rattlesnake predation lies in the fact that chemosensory searching is
released by striking. Hence, this relationship imposes a sequential
dependency upon the responses such that it is extremely unlikely
(unless snakes are explicitly trained to take dead prey and, thus
bypass the need to strike) for searching to occur prior to, or in
the absence of, striking. Secondly, some snakes that refuse to
feed in captivity can be induced to do so if they can be "teased"
into striking. This technique is commonly employed by reptile
curators (C.W. Radcliffe, personal communication). Perhaps the
technique works because striking, even if it is induced by
"teasing", activates chemosensory searching. If so, we have an
example of a response chain which will run to completion (i.e.,
swallowing) even if it is initially activated in an artificial
manner.

Another point of interest is the fact that chemosensory searching occurs at a high rate after striking even though we took great
care not to place mouse odors in the post-strike environment (mice
were never allowed to touch the floor or walls of the snake's cage
and were removed immediately after the strike). Nonetheless, TFRs
increased consequent to the strike and remained at asymptotic levels
for at least 30 minutes (Scudder, in preparation). This does not
mean that TFRs would not be still higher if prey odors were available; rather, it means that a strike will induce high TFRs even in
the absence of such cues. The next section will explore this point
further.

Finally, all of our experiments involved successful strikes
(i.e., the snake actually contacted and envenomated the mouse). We
have not examined the extent to which chemosensory searching can be
activated by an unsuccessful strike (e.g., a "miss"). However, it
is our impression based on casual observations that an unsuccessful
strike will not induce high TFRs. We believe that feedback mechanisms exist which inform the predator of the success of the strike
and that these mechanisms are tied into the chemosensory searching
phenomenon. More specifically, we envision a physiologically-
based "if statement": if the strike is successful, search for the
trail and follow it; otherwise, strike again.

Effects of Prey and Other Odors on TFR After Striking Prey

Having satisfied ourselves that striking is a prerequisite for
chemosensory searching in rattlesnakes, our next step was to determine if the chemical composition of the post-strike environment
would influence TFRs. The following procedures were used: Rattlesnakes of several taxa were exposed to the experimental (strike)
and control (no-strike) conditions described above. However, just
after the mouse was removed from the snake's cage, a petrie dish
containing cedar shavings was placed into the snake's cage. The
mouse was never allowed to touch the snake's cage, hence, no odors
were deposited by the prey. The cedar shavings were either clean,
or soiled by mice which had lived on them for at least three days,
or lightly doused with perfume.[3] All tongue flicks emitted by the
rattlesnakes in all conditions were recorded by hand-held counters
for the following ten minutes.

Mean TFRs for three taxa of rattlesnakes exposed to these
contingencies are shown in Figure 1.

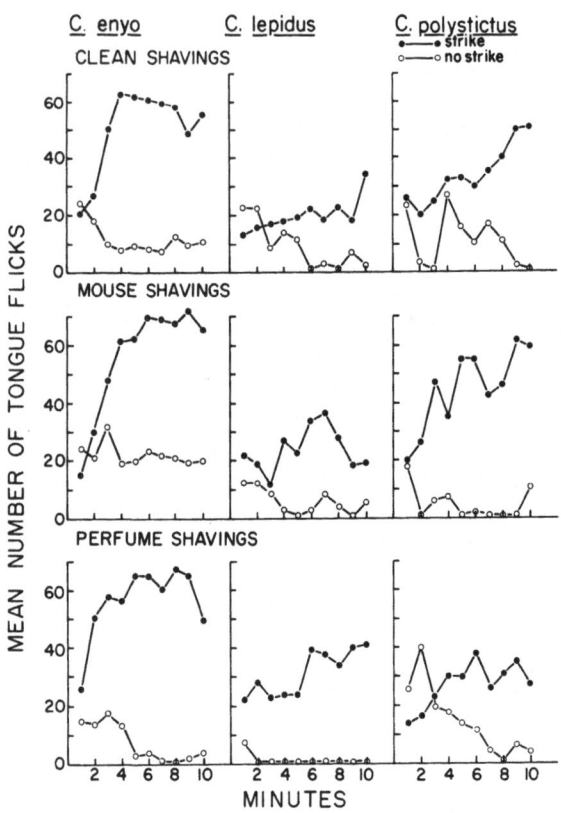

Figure 1.   Mean TFRs during 10 min. after striking a mouse and
during 10 min. after seeing, smelling and detecting
thermal cues arising from a mouse.   Subjects were
four C. enyo, two C. lepidus and two C. polystictus.
Each animal was observed six times: three times
under the strike (experimental) condition and three
times under the no-strike (control) condition .
During these runs petrie dishes containing clean
cedar shavings or shavings with mouse or perfume odors
were placed into the snakes' cages.   Each snake
received the six conditions in a different randomly
determined order.

First, it is clear that the strike vs. no-strike effect occurred in all conditions. Second, the taxa differ in the magnitude of this effect; C. enyo showed a stronger effect than did C. lepidus or C. polystictus. We suspect that this difference resulted from the fact that our specimens of C. enyo were calmer and better adjusted to captivity than were the C. lepidus and C. polystictus. Third, and most important, the TFR after striking was not significantly influenced by chemical composition of the post-strike environment. Once chemosensory searching was released by striking, it rose to a maximal level and sustained itself regardless of the type of odor we placed into the cage via the petrie dishes. It is for this reason that we (Chiszar, Radcliffe and Scudder, 1977) have suggested that chemosensory searching after striking is a fixed action pattern. (The term "modal action pattern" is preferred and we will use it in the future; c.f., Barlow, 1968, 1977).

Since the completion of the study whose results are shown in Figure 1, we have run a variety of additional taxa through the same paradigm. In most cases, the results were like those in Figure 1; namely, post-strike TFR always rose to a maximum independent of the odor introduced into the snake's cage. However, one exception has so far come to light. Specimens of C. d. terrificus showed higher TFRs during the post-strike period when mouse odors were present than when clean or perfumed shavings were present. This may represent a greater dependence on prey chemicals in these neo-tropical snakes than in the dry-land forms. Speculation concerning this matter is forthcoming (Scudder, in preparation; see the Discussion section of this paper).

How Does Striking Release Chemosensory Searching?

Another question of some importance concerns the mechanism(s) by which the act of striking results in the activation of chemosensory searching. At least two possibilities have occurred to us. First, striking is a vigorous muscular response which undoubtedly has proprioceptive consequences. Perhaps the proprioceptive afference arising from striking somehow activates the vomeronasal system. Although we are unable to specify the specific neural pathways through which this effect might occur, there is nothing against the principle involved. Second, striking brings the roof of the snake's mouth into contact with the prey's integument. Since the ducts to the vomeronasal organs open through the roof of the mouth, it is easy to hypothesize that the snake receives a sudden burst of chemical input directly into the vomeronasal system as a consequence of striking. Perhaps this input somehow primes or activates the vomeronasal system, thus producing elevated TFRs immediately after the strike.

To separate these two potential consequences of striking and
to evaluate their relative contributions to chemosensory searching,
Scudder (in preparation; see also Scudder, Short and Chiszar, 1978)
utilized a method which has long been a mainstay of ethological
research (Tinbergen, 1948, 1951). Specifically, he constructed
cotton models of mice which were effective in releasing strikes in
our rattlesnakes. Some models were kept sterile, others were
doused with perfume and others were invested with mouse odors.
We then recorded TFRs in a variety of rattlesnakes after they
struck each kind of model. If striking per se (i.e., proprioception)
is the causative factor in chemosensory searching, then high TFRs
ought to be observed after any of the models have been struck.
However, if input of prey chemicals to the vomeronasal organs
concomitant with striking is the critical element, then high TFRs
ought to be seen only after striking the mouse-odored model.

Figure 2 presents the data from six S. c. tergeminus which
have been run through all of the above conditions. Six specimens
from each of three additional taxa have also been run through these
conditions; but these data have not yet been statistically analyzed.
The basic pattern of results shown in Figure 2, however, appears to
be representative of the results found with the other species (S.
m. barbouri, C. v. viridis and C. d. terrificus). The strike –
no-strike effect is again obvious, indicating that biomechanical
consequences of striking do, in fact, contribute to the activation
of chemosensory searching. This is true even when the snakes have
struck sterile or perfumed models. Yet, it is also clear that
striking the mouse-odored model generated a higher TFR than was
seen after the other two models were struck. This reveals that the
chemical concomitants of striking also contribute to chemosensory
searching.

At this point, we conclude that both proprioception and chemical input to the vomeronasal organ are important. In fact, it
appears that the contributions of these two factors are additive
since together (i.e., striking a model that has been mouse-odored)
they produce a TFR which is comparable to that seen after striking
a real mouse.

Chemosensory Searching and Prey Selection

If rattlesnakes have always been offered live prey in captivity,
they will be unlikely to swallow unless they have just struck prey.
This does not mean that rattlesnakes cannot be trained to accept
dead food without striking; rather, it means that under normal
conditions, strike-induced chemosensory searching is important for
activating or priming ingestive responses. To document this
generally believed phenomenon, we offered dead prey to rattlesnakes
who had recently struck (condition S) a mouse, and to rattlesnakes
that had not (condition NS). Of 94 observations made under condi-

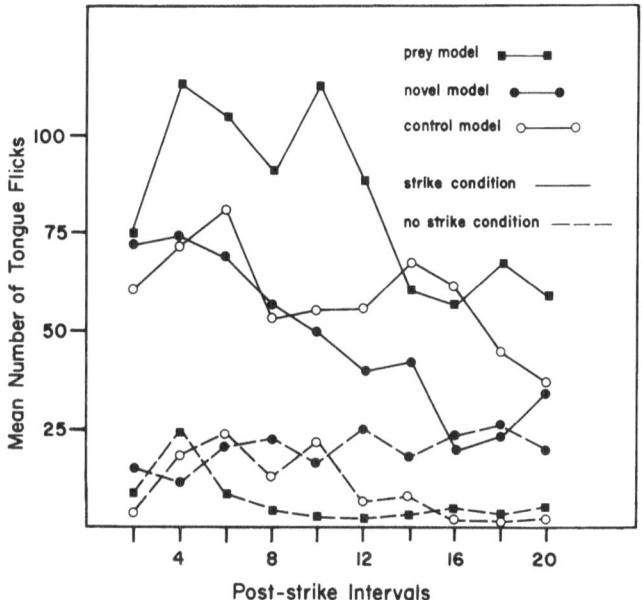

Figure 2.   Mean TFRs for six <u>Sistrurus</u> <u>c</u>. <u>tergeminus</u> observed under
six conditions.   In each condition a cotton mouse model
was suspended into the snake's cage for three sec.   In
three conditions the snake was then allowed to strike
the model; in the remaining conditions the model was
removed without giving the snake the opportunity to
strike.   During these runs the model was either sterile
or contained mouse (prey) or perfume (novel) odors.

tion S, 67 (71.3%) resulted in grasping and/or swallowing within
20 minutes. However, of 24 observations made under condition NS,
only 5 (20.8%) resulted in grasping and/or swallowing. This
difference was statistically significant and thus confirms the
notion that striking induces a preparedness to accept food.

At this point, David Duvall hypothesized that rattlesnakes
might prefer an envenomated mouse if given a choice between two
mice (one killed by envenomation and one killed by other means
such as cervical dislocation or asphyxiation). To test this idea,
Duvall allowed a variety of rattlesnakes to strike mice which were
then removed from the snakes' cages. Upon the death of the envenomated mouse, another of the same inbred strain, sex and age
was killed manually. Then the two mice were placed side by side
into a snake's cage, equidistant from the predator's head. Of
67 cases in which choices were made, 46 involved selection of the
envenomated mouse (Duvall, Chiszar, Trupiano and Radcliffe, 1978).

This finding has since been replicated with refined procedures
(Duvall, Scudder and Chiszar, in press) which indicated (1) that
envenomation alters chemical properties of oral and/or nasal
tissues of the prey, and (2) that rattlesnakes are particularly
sensitive to these alterations and may, in fact, use them to
identify the head-end of the dead mouse.

Strike Induced Inhibition of Subsequent Strikes

Once a rattlesnake has struck a mouse and has begun the
chemosensory searching process, the predator is unlikely to strike
a second live mouse. Our first experiment on this topic involved
presentation of two mice. The first was removed after being struck,
and the second was introduced either immediately after the first was
removed, 3 minutes later, 5 minutes later, 10 minutes later, 15
minutes later or 20 minutes later. The first mouse was always
struck with 30 seconds. The second mouse was usually not struck
at all (within a 300 second trial) if it was introduced within 5
minutes after the first mouse was struck. After the passage of
20 minutes, the second mouse was struck with a latency that no
longer differed from that involved in striking the first mouse
(Chiszar, Radcliffe and Scudder, 1977). If a rattlesnake is
allowed to ingest the first mouse, the second one will be struck
immediately (Chiszar, Radcliffe and Smith, 1978), indicating that
signals arising from the act of swallowing are able to disinhibit
the strike-induced inhibition of subsequent strikes (SIISS).

We are viewing SIISS as an attentional phenomenon. During
the time that the snake is engaged in chemosensory searching,
visual and/or thermal cues from a second live mouse are ignored.
There is no doubt that the snake can detect the second mouse; in
fact, it is not unusual for the snake to move away from the second

mouse. Hence, SIISS is most likely based on changes in attentional processes rather than fluctuation in sensory thresholds consequent to striking the first mouse. Perhaps SIISS somehow contributes to the utilization of chemical cues associated with the first mouse's trail by preventing the snake from being distracted by the second mouse. Another (not mutually exclusive) possibility derives from Duvall's work. We know that odors arising from an envenomated mouse are different from those arising from a nonenvenomated one, and we know that the snakes prefer the former. Hence, SIISS may be based on the fact that a rattlesnake simply prefers to attend to chemical cues associated with a struck prey than to chemical, visual or thermal cues associated with a nonenvenomated one. If the snake's attentional registers are limited, then attending to the former might preclude attending to the latter. Failure to attend to the latter should eliminate striking the second live mouse as long as cues deriving from the envenomated mouse are available.

We have recently completed an experiment that supports this interpretation. Rattlesnakes under prolonged food deprivation will accept nonenvenomated dead prey even without striking. Increased hunger seems to reduce or eliminate the strong preference for chemical cues associated with envenomated rodents. Moreover, increased hunger (8 weeks of deprivation) eliminated SIISS: the second mouse was struck immediately even though the first one had not been swallowed (Miller and Scudder, 1979). Although several interpretations of SIISS have suggested themselves to us, we consider the "competing stimulus" hypothesis to be the most parsimonious one at present.

DISCUSSION

Our experiments have shed some light on the mechanisms which mediate sequential dependencies in rattlesnake predation. Striking is first released by visual (movement) and thermal cues arising from prey. Chemosensory searching is caused to occur by proprioceptive and vomeronasal consequences of striking. Chemosensory searching initially facilitates prey tracking and then location of the dead prey's head. Also, chemosensory searching and/or striking primes the snake to begin swallowing once the prey is located. This collection of effects explains how it happens that individual components of the response chain rarely occur out of the modal order.

Research currently in progress is exploring the factors which give rise to variations of the modal order. For example, we are examining the effect of hunger level on the rattlesnake's (and other species) willingness to accept dead prey without first striking it. Our initial results suggest that increasing hunger

produces reductions in sensory thresholds for detection of prey
odors. Hence, a sufficiently hungry snake may be able to attend
to odor trails even without making a priming contact with prey.
This may be the mechanism for carrion feeding in snakes which
practice this behavior on a facultative basis. Although mechanical
contact with prey may be necessary to prime trailing and ingestive
responses in mildly hungry snakes, increased hunger may cause the
vomeronasal system to become extremely sensitive such that priming
by actual contact with prey is no longer necessary. In such cases,
snakes may be induced to trail and swallow entirely on the basis
of chemical cues such as might arise from carrion (Chiszar,
Lipetz, Scudder and Pasanello, in preparation; Cowles and Phelan,
1958). This line of inquiry has been particularly exciting because it suggests the existence of multiple feeding strategies
that can be triggered by stimulation arising from and correlated
with the degree of food deprivation. At low levels of deprivation,
many snakes might be called ambushers in that they position themselves and literally wait for prey to pass by.

        At very high levels of deprivation, the same snake may become
a carrion feeder. At intermediate levels of hunger a mixture of
both strategies may occur such that snakes will seek out (presumably
on the basis of chemosensitivity) live prey and then engage in a
more or less modal sequence. This could be called foraging. Perhaps these multiple strategies should be conceptualized as a
"behavioral scale" (Wilson, 1975) along which the snake can move as
a consequence of hunger.

        Rattlesnakes appear to possess an atavistic sensitivity to
odors associated with fish mucus (Chiszar, Scudder and Smith, in
press). Perhaps this is a remnant of the presumed Agkistrodon
ancestry of rattlesnakes (Gloyd, 1940, p. 249). In any case, we
are currently exploring the possibility of piscivorous feeding
behavior in rattlesnakes,and the extent to which the modal response sequence might be altered in this context. Related to this,
we have extended our observations to water moccasins with very
interesting initial results. When water moccasins are exposed
to fish mucus odors, they behave like garter snakes and other
natricines. That is, the moccasins exhibit far higher TFRs than
they do under appropriate control conditions. However, when they
are exposed to rodent odors they show no elevation in TFR. It is
well known that water moccasins will accept rodent prey, however,
and ours do so quite readily. Accordingly, we performed a strike -
no-strike experiment with moccasins and found results much like
those discussed above for rattlesnakes (Chiszar, Simonsen,
Radcliffe and Smith, in press). Hence, water moccasins, like
rattlesnakes, do not respond with increased TFR to rodent odors
if they have not previously struck a rodent. But, if they have
recently struck a mouse, they exhibit a strong TFR elevation
(i.e., strike-induced chemosensory searching). Thus, moccasins

reveal the natricine strategy (immediate response to chemical cues) when offered fish and the rattlesnake strategy when offered rodents. We are now working on the possibility that increased hunger will cause water moccasins to employ the natricine strategy even when rodent prey are offered. Such an outcome might confirm our idea about the dependence of sensory thresholds on hunger level.

In any case, water moccasins seem to be the most complex snakes we have so far examined. Their eclectic feeding preferences and multiple strategies indicate the existence of sophisticated perceptual and integrative processes. This raises some very intriguing questions about the evolution of feeding mechanisms in the Crotalidae. If contemporary Agkistrodon species are at all representative of Agkistrodon ancestors, then we must assume that the latter were not only more generalized than their contemporary rattlesnake descendants but that they were also more complex in terms of perceptual and central-processing capacities. As rattlesnakes adapted to their current environmental nitches, they must have become increasingly specialized to deal with a reduced variety of food types. This, in turn, resulted in a loss of some ancestral capacities and/or sensitivities. Accordingly, in some respects, we might consider at least some species of rattlesnakes to have undergone a secondary simplification. To examine this possibility, we are beginning to compare dry-land and wet-land rattlesnake species in order to measure the extent to which the latter have retained Agkistrodon feeding eclecticism and, hence, continue to possess more complex perceptual and central processes than dry-land forms.

REFERENCES

Barlow, G.W. 1968. Ethological units of behavior. In D. Ingle (Ed.) The General Nervous System and Fish Behavior. Chicago, University of Chicago Press, 217-232.

Barlow, G.W. 1977. Modal action patterns. In T. Sebeok (Ed.) How Animals Communicate. Bloomington, Indiana, Indiana University Press, 98-134.

Baumann, F. 1927. Experimente uber den Geruchssinn der Viper. Rev. Suisse Zool. 34, 173-184.

Baumann, F. 1928. Uber den Nahruhgserwerb der Viper. Rev. Suisse Zool. 35, 233-239.

Burghardt, G.M. 1967. Chemical cue preference of inexperienced snakes: comparative aspects. Science 157, 718-721.

Burghardt, G.M. 1969. Comparative prey-attack studies in newborn snakes of the genus Thamnophis. Behaviour 33, 77-114.

Burghardt, G.M. 1970. Chemical perception in reptiles. In J.W. Johnston Jr., D.G. Moulton and A. Turk (Eds.) Communication by Chemical Signals. New York, Appleton-Century-Crofts, 241-308.

Chiszar, D. and Radcliffe, C.W. 1976. Rate of tongue flicking by
    rattlesnakes during successive stages of feeding on rodent
    prey. Bull. Psychon. Soc. 7, 485-486.
Chiszar, D. and Radcliffe, C.W. 1977. Absence of prey-chemical
    preferences in newborn rattlesnakes (Crotalus cerastes, C.
    enyo, and C. viridis). Behav. Biol. 21, 146-150.
Chiszar, D., Radcliffe, C.W. and Scudder, K.M. 1977. Analysis of
    the behavioral sequence emitted by rattlesnakes during feeding
    episodes. I. Striking and chemosensory searching. Behav.
    Biol., 21, 418-425.
Chiszar, D., Radcliffe, C.W. and Smith, H.M. 1978. Chemosensory
    searching for wounded prey by rattlesnakes is released by
    striking: a replication report. Herp Review 9, 54-56.
Chiszar, D., Scudder, K.M. and Knight, L. 1976. Rate of tongue
    flicking by garter snakes (Thamnophis radix haydeni) and
    rattlesnakes (Crotalus v. viridis, Sistrurus catenatus
    tergeminus and S. c. edwardsi) during prolonged exposure to
    food odors. Behav. Biol. 18, 273-283.
Chiszar, D., Scudder, K.M. and Smith, H.M. in press. Chemosensory
    investigation of fish mucus odor by rattlesnakes. Bull.
    Maryland Herp. Soc.
Chiszar, D., Simonsen, L., Radcliffe, C.W. and Smith, H.M. in press.
    Rate of tongue flicking by water moccasins (Agkistrodon
    piscivorous) during prolonged exposure to various food odors;
    and, strike-induced chemosensory searching in the cantil
    (Agkistrodon bilineatus). Trans. Kan. Acad. Sci.
Duvall, D., Chiszar, D., Trupiano, J. and Radcliffe, C.W. 1978.
    Preference for envenomated rodent prey by rattlesnakes. Bull.
    Psychon. Soc. 11, 7-8.
Duvall, D., Scudder, K.M. and Chiszar, D. in press. Rattlesnake
    predator behavior: Mediation of prey discrimination, and
    release of swallowing by odors associated with envenomated
    mice. Animal Behavior.
Dullemeijer, P. 1961. Some remarks on the feeding behavior of
    rattlesnakes. Kon. Ned. Akad. Wetensch. Proc. Ser. C 64,
    383-396.
Gloyd, H.K. 1940. The Rattlesnakes, genera Sistrurus and Crotalus.
    A study in zoogeography and evolution. Chicago, The Chicago
    Academy of Sciences, special publication number four.
Kubie, J. and Halpern, M. 1975. Laboratory observations of trail-
    ing behavior in garter snakes. J. Comp. Physiol. Psychol.
    89, 667-674.
Losey, G.S. 1976. Animal models: do they fool them or us?
    Paper presented at meetings of the Animal Behavior Society,
    Boulder, Colorado.
Miller, T. and Scudder, K.M. 1979. Prolonged food deprivation
    decreases strike latency and increases strike accuracy in six
    species of Crotalus (Reptilia: Serpentes). Presented at meet-
    ings of the Colorado-Wyoming Academy of Science, Durango,
    Colo. (title #233).

Naulleau, G. 1964. Premieres observations sur le comportement de
    chasse et de capture chez les vipers et les couleuvres.
    Terre Vie 1, 54-76.
Naulleau, G. 1966. La Biologie et le Comportement Predateur de
    Vipera aspis au Laboratoire et dans la Nature. These, Paris:
    P. Fanlac.
Scudder, K.M. 1976a. Responses of garter snakes (Thamnophis
    radix haydeni) during prolonged exposure to fish-permeated
    water. The Colorado Herpetologist 2, 6-10.
Scudder, K.M. 1976b. "Perceptual switching"in rattlesnakes.
    Paper presented at meetings of the Animal Behavior Society,
    Boulder, Colorado.
Scudder, K.M. In preparation. Mechanisms mediating the sequential
    aspects of predatory episodes in Crotalid snakes. Ph.D.
    dissertation, University of Colorado, Boulder.
Scudder, K.M., Short, J.S. and Chiszar, D. 1978. Mechanisms
    activating chemosensory searching in rattlesnakes during
    predatory episodes. Paper presented at meetings of the
    Animal Behavior Society, Seattle, Washington.
Tinbergen, N. 1948. Social releasers and the experimental method
    required for their study. Wilson Bulletin 60, 6-51.
Tinbergen, N. 1951. The Study of Instinct. Oxford: Clarendon
    Press.
Wiedemann, E. 1932. Zur Biologie der Nahrungsaufnahme der
    Kreuzotter, Vipera berus L. Zool. Anz. 97, 278-286.

FOOTNOTES

1. The research here reported was made possible by three grants
from the M.M. Schmidt Foundation. We would like to thank the
following people for their friendship and collaboration during
various phases of the work: T. Carter, D. Duvall, F. Igelhart,
L. Knight, V. Lipetz, L. Pasanello, C.W. Radcliffe, J.S. Short,
L. Simonsen, H.M. Smith, J. Trupiano, M. Wand, and S. Wellborn.
Finally, we wish to thank J.L. McGaugh, editor of Behavioral
Biology, for permission to reproduce Fig. 1. Requests for reprints
should be sent to D. Chiszar, Department of Psychology, University
of Colorado, Boulder, Colorado, 80309.

2. It is important to keep in mind that all of our snakes were
individually housed and that all snakes were always fed live rodent
prey. Results may be quite different if either of these setting
conditions (especially the latter) is altered.

3. Mouse odors are presumed to be biologically significant to
rattlesnakes; but, such odors are also pungent (at least to human
observers). Perfume is considered to be pungent but biologically
irrelevant. Hence, we use the perfume condition as a pungency
control (see Losey, 1976).

WHAT THE NOSE LEARNS FROM THE MOUTH

John Garcia and Kenneth W. Rusiniak

Department of Psychology and Mental Retardation
Research Center, University of California
Los Angeles, CA 90024

The mouth is a border zone dividing the feeding domain into two separate regions, each with its own empirical laws. The first region is the external arena, in which terrestrial animals cope with signals precisely located in space and time to obtain food or drink. The second is the internal homeostatic pool, which places demands for food or drink upon the animal. Pavlovian laws of conditioning were formulated in the external environment, where the flavor of food acts like an unconditioned stimulus (UCS) which, if applied promptly, can be used to modify a vast array of coping behaviors in man, rat or pigeon. These same laws do not apply to the internal environment, where the flavor of food acts more like a conditioned stimulus (CS) and a specialized visceral feedback acts as the UCS to modify the specific flavor incentive for which man, rat or pigeon strives.

TASTE IS A STRONGER CS THAN ODOR FOR A POISON UCS

The nose is situated just above the mouth in most terrestrial animals. From this vantage point, the nose sniffs food or drink as it approaches and before it enters the mouth to be tasted. Thus, odor is the first component and taste is the second component of the sequential compound CS called flavor. However, when the role of odor and taste in toxiphobia are independently evaluated, taste proves to be a much more effective CS than odor. For a flavored fluid followed by delayed poison effects, the taste alone is a sufficient CS for establishing a conditioned flavor aversion. Anosmic animals condition as readily as intact animals, indicating that taste is sufficient and odor is not necessary. In addition, when an odorant is placed on a paper disc surrounding a water spout, the odor alone is a relatively ineffective CS for intact animals conditioned with delayed illness (Hankins et al., 1973). This differential effectiveness cannot be explained by the relative salience of the odor and taste components, because when the same flavor is paired with immediate foot-shock, odor is much more effective than taste. Anosmic animals are severely handicapped in flavor-shock conditioning; so are intact rats when the odor component of the flavored fluid is masked (see Figure 1; Hankins, Rusiniak and Garcia, 1976).

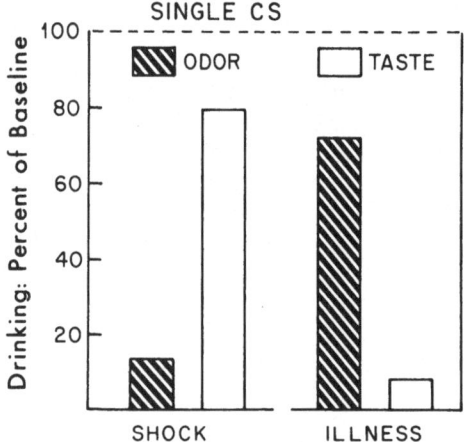

Figure 1.   Odor alone versus taste alone as single CSs for shock or lithium-induced illness. For independent groups of rats (n = 10), on the left, almond-scented tap water or 0.1% saccharin water signalled immediate shock on 5 trials randomly interspersed with 5 non-shock water trials each day. After forty trials, training was terminated when the almond odor group reached asymptote. On the right, the same stimuli signalled 190 mg/kg lithium chloride intubation 30 min later every third day for 3 pairings, and training terminated when the saccharin taste group reached asymptote.

The selective nature of this conditioning is clear-cut but relative. Taste-shock conditioning can be established with intense stimulation, promptness and practice, but never approaches the speed and efficiency of **odor-shock** conditioning. Similarly, odor-illness conditioning is possible with immediate illness and repeated trials but it never approaches the robust one-trial potency of taste-illness conditioning (e.g., Hankins et al., 1973; Hankins et al., 1976). In any case, under ordinary circumstances in which danger must be actively sniffed out by the rat, a scent which is a perfectly good cue to learn about the painful insults of the external world is useless to avoid the noxious effects of slow-acting poison upon the homeostatic environment (Garcia and Brett, 1977). Odor is an external CS designed to serve appetitive coping behaviors more efficiently than consumption and homeostatic balance.

TASTE POTENTIATES ODOR, THE OPPOSITE OF OVERSHADOWING

An associative analysis of the rat feeding on a poisoned meal leads to this paradox. On the one hand, given that the nose sniffs the food before the mouth tastes it, odor should be the better CS because taste is a redundant CS, according to the information hypothesis of Egger and Miller (1962). Thus, odor should block or overshadow conditioning to taste. On the other hand, given that taste is a stronger more effective CS than odor when tested alone, taste should overshadow conditioning to odor, according to Pavlovian laws (Kamin, 1968; Mackintosh, 1974, pp. 48-51). Thus, a simple information hypothesis will not do.

Guided by conditioning principles, we formulated a two-phase hypothesis of higher order conditioning to resolve this paradox. In the first phase, taste associated with toxic illness becomes aversive, and in the second phase, odor associated with the aversive taste becomes secondarily aversive (Garcia, Rusiniak and Brett, 1977). But the two-phase hypothesis was proved inadequate by tests on laboratory rats. When odor was repeatedly paired with an aversively conditioned taste in a second-order conditioning paradigm, odor did not readily acquire the capacity to delay or depress consumption (Rusiniak, Hankins, Garcia and Brett, 1979). Such second-order effects may be obtained under other circumstances (Rusiniak, Gustavson, Hankins and Garcia, 1976), but seem too weak to account for the way rats and coyotes learn to avoid poison with a sniff in one or two trials.

What emerged from these studies was a robust effect quite the opposite of overshadowing. When taste and odor were combined to form a compound CS for a delayed poison UCS and then tested separately in extinction, odor alone proved to be a more powerful deterrent than taste alone. In fact, odor continued to exert a depressive effect upon consumption even after the aversion to taste was completely extinguished. In another compound CS experiment, the odor component was held constant and the taste component was varied. Odor at zero taste concentration was completely ineffective when paired with a single toxic lithium treatment delayed thirty minutes. However, the strength of the odor component tested alone in extinction, increased as a direct function of the taste concentration used in acquisition. Interestingly, a taste too weak to be effective when used alone, was sufficient to potentiate the attendant odor cue (see Figure 2). Potentiation may be a two-way effect; odors may potentiate weak tastes but the effect is not as dramatic as the potentiation of odor by taste. As might be expected, the combination of an ineffective taste and an ineffective odor add up to a potent compound CS (Rusiniak et al., 1979). Furthermore, it appears that the potentiation of odor by taste may be limited to internal homeostatic conditioning. When immediate foot-shock is the UCS, odor is the strongest CS, and it does not effectively potentiate the weaker taste component (see Figure 3; Rusiniak, Lopez and Rice, 1979).

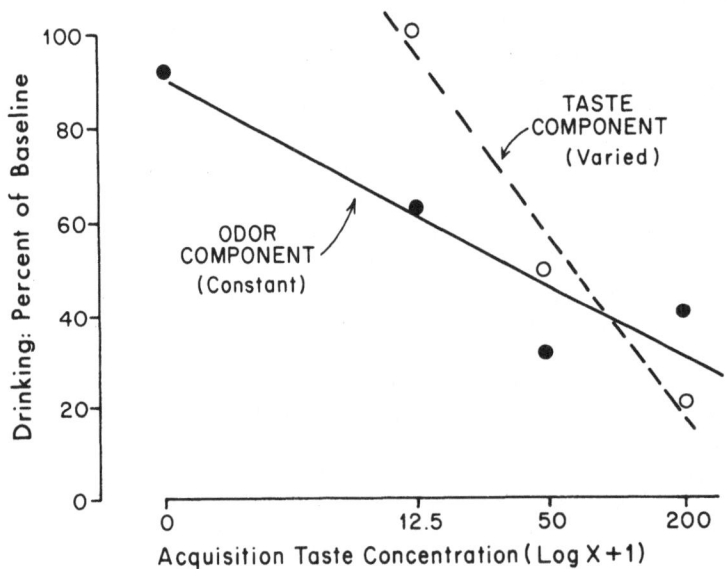

Figure 2. Synergistic potentiation of odor by taste. The almond scent concentration was held constant at 2% and the concentration of saccharin was varied from 0 to 200 milligrams of saccharin per liter of water. Four groups of rats (n = 16) were given a single pairing of the specified compound flavor and then the odor and taste components were tested separately, as illustrated here (Replotted from Rusiniak et al., 1979).

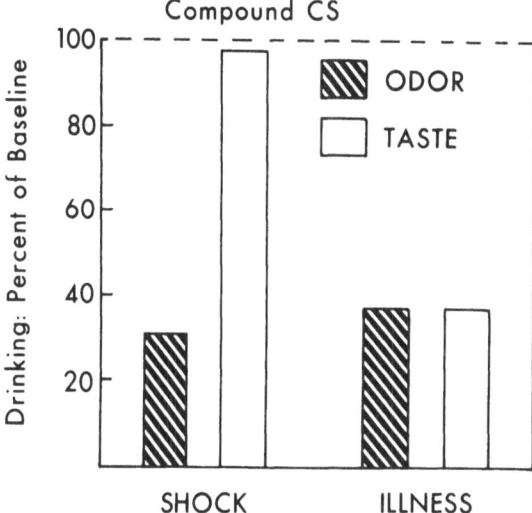

Figure 3. Potentiation of odor by taste after compound conditioning with lithium illness, but not shock. On the left,
compound almond-scented saccharin signalled immediate
shock, as in Fig. 1, for one group of rats (n = 10).
After 100 trials, almond odor and saccharin taste were
tested alone in extinction. On the right, compound
almond-scented saccharin signalled 60 min delayed
lithium illness on one trial for one group of rats (n =
10); almond odor alone and saccharin taste alone were
then tested in extinction.

THE BIRD'S EYE LEARNS ABOUT POISON FROM ITS BEAK

A similar phenomenon was discovered in hungry red-tailed hawks.
For some of these raptors, consumption of black laboratory mice was
repeatedly paired with lithium illness; subsequently, they developed
an aversion for all laboratory mice including white mice, a standard
dietary item never paired with illness. Presumably, black and white
mice taste alike even though they look different. For other hawks,
the black dead mouse was flavored with a mild quinine solution but
was, nevertheless, avidly consumed by the carnivorous birds. After
this black-bitter mouse was paired with the toxic UCS in a single
trial, the hawks specifically avoided black dead mice by visual
inspection alone, as if the taste component had potentiated the
visual cues (Brett, Hankins, and Garcia, 1976). A similar, but
more complete study has been conducted on thirsty pigeons. First,
blue water alone was demonstrated to be a weak CS and salty water

alone was demonstrated to be a strong CS for a toxic lithium UCS. When blue and salty were combined into a compound CS, the taste did not overshadow color. Instead, the stronger component potentiated the weaker component so that color appeared to be even stronger than taste in extinction trials (Clarke, Westbrook and Irwin, 1979). Figure 4 compares this visual potentiation in pigeons with odor potentiation in rats.

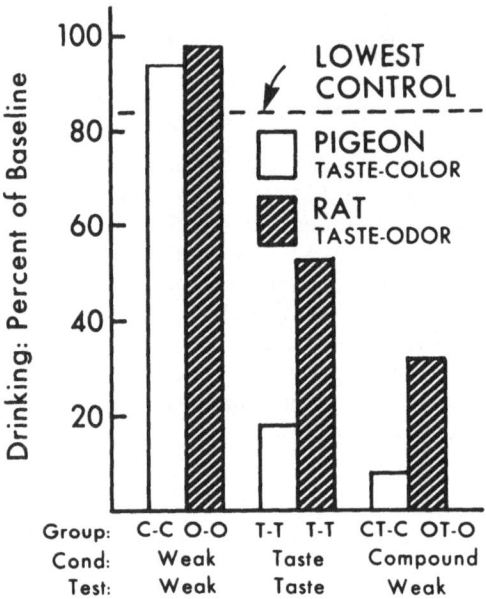

Figure 4.   Synergistic potentiation in pigeons and rats. After a single pairing with delayed lithium illness, blue color alone for pigeons, and almond odor alone for rats were weak cues (left); salt and saccharin tastes were effective, respectively (center). When compounded with taste in a single poison trial, weak cues became strong when tested alone (right). (Pigeon data replotted from Clarke et al., 1979).

Let us summarize the evidence presented thus far. Poisoning produces a form of internal conditioning which is quite different from Pavlovian conditioning. Taste is the most effective CS. Conditioning can occur in one or two trials even when the poison UCS is presented a half hour or more after the taste CS. Moreover, a strong taste CS will not necessarily overshadow a weaker CS, it may potentiate the weaker CS. In mammalian species, taste potentiates odors and in avian species, taste potentiates visual food cues. Tests with rats indicate that when shock is the UCS, odor is a more effective CS than taste and that a strong CS will overshadow rather than potentiate a weaker CS. Therefore the unique features of internal conditioning are a function of the poison UCS.

NEUROVASCULAR FEEDBACK IS THE UCS FOR CONDITIONING THE PALATE

There has been much confusion regarding the nature of the UCS, as investigators searched for a common factor among the various agents used to produce a conditioned taste aversion (see Coil and Garcia, 1978 for a review). The UCS is neither toxin, nor toxicosis per se, it is a stimulus. And the common factors of stimulation lie not in the external agents, but within sensory neuronal systems, as Johannes Muller pointed out over a century ago.

Under natural feeding conditions, the meal enters the mouth and provides the CS by stimulating the chemoreceptors of taste which send fibers directly to the brain stem area controlling emesis. The UCS effects of the ingested meal converge directly to the same area via many other neural and vascular routes. If the toxic irritation is limited to the stomach, the vagus carries the noxious message to the emetic system; severing the vagus blocks the conditioned taste aversion in rats (Coil, Rodgers, Garcia and Novin, 1978) and also blocks vomiting in dogs (Borison and Wang, 1953). If the toxic chemical is introduced into the blood, it circulates to the area postrema where internal monitors relay the noxious stimulation to the adjacent emetic system. When this vascular UCS route is employed, severing the vagus blocks neither conditioned aversions in rats nor vomiting in dogs. However, extinction is rapid in vagotomized rats, even when an intravenous toxic UCS is employed, probably because the vagus is a mixed nerve with sensory and motor fibers serving the conditioned response (CR). In the intact rat, when the aversive stimulus contacts the tongue during extinction, efferent fibers "turn the stomach". These gastric CRs are then fed back to the CNS via the sensory fibers of the vagus to maintain the strength of the aversion (see Figure 5). Furthermore, if the emetic system is depressed by anti-emetic drugs, conditioned flavor aversions are attenuated (Coil, Hankins, Garcia and Jenden, 1978).

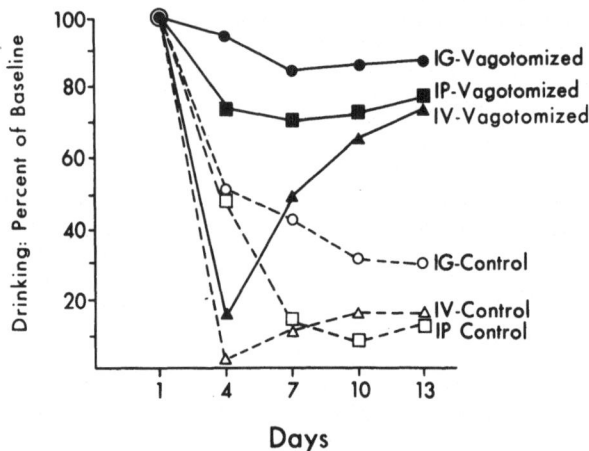

Figure 5.   The effect of vagotomy on taste aversions.   Vagotomy
blocked the acquisition of an aversion for 0.1%
saccharin induced by repeated applications of 5 mg/kg
copper sulphate in the stomach (IG) or the peritoneal
cavity (IP), but not when the poison (10 mg/kg) was
introduced directly into the blood (IV).   Note that
extinction was rapid in the IV-Vagotomized group,
indicating that the vagus contributes to the maintain-
ence of the aversion.   (Replotted from Coil et al.,
1978).

        The action of this brain stem area extends far beyond the
emetic response to influence the entire feeding system.
Palatability can be raised to facilitate sustenance as well as
lowered to avoid overindulgence.   Flavors gain palatability when
paired with beneficial UCS injections ending a thiamine deficiency
or recuperation from an emetic injection (Garcia, Hankins, and
Rusiniak, 1979).   Rats depleted of salt seek flavors previously
associated with salt (Fudim, 1978).   Glucose is pleasant to the
tongue when a person's blood sugar is low, but unpleasant when
the blood sugar is restored (Cabanac, 1971).   Gastric receptors
signal the commencement and termination of eating (Deutsch, J.A.,
Young, W.G. and Kalogeris, T.J., 1978), apparently assessing the
nutritional value of the stomach load before absorption restores
the cellular deficits.   Severing the vagal route to the brainstem
disrupts modulation of feeding behavior (Novin, D., 1976).   Thus,
the vascular route and the vagal route to the brainstem are involved in both positive and negative modification of the incentive
value of flavor.

INDEPENDENCE OF THE INTERNAL AND EXTERNAL CONDITIONING DOMAINS

The internal domain, where the palate is conditioned by visceral feedback, is independent in many ways from the external domain, where behavior is conditioned by surface contact reinforcement (see Garcia et al., 1977 for a review). For example, a rat can be trained to make an instrumental response, to press a lever in a box or turn left in a spatial maze, to obtain a sweet reward. That sweet reward can then be made aversive with contingent toxicosis in another place at another time. But when the rat is returned to the original situation under extinction conditions, it will continue to make the instrumental response, as if it is not aware that the sweet reward for which it labors in vain has been changed into a disgusting fluid.

The independence of external behavior and internal regulation is also seen in the behavior of wild mammalian predators. A coyote accustomed to preying on rabbits and then given a meal of rabbit flesh laced with lithium will subsequently charge eagerly towards a rabbit seen in the distance. The odor of rabbit in its nostrils and/or the taste of rabbit in its mouth will cause the coyote to break off its attack and slink away, ears back and tail down in disgust and confusion. During a series of extinction trials, the coyote will often approach and sniff the aversive rabbit carefully. Ultimately, it may again kill the rabbit and gingerly eat a small bit, returning after a while to sample rabbit again and again, thus regaining its taste for its natural prey through nutritional feedback. Hungry wolves averted to the taste of mutton will display social pack behavior toward living sheep. The prey is still visually attractive but gastronomically repulsive. Faced with conflict, the social wolves orient toward their former prey, but if they approach too near, the defenseless sheep bluffs and the wolves give way, bowing in submission like puppies. After feeding on rabbit, the wolves completely ignore the sheep, as if satiation has dissolved the curious and artificial social contract between prey and predator (Gustavson et al., 1976).

Furthermore, the internal incentive modification process and the external behavior modification process can change in opposing directions in the same animal without undue conflict. Using a conditioned emotional suppression (CER) paradigm, Brett (1977) presented anosmic rats with bitter harmless water accompanied by a single pulse of foot shock at the midpoint of the brief (60 sec) drinking interval. With repeated trials, the bitter water gradually became a danger signal eliciting the CER just prior to the time of shock, much as an audiovisual signal elicits a precise CER. At the same time, bitter water was satisfying thirst. Therefore bitter gained incentive value and drinking gradually increased in the post-shock period (see Figure 6). Similar opposing effects

were obtained in a shuttle-box where an aversive solution of
saccharin was the only cue by which anosmic rats could avoid foot
shock. Ultimately, they learned to run away from the taste CS,
but when they were offered that same flavor in another context,
their preference for the taste CS had increased twofold (Garcia
and Hankins, 1977). Thus, when a taste CS is followed by an external shock UCS, the avoidance responses are specific to the time
and place of shock UCS, indicating the UCS is feared but the CS is
not. Other studies indicating that signalled shock (UCS + CS) is
preferred to unsignalled shock (UCS - CS) also indicate that the
CS does not become aversive in shock studies (Weiss, 1972). In
contrast, when a taste CS is followed by a toxic UCS, the taste CS
is rejected at other times and in other places, indicating that the
CS has become distasteful in and of itself.

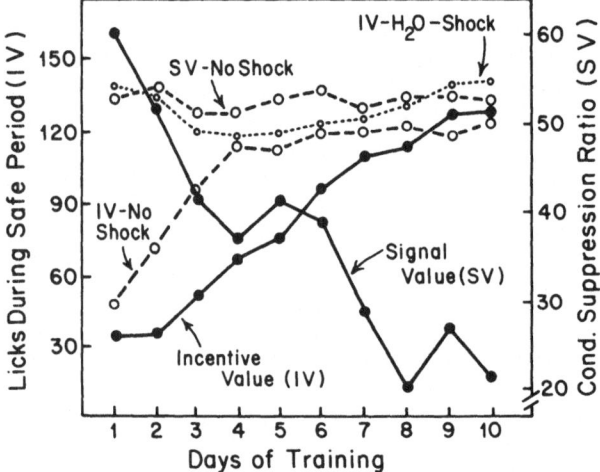

Figure 6.   Signal versus incentive modification in the same animals.
            An aversive bitter taste (.15 g/L) signalled a single
            pulse of shock delivered at the midpoint of a 1 min
            drinking period. With practice on 10 daily trials, rats
            (n = 8) learned to use the aversive taste to defend
            against shock (signal value), but drank more of the
            aversive shocked fluid (incentive value) after shock be-
            cause it reduced thirst without ill effects. Control
            groups (n = 5) drank bitter without shock (no shock),
            or water followed by shock (H2O-shock). (Replotted
            from Brett, 1977).

INDEPENDENCE OF INTERNAL CONDITIONING AND COGNITIVE AWARENESS

In humans, flavor-illness conditioning appears to be relatively independent of cognitive associations relating cause to effect. During motion illness caused by a tumultuous sea, the emetic reflex is apt to expel innocent food and drink from the body and establish a detrimental aversion for nutritious food despite the awareness that sea-sickness, not food, is the true source of his discomfort. Children under treatment for neoplastic disease develop taste aversions for the specific flavor of ice cream presented before chemotherapy, despite their knowledge gained from prior experience that the nausea was caused by the medication, not the ice cream (Bernstein, 1978).

In some respects, the rat acts as if it is attributing its visceral state to appropriate causes in a cognitive, almost human way. It develops an aversion for the flavor of food or drink more readily than it does for the food dish or for the sickness chamber. All things being equal, it will develop a stronger aversion for the last flavor consumed. But, if the last meal was a familiar, safe food, and the prior meal was a novel previously untested food, it will develop a stronger aversion for the novel food despite the longer CS-UCS interval. During its illness and recovery, the rat is careful of its diet, rejecting strong novel tastes. After recovery, the rat retains an enduring suspicion of any food similar to the poisoned food (see Garcia et al., 1977 for a review).

Moreover, a clear appreciation or "awareness" of the toxic UCS reaction may not be necessary for a conditioned taste aversion in animals. In fact, if the rat is allowed to drink a flavored solution and remain alert for about fifteen minutes, perhaps to consolidate the CS memory trace, the delayed nauseous UCS is effective even when presented after cortical functioning has been degraded by anesthesia (Roll and Smith, 1972) or by potassium chloride applied to the cortex (Buresova and Bures, 1973). Since the emetic UCS is effective when the subject appears partly or wholly unconscious, it is entirely possible that many "irrational" food aversions may result from adventitious conditioning by a delayed UCS of which the subject has no memory.

UNION OF THE INTERNAL AND EXTERNAL DOMAINS BY THE PERIPHERAL SENSORS

It is the peripheral taste receptors in the mouth which unite the external hunting region of the feeding domain with the internal regulatory region. By synergistic potentiation, taste directs olfaction's role in feeding, for the nose is a busy organ performing a variety of other functions in the external world. Consider

the rat who sniffs out and eats a tainted meal without detecting
the slow-acting poison. The rodent then turns homeward, avoiding
the odor of predators and the scents marking the territory of its
conspecifics. Safe in its own lair, the rat greets its receptive
mate with much sniffing and copulates. Then it becomes ill. If
the rat obeys Pavlovian laws for external conditioning, it would,
of necessity, avoid the last salient odor lingering in memory.
This would hardly be adaptive. Instead, guided by a visceral consummatory UCS specific to the feeding system, the rat retrieves the
odor memory that was compounded with the peripheral taste CS used
to monitor the consummatory act of feeding. Since the visceral
effect of poison is noxious, taste and its attendant odor become
noxious and subsequently odor, an external CS blocks the approach
to food. Thus, the appetitive and consummatory phases of feeding
are united into one smooth adaptive sequence. Other odors do not
interfere, because each odor is indexed to its own respective
behavioral system by the peripheral sensors which guide the consummatory act ending that sequential pattern, be it defense against
predators, recognition of territory, or courtship of a mating
partner.

Like the mammalian nose, the avian eye is a busy information
channel, seeking food, detecting predators and identifying mates.
When a food-related nausea strikes, conditioning accrues to the
visual cues which attended the taste of the meal. Other visual
cues in spatiotemporal position to interfere with visual taste-
illness learning do not do so, because each visual cue is
associated to the behavioral situation for which it is relevant.
There is some evidence that taste can synergistically potentiate
visual food stimuli in the rat, but visual potentiation in the rat
(Galef and Osborn, 1978) is probably not as powerful as odor
potentiation (Rusiniak et al., 1979) since it requires immediate
illness and extinguishes rapidly. Furthermore, when an audio-
visual CS and a taste CS were combined and followed by an illness
UCS, the taste CS did not effectively potentiate the audio-visual
CS (Garcia and Koelling, 1966; Domjan and Wilson, 1972). To
observe a truly robust visual potentiation, one must go to the
birds.

CONDITIONING AND LEARNING WITHIN A DARWINIAN FRAMEWORK

Appropriately enough, flavor-illness conditioning was an
integral part of evolutionary theorizing from the beginning. For
example, Darwin (1871) explained some apparent exceptions to
adaptive coloration in terms of contingent associations. Gaudy
caterpillars regularly flaunt their intense color patterns on exposed twigs, catching the eye of every passing bird and violating
the rule of camouflage. These tender larval creatures prove to

have a nauseous taste, but this is not enough, for the slightest peck by the bird would be invariably fatal to the caterpillar. However, after one or two such pecks, the bird associates the conspicuous colors with the nauseous taste and thus avoids other members of the same species promoting their survival.

The role of learning within the evolutionary process has been put forth in modern times by Konrad Lorenz (1965) in his essay "Evolution and the Modification of Behavior". Over succeeding generations, each species interacts with its environment by mutation and selection, by the trials and the successes of varying individuals. This long-term process, analogous to induction, encodes information into the genome of the species which, in turn, directs the construction of individuals out of the environmental substrate. The short-time vagaries of the niche are such that learning, an inductive process of information gathering within the niche by each individual, is indubitably adaptive. Therefore, learning is best viewed as plastic behavioral components inter-calated within phylogenetically determined action patterns, like cartilaginous discs between bony vertebrae.

In summary, this coordination of nature and nurture is re-vealed by the marvelous defense in depth protecting both mammals and birds from poison. Each individual comes equipped with an inborn aversion for bitter, which shields it from the toxic plants in the natural world where it evolves. The CNS directs the vomiting reflex to eject the remnants of any meal which disagrees with the gastrointestinal system; it also records a specific learned aversion, adding to the individual's store of species-specific wisdom. The rat's mouth teaches its nose, by contiguity, which odor is to be associated with long-delayed poisonous effects, because the specialized rat is designed by nature to sniff for food in dim rough terrain with its gaze fixed upward, providing a pro-tective canopy against shadowy attacks from above. And by the same rule, the pigeon's beak teaches its eye to avoid the visual aspects of poisoned food because the specialized pigeon is designed to operate in a bright spacious ocean of shifting air where vision is supreme.

Finally, a simple description of the stimuli in the niche is not an explanation. The coyote and the red-tailed hawk both hunt in the bright prairie, visually directing their attacks on rabbits and rodents. But when the coyote is poisoned, taste potentiates food odor, revealing its kinship to the rat, and when the red-tailed hawk is poisoned, taste potentiates visual food cues, revealing its kinship to the pigeon. Thus, learning, like fixed action patterns and morphological structure, is also a matter of taxonomy. And a complete explanation of learning requires a description of the phylogenetic and ecological context in which it occurs.

REFERENCES

Bernstein, I.L. 1968. Learned taste aversions in children re-
    ceiving chemotherapy. Sci. 200: 1302.
Borison, H.L. and Wang, S.C. 1953. Physiology and pharmacology
    of vomiting. Pharmacol. Rev. 5: 193.
Brett, L.P. 1977. Experimental extensions of the cue-consequence
    aversive conditioning paradigm. Ph.D. Dissertation, UCLA.
Brett, L.P., Hankins, W.G. and Garcia, J. 1976. Prey lithium
    aversions III: Buteo hawks. Behav. Biol. 17: 87.
Buresova, O. and Bures, J. 1973. Cortical and subcortical com-
    ponents of the conditioned saccharin aversion. Physiol.
    Behav. 11: 435.
Cabanac, M. 1971. Physiological role of pleasure. Sci. 173: 1103.
Clarke, J.C., Westbrook, R.F. and Irwin, J. 1979. Potentiation
    instead of overshadowing in the pigeon. Behav. Neur. Biol.
    25: 18.
Coil, J.D. and Garcia, J. 1978. Conditioned taste aversions: The
    illness US, In Biological Approaches to Learning" papers
    presented at the Northeastern Regional Meeting of the Animal
    Behavior Society, St. Johns, Newfoundland, Canada, pp. 1-28.
Coil, J.D., Hankins, W.G., Jenden, D.J. and Garcia, J. 1978. The
    attenuation of a specific cue-to-consequence association by
    antiemetic agents. Psychopharmacol. 56: 21.
Coil, J.D., Rogers, R.C., Garcia, J. and Novin, D. 1978. Con-
    ditioned taste aversions: Vagal and circulatory mediation of
    the toxic unconditioned stimulus. Behav. Bio.. 24: 509.
Darwin, C. 1871. The Descent of Man and Selection in Relation to
    Sex. Murray, London.
Deutsch, J.A., Young, W.G., Kalogeris, T.J. 1978. The stomach
    signals satiety. Sci. 201: 165.
Domjan, M. and Wilson, N.E. 1972. Specificity of cue to conse-
    quence in aversion learning in the rat. Psychon. Sci. 26: 143.
Egger, M.D. and Miller, N.E. 1962. Secondary reinforcement in rats
    as a function of information value and reliability of the
    stimulus. J. Exp. Psychol. 64: 97.
Egger, M.D. and Miller, N.E. 1963. When is a reward reinforcing?
    An experimental study of the information hypothesis. J. Comp.
    Physiol. Psychol. 56: 132.
Fudim, O.K. 1978. Sensory preconditioning of flavors with a
    formalin-produced sodium need. J. Exp. Psychol: An. Behav.
    Proc. 4: 276.
Galef, B.G. and Osborn, B. 1978. Novel taste facilitation of the
    association of visual cues with toxicosis in rats. J. Comp.
    Physiol. Psychol. 92: 907.
Garcia, J. and Brett, L.P. 1977. Conditioned responses to food odor
    and taste in rats and wild predators, In The Chemical Senses
    and Nutrition. (M.R. Kare, and O. Maller, Eds.). pp. 277-290,
    Academic Press, New York.

Garcia, J. and Hankins, W.G. 1977. On the origin of food aver-
    sion paradigms, In Learning Mechanisms in Food Selection.
    (L.M. Barker, M.J. Best and M. Domjan, Eds.), pp. 3-19, Baylor
    University Press, Waco.
Garcia, J., Hankins, W.G. and Rusiniak, K.W. 1974. Behavioral
    regulation of the milieu interne in man and rat. Sci. 185: 824.
Garcia, J. and Koelling, R.A. 1966. Relation of cue to consequence
    in avoidance learning. Psychon. Sci. 4: 123.
Garcia, J., Rusiniak, K.W. and Brett, L.P. 1977. Conditioning
    food illness aversions in wild animals: Caveant Cononici,
    In Operant-Pavlovian Interactions (H. Davis and H.M.B. Hurwitz,
    Eds.), pp. 273-316, Lawrence Erlbaum Press, New Jersey.
Gustavson, C.R., Garcia, J., Hankins, W.G. and Rusiniak, K.W. 1974.
    Coyote predation control by aversive conditioning. Sci. 184:
    581.
Gustavson, C.R., Kelly, D.J., Sweeney, M. and Garcia, J. 1976.
    Prey-lithium aversions. I: Coyotes and wolves. Behav. Biol.
    17: 61.
Hankins, W.G., Garcia, J. and Rusiniak, K.W. 1973. Dissociation
    of odor and taste in baitshyness. Behav. Biol. 8: 407.
Hankins, W.G., Rusiniak, K.W. and Garcia, J. 1976. Dissociation
    of odor and taste in shock-avoidance learning. Behav. Biol.
    18: 345.
Kamin, L.J. 1969. Predictability, surprise, attention and condi-
    tioning, In Punishment and Aversive Behavior (B.A. Campbell
    and R.M. Church, Eds.), pp. 279-296, Appleton-Century-Crofts,
    New York.
Lorenz, K. 1965. Evolution and Modification of Behavior.
    Univeristy of Chicago Press, Chicago.
Mackintosh, N.J. 1974. The Psychology of Animal Learning,
    Academic Press, New York.
Novin, D. 1976. Visceral mechanisms in the control of food
    intake, In Hunger: Basic Mechanisms and Clinical Applications,
    (D. Novin, W. Wyrwicka and G.A. Bray, Eds.) pp. 357-367,
    Ravin Press, New York.
Roll, D.L., Smith, J.C. 1972. Conditioned taste aversion in
    anaesthetized rats, In Biological Boundaries of Learning
    (M.E.P. Seligman and J.L. Hager, Eds.) pp. 98-102, Prentice-
    Hall, Englewood Cliffs, New Jersey.
Rusiniak, K.W., Gustavson, C.R., Hankins, W.G. and Garcia, J.
    1976. Prey-lithium aversions. II: Laboratory rats and
    ferrets. Behav. Biol. 17: 73.
Rusiniak, K.W., Hankins, W.G., Garcia, J. and Brett, L.P. 1979.
    Flavor-illness aversions: Potentiation of odor by taste in
    rats. Behav. Neur. Biol. 25: 1.
Rusiniak, K.W., Lopez, S.C. and Rice, A.G. 1979. Odor and taste:
    Potentiation in illness aversion, overshadowing in shock
    avoidance, Paper presented at the 59th Annual Western
    Psychological Association Meeting, Abs. pg. 50, San Diego.

Weiss, J.M.   1972.   Psychological factors in stress and disease.
     Sci. Am. 226: 104.

ACKNOWLEDGEMENT

     This work was supported by USPHS NIH Research Grant NS 11618
and Program Project Grants HD 05958 and AAO 3513.

RAT PUP'S FOOD CONSUMPTION AS A FUNCTION OF PREWEANING TASTES AND

ODORS*

Paul M. Bronstein[1] and David P. Crockett[2]

[1]Department of Psychology, Trenton State College
Trenton, NJ 08625
[2]Department of Psychology, Hunter College
New York, NY

The exposure of immature rodents to distinctive foods or odors can cause these animals to subsequently recognize and orient towards the previously experienced stimulus. Cheal (1975) and Leon (1975) have provided excellent reviews of this literature. One method of undertaking such treatment is by feeding maternal rats some specially scented or uniquely flavored substance. Pups reared by such dams show a preference or tolerance for the distinctive cue which had been added to the food or water supply of their maternal caretaker (Bronstein and Crockett, 1976a, b; Capretta and Rawls, 1974; Galef, 1979; Galef and Henderson, 1972; Leon, 1977). The mechanisms responsible for these effects have received little atten- tion, although vague allusions to an imprinting-like process have appeared (Burghardt and Hess, 1966; Porter and Etscorn, 1974). The present investigations are attempts to determine the processes em- ployed by rat pups in establishing a preference for their mother's rations.

There appear to be two general strategies which might lead an isolated rat pup to identify its mother's rations at the time of weaning. First, a juvenile could employ a peripheral (perhaps gustatory) trace of maternal foods and match this cue to other, similar stimuli encountered when independent feeding is first initiated. Second, the pups might form a maternal-food preference as a result of some centrally mediated learning.

EXPERIMENT I

Evidence concerning the role of learning in rat pups' identifi- cation of foods is circumstantial and inconclusive. Bronstein,

Levine, and Marcus (1975) made several suggestions leading to the tentative conclusion that rat pups use some peripheral trace of the maternal rations, picked up in the nest, in making their initial food choices. First, preweaning exposure to foods or odors usually has only a weak or short-lived effect (e.g., Bronstein, et al., 1975). Second, when pups have shown a preference for their mother's food for several consecutive tests (Galef and Henderson, 1972), those trials were very brief (20 min) and were also interrupted by pro- longed exposures of the pups to their dams. A peripheral trace of the maternal diet could have been replenished during the day-long intertrial intervals used by Galef and Henderson. On the other hand, Cheatle and Rudy (1979), Rudy and Cheatle (1977), Smith and Spear (1978), and Thoman, Wetzel and Levine (1968) have all demonstrated some food- or odor-related learning in rats less than one week old. Those findings clearly dictate the possibility that food-related learning might normally occur when rat pups are reared by a dis- tinctively fed female.

The working assumption of the first study was that if rat pups do learn something about their mother's foods during the suckling period, they might be informed by their dam that her ingesta are safe--i.e., unlikely to be followed by malaise (Kalat and Rozin, 1973). This hypothesis was derived from several sources of in- direct evidence. First, there is a formal similarity between the suckling period and the procedures normally used to induce safety learning: both treatments seemed to involve the exposure of animals to tastes in the absence of malaise (cf. Bolles, Riley and Laskowski, 1973; Galef and Sherry, 1973). Second, pups behave as if they had learned that the food consumed by their maternal caretaker was safe. That is, weanling rats will consume relatively large quanti- ties of their mother's food while rejecting or being suspicious of nonmaternal rations (Bronstein, et al., 1975). Third, it seemed plausible that relatively inexperienced omnivores such as weanling rats would have the need to rapidly assess the dangers associated with novel food sources. A capacity for learning about the safety of maternal foods prior to weaning would, we conjectured, reduce the inherent danger for pups when starting to forage relatively independently.

The question of whether rat pups learn not to associate their mother's rations with malaise is addressed in the first experiment. The design used is a variant of a paradigm employed by Revusky and Bedarf (1967). Those investigators showed that in adult rats a prior history of ingesting a flavored solution would hinder the association of that taste with a malaise induced by X-rays. The specific focus of the current inquiry was to learn whether a 2-to-3- week suckling period with a distinctively fed dam reduces the tendency of weanlings to associate an experimenter-induced illness with the mother's rations.

Method

Subjects. Albino rats of Sprague-Dawley descent, randomly bred in our colony for 3 yr, were used. All animals were maintained solely on ad lib Purina lab chow and water unless otherwise noted. The laboratory was heated to approximately 72° F (22.2°C) and lighted 18 hr/day (0800 to 0200).

Sixteen adult females gave birth in individual, plastic delivery cages (Carworth) on a bed of Sanicel and newsprint. Litters were culled to 10 pups within 24 hr of parturition; and birth dates are designated "Day 0" throughout this paper. On day 2 postpartum all litters were culled to six pups (three males per litter), crossfostering accomplished, and finally, dams and foster pups were rehoused with restricted access to the experimental diets. These foods were (a) Purina lab chow powder, and (b) Turtox fat sufficient diet. Eight females were maintained on each of these foods after crossfostering. Galef and Henderson (1972) have shown that these two foods facilitate the formation of preferences for maternal foods among rat pups.

The 16 females and their litters were divided into two groups of eight dams, with crossfostering taking place within each of these subgroups. The shifting of pups was done randomly, but with the following restrictions: (a) Half of each foster litter (i.e., three pups) would be comprised of males. (b) No female nursed any of her own progeny. (c) Half of the pups in each foster litter were born to females who continued to eat only Purina lab chow after crossfostering.

Immediately following crossfostering, each dam and her foster litter were transferred to 2/3 of Wahmann gang cage (model LC75/SC). This litter compartment (44 x 18 x 24 cm) was provided with ad lib water, newsprint for nesting material, and was underlined with Plexiglas to prevent the loss of pups. No food was ever offered to the animals in the litter compartment. The remaining portion of each cage (22 x 18 x 24 cm) served as the dam's feeding compartment and was isolated by an opaque, Plexiglas barrier. Within this latter area, containing both food and water, each dam was fed three 90-min meals daily. Thus, all pups were prevented from having any direct contact with food prior to testing.

Apparatus. The testing chambers have been described in detail by Bronstein, et al. (1975). Briefly, each pup was tested individually in a cylindrical, Plexiglas arena (24 cm in diameter and 7.2 cm deep). Each arena was fitted with two plastic food cups so that every animal could be offered a choice between Purina and Turtox powders.

Procedure. On days 20 and 21 each litter was placed en masse
into one of the testing arenas with both of the food cups in place
and empty for 1-hr habituation trials.

On day 21, following the second exposure to the apparatus and
when the pups had been away from their foster dams for approximately
5 hr, each litter was divided into three pairs. Each pair of pups
(one male and one female) then received a different treatment.

Two animals from each litter (Group A) were syringe fed 0.25 cc
of a wet mash made from water mixed with a quantity of their foster-
mother's diet. Ten min following this initial feeding the Group A
pups were given a second 0.25 cc of mash made from a mixture of water
and the powdered food (either Turtox or Purina) not eaten by their
maternal caretaker. The two food-and-water mixtures were of approximately the same viscosity. They were administered by placing a
needleless syringe to the mouth of the animals; all pups readily
accepted both mixtures as they were slowly expressed. The Purina
mash was made from one part powder and six parts tap water (by
weight); the Turtox mixture was made from four parts food and three
parts water. Twenty min following their second syringe feeding, all
Group A animals were given an ip injection of 0.15 M LiCl (2.0 ml/
100 g body weight).

The second pair of pups in each litter was treated identically
to the Group A animals, except that these Group B pups received their
syringe feedings in counterbalanced order. That is, the Group B
animals were exposed to 0.25 cc of nonmaternal mash prior to ingesting 0.25 cc of the mixture made from water plus some of the food
eaten by their foster mother during the suckling period. Group C,
the third pair of animals from each litter, served as controls and
were merely handled while the other pups were either fed or injected.

Between the day 21 operations (feedings, injections, or hand-
lings), each pup was housed individually in a wire-mesh cage (Wahman,
model LC75/SA). Following the day 21 manipulations, the animals were
again housed individually in this cage (now with water available ad
lib) for 18 hr after which testing began. Two tests, each of 60 min
duration, were administered, and each rat could choose between Purina
and Turtox powders on each test. The positions of the two foods
were alternated at the midpoint of each trial. There was a period
of 7 hr between tests, and the pups were rehoused in their individual
cages with ad lib water during this interval.

Results and Discussion

Eating was measured by weighing the food cups both at the start
and at the conclusion of each testing period. The balance used was

accurate to 0.0001 g, and a subject was required to consume at least
0.025 g of food (Turtox + Purina) for inclusion in the study. During
the first trial one Turtox-reared pup failed to reach this ingestion
criterion. Also, excessive spillage caused the removal of one pup
during the first 60-min test, as well as the elimination of seven
subjects during the second trial.

   Throughout this paper only two-tailed probabilities are used,
with alpha set at 0.05 unless otherwise noted. We also followed the
conservative statistical procedures suggested by Abbey and Howard
(1973). Accordingly, only one degree of freedom was associated with
each of the 16 foster litters. Within each litter, scores for like-
treated subjects were pooled, and that mean was taken to represent
the within-litter group. Even after the elimination of a few pups
due to spillage, usable data were always collected from at least one
weanling per treatment condition in each litter. Also, within each
maternal-feeding condition there were no differences between Group A
and Group B. Therefore, these two groups were combined and only
statistics comparing poisoned and unpoisoned subgroups will be detailed.

   For both Purina- and Turtox-reared animals, the administration
of LiCl significantly reduced the total amount of food eaten. On
each trial, Group C pups were found to eat more than their poisoned
littermates (all $ps <.05$). Regardless of their foster mother's diet,
each of the uninjected control groups ingested approximately the
same total amount of food during each test, ts (14)$< .71$; $ps > .50$).

   On both tests each subject's relative preference for Purina
was calculated as that percentage of its total intake consumed
from the Purina trough--i.e., (grams of Purina/total grams of
Purina + Turtox) x 100. A comparison of the two uninjected groups
revealed that for the first hr of testing the pups reared by Purina-
fed dams ate a significantly greater proportion of their food from
the Purina cup than the Group C rats nursed by Turtox-fed foster
mothers, t (14) = 3.88; $p < .01$). Thus, Experiment I is a further
demonstration of the transfer of traces of Purina and Turtox from
rat mothers to their foster young. The behavior of our uninjected
controls during the initial preference test reconfirms the findings
of Galef and Henderson (1972) and Levine and Bronstein (1976); and
the results of Trial 1 are documented in Figure 1. However, the
differences between control groups was no longer apparent during
the second hr of feeding, t (14) = 1.68; $p < .1$. As previously
shown (Bronstein, et al., 1975; Galef and Henderson, 1972), the
exposure of pups reared by Purina-fed dams to Turtox reduces their
preference for the maternal diet. Exactly this change in eating
was noted among Purina-reared controls (see Figure 2).

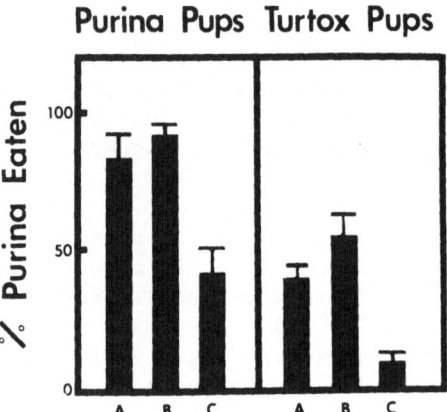

Fig. 1. Mean percent Purina consumed during the first trial of
         Experiment I. The behavior of pups raised by Purina-fed
         dams is depicted in the left-hand panel, and the eating of
         weanlings whose fostermothers ate Turtox is shown at the
         right. In all figures, the vertical lines indicate one
         standard error about each mean. The letters on the hori-
         zontal axis designate the treatments of the groups as de-
         scribed in the Method section.

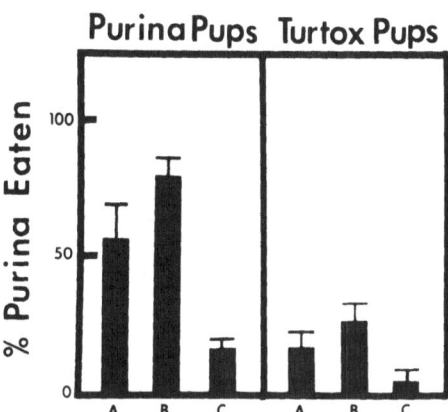

Fig. 2. Mean percent Purina consumed by each group during the second
trial of Experiment I.

        In comparison to their Group C littermates, Purina-reared pups
that were poisoned on day 21 showed a reliable increase in their
preference for Purina. This effect was noted during both eating
tests, ts (7)> 5.14; ps< 0.01. Likewise, the poisoned animals
reared by Turtox-fed dams also preferred Purina reliably more than
their Group C littermates; this effect also occurred during each
test, ts> 3.50; ps< 0.01.

        Regardless of their mother's diet, an injection of LiCl re-
sulted in a reliable increase in pups' preference for Purina. It
may be that for our animals Turtox is simply more salient (i.e.,
more associable with illness) than Purina chow. This simple conclusion may be rejected, however, by comparison of the eating behaviors of the poisoned rats from the two maternal-diet conditions.
During each test, poisoned, Purina-reared weanings preferred Purina
significantly more than pups that were poisoned after being raised
by Turtox-fed dams, ts (14)> 5.22; ps< 0.01. Poisoning with LiCl
resulted in a heightened avoidance of Turtox in both of the maternal-
food conditions. However, by comparing the magnitudes of these

aversions, it is clear that the Purina-reared animals had been inoculated against the malaise becoming associated with their mother's food.

Furthermore, the order in which tastes were offered to pups on day 21 was not a relevant factor in determining subsequent behavior. Subjects were not simply associating the noxious consequences of their being injected with that taste most recently sampled. It follows, therefore, that weanlings are able to choose foods on the basis of a memory of the mother's diet. They need not employ their most recent mouth trace when selecting foods. One explanation for these results is that pups associate an illness preferentially with foods which their mothers had not recently consumed. A second possibility is that pups reared by Purina-fed dams react more strongly to LiCl than do pups raised by Turtox-fed foster mothers. Yet another complicating factor is the possibility of differential associability of Turtox and Purina tastes with the malaise induced by LiCl--i.e., differential salience. It might be, for instance, that of the two flavors we used, Turtox has the greater affinity for the toxicosis. The reduction in novelty of Turtox among pups whose dams ate that food could have led to the current findings.

It is not possible at this time to select the most appropriate explanation or interaction. That decision will probably require the use of foods of known salience which can also be passed from mothers to pups. Nevertheless, Experiment I does show that pups, like adults, do store memories of prior taste and odor experiences. This information was used by the weanlings in selecting their first meal of solid food following toxicosis.

EXPERIMENT II

The first study showed the avoidance of a novel (nonmaternal) food by pups reared by Purina-fed dams after the weanlings had been poisoned on day 21. The second investigation attempted to examine which specific events occurring at weaning are responsible for the pups' bait shyness.

Method

Subjects. Six female rats and their offspring were used. These animals were drawn from the same stock and maintained under the same colony conditions as used in Experiment I. Each litter was culled to 10 pups on the day of parturition, and each dam was permitted to rear her litter in the plastic delivery cage (Carworth) until they were 10 days old. Water and Purina lab chow were continuously available to the mothers during this period.

On day 10 each female and her litter were transferred to the divided Wahmann gang cage described in the first study. The mothers then reared their progeny for 11 additional days. The pups were separated from their caretaker's feeding compartment during this phase of the study, and the dams were fed four 1-hr meals daily.

Apparatus. The testing apparatus was identical to that used in Experiment I.

Procedure. Repeating the procedures of the first study, each litter was given a 1-hr habituation trial on days 20 and 21. Following the last of these exposures, eight pups were randomly selected for testing from each litter. These animals were returned to their rearing cage (containing ad lib water, but no food) for 5 hr during which time they were isolated from their dams.

Following this maternal deprivation, every litter was randomly divided into four treatment groups, each containing one male and one female pup. The four groups were treated as follows on day 21: (a) The first pair of weanlings received sequentially the two types of wet mash used in Experiment I followed by an ip injection of LiCl. (b) The second pair received the two tastes, but were then injected with isotonic saline. In every litter, one pup in each of the first two groups received the Turtox mash prior to Purina mixture. The counterbalanced order was used for the other member of each couple. (c) Group 3 animals received no mash, but had empty, clean syringes placed in their mouths to simulate the handling used during administration of the tastes to their littermates. These pups later received a LiCl injection. (d) The final pair of subjects was stimulated with empty syringes following which an injection of isotonic saline was administered.

A 30 min delay elapsed between the final presentation of a taste (or an empty syringe) and the administration of an injection. All other details concerning the injection procedures replicated the methods of Experiment I.

Each pup was pair housed with the littermate of the same treatment group both for the 17.5 hr following its injection as well as during the 6-hr intertrial interval. Animals were maintained in Wahmann cages (model LC 75/SA) without food, but with ad lib water, during these times. Preference tests were identical to those outlined in Experiment I.

Results

The statistical procedures and eating criterion (0.025 g/test) of the first study were again in effect. One pup failed to consume enough food in each trial, and its data were not examined. Also, two

animals were eliminated from Trial 1 and six subjects were removed
from Trial 2 as a result of their spillage of food. Even after the
removal of some animals due to spillage, usable data were always
collected from at least one weanling per treatment condition in each
litter. Furthermore, within each of the groups receiving the Purina
and Turtox tastes, no differences emerged as a function of the order
in which the mixtures were given. Data collected from the two taste-
order subgroups were, therefore, combined prior to analysis.

An analysis of variance was first used to determine whether any
of the treatments altered the total amount of foods ingested. As
noted in Figure 3, such an effect was noted on each trial--Trial 1:
$\underline{F}$ (3, 15) = 30.97, $\underline{p}$< 0.01; Trial 2: $\underline{F}$ (3, 15) = 5.31, $\underline{p}$< 0.05.
Post hoc analysis of the first trial revealed the following: Pups
experiencing both tastes and LiCl ate significantly less food than
any of the other groups, 2.84<$\underline{t}$s (5)<17.05; $\underline{p}$s< 0.05. Animals receiving the toxin unpaired with tastes consumed reliably less food
than littermates receiving the same pair of tastes followed by $\underline{ip}$
NaCl, $\underline{t}$ (5) = 5.51; $\underline{p}$< 0.01. Finally, those weanlings presented with
the two tastes paired with saline ate more food than pups receiving
the NaCl without the prior exposure to wet mash, $\underline{t}$ (5) = 9.67; $\underline{p}$< 0.01.
Post hoc analysis of Trial 2 showed that pups given samples of mash
paired with LiCl ate reliably less food than either group receiving
an injection of saline, $\underline{t}$s (5)> 3.82; $\underline{p}$s< 0.02. No other differences
were significant during the second trial.

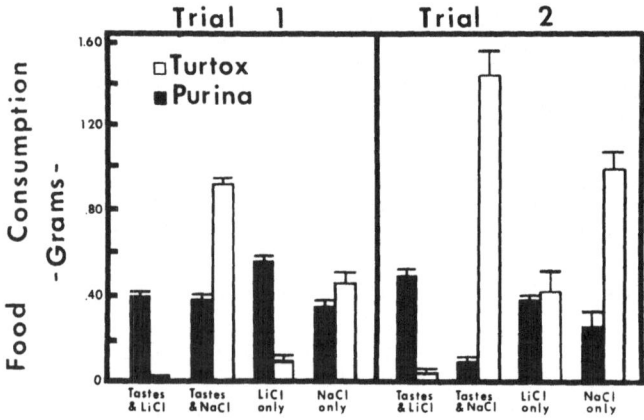

Fig. 3.   Mean grams of Purina and Turtox eaten as a function of
          treatment and testing trials in Experiment II.

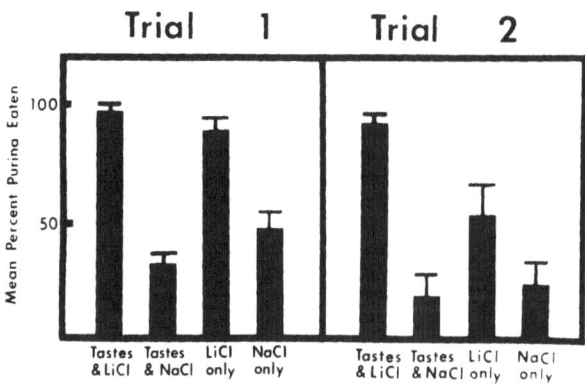

Fig. 4.   Mean percent Purina consumed as a function of treatment and testing trials in Experiment II.

Analyses of variance also showed that our treatments modified food preferences reliably during both tests--Trial 1: $\underline{F}$ (3, 15) = 21.67, $\underline{p} < 0.01$; Trial 2: $\underline{F}$ (3, 15) = 6.96, $\underline{p} < 0.01$. These data are presented in Figure 4.

On Trial 1, each group receiving LiCl ate a greater proportion of food from the Purina cup than either of the NaCl-injected groups, $\underline{t}s$ (5) > 2.84; $\underline{p}s < 0.05$. Pups poisoned following food administration did not prefer Purina any more strongly than animals injected with LiCl after 5 hr isolation from tastes. Also, within each injection condition, the two taste-order groups were not differentiable, $\underline{t}s$ (5) < 1.51; $\underline{p}s > 0.20$. During the second preference test, pups that had been given LiCl following tastes of the maternal rations showed a stronger aversion than either of the groups injected with NaCl, $\underline{t}s$ (5) > 3.75; $\underline{p}s < 0.02$. No other intergroup differences were reliable.

It is clear that the pairing of tastes with illness at weaning was not necessary for pups to increase their avoidance of the non-maternal Turtox on Trial 1. Unpaired LiCl was no less effective than the taste-illness combination. In comparison to the pups receiving LiCl only, the pairing of tastes with illness resulted in significantly more suppression of total eating on Trial 1 ($p < 0.05$; see Figure 3). However, the preference scores for these two groups were statistically undifferentiable (Figure 4).

Some consequences of pairing LiCl with tastes were seen on Trial 2. First, animals receiving the conjunction of tastes and illness ate reliably less Turtox than animals receiving only LiCl ($p < 0.01$; see Figure 3). Second, while the preferences for Purina were not reliably different in the two poisoned groups, the pairing of LiCl and tastes caused animals to maintain a higher Purina preference than either of the NaCl-injected control groups. Pups that received unpaired LiCl could not be differentiated from pups injected with NaCl. Thus, an injection of LiCl at weaning will heighten the maternal-food preference of Purina-reared pups. Furthermore, if the illness was preceded by tastes, the pups (a) showed a significantly greater reduction in their total food consumption, and (b) persisted in their Turtox avoidance longer than weanlings poisoned in the absence of taste cues.

GENERAL DISCUSSION

Experiment I reconfirmed prior findings that rat pups are capable of establishing a preference for foods eaten by their maternal caretakers. This effect was quite robust, even with the rather stringent statistical criterion of the current work. Furthermore, several aspects of the present studies indicate that the weanlings chose their mother's diet on the basis of some learning.

Three kinds of evidence support the conclusion that pups' learning mediates food choices at the time of weaning. First, the order of taste presentations at weaning was, in both studies, without effect in altering eating. The recency with which a wet mash was experienced failed to affect eating behavior; this finding would be improbable if the pups had not done some learning prior to weaning. Second, the magnitude of Turtox avoidance in Experiment I was predicted by the preweaning experiences of the animals. Heightened avoidance occurred when foster mothers had not themselves eaten Turtox. Third, the behavior of Experiment II animals that were poisoned without previously eating any wet mash also favors a learning interpretation. One might hypothesize that at weaning these pups had not learned about their maternal diet, and carried with them only a recently acquired, peripheral trace of their mother's food. However, since weanlings are quite able to acquire a toxiphobia (Galef and Sherry, 1973; Grote and Brown, 1971), one would then expect that this peripheral trace of Purina would become associated with the experimenter-induced illness. This line of reasoning predicts that the pups of Experiment II that were treated with LiCl in the absence of any wet mash would show a reduction in their intake of Purina.

Our actual findings contradict this prediction. As seen in Figure 4, subjects receiving only LiCl increased their preference for Purina in comparison to littermate controls. Buresova and Bures (1975) speculated that prior to weaning rat pups learn that their maternal diet is safe--i.e., unlikely to be associated with illness. The current results conform to that suggestion.

Previous data on the start of independent feeding in weanling rats suggest that while pups might initially establish a preference for their maternal diet, this first orientation is rapidly subordinated to gustatory preferences. Sweet tastes such as the Turtox diet become attractive and overcome the maternal-food preference if an hour of ad lib sampling is permitted (Bronstein et al., 1975). Prior studies have shown that pups' preference for their mother's food is a short-term bias in the feeding behavior of the juveniles.

The current data, however, point to an increased reliance among pups on memories of their mother's rations following illness. Moltz (1979) has hypothesized that through suckling and coprophagia the rat pup acquires inoculations against enteric infections from its mother. In learning about maternal rations pups are acquiring a second behavioral line of defense against illness.

Footnote

*This research was supported by Grants MH 22027-01 and MH 26372-01 from the National Institute of Mental Health. We were greatly distressed by the unscientific criticism of Lester Aronson's research tactics. Aronson's writings have inspired interest in the processing of olfactory signals by animals. This paper is dedicated to Dr. Aronson on the occasion of his retirement.

REFERENCES

Abbey, H. and E. Howard. 1973. Statistical procedure in developmental studies with multiple offspring. Dev. Psychobiol. 6 329.
Bolles, R.C., A.L. Riley and B. Laskowski. 1973. A further demonstration of the learned safety effect in food aversion learning. Bull. Psychonom. Soc. 1 190.
Bronstein, P.M. and D.P. Crockett. 1976a. Maternal rations affect the food preferences of weanling rats: II. Bull. Psychonom. Soc. 8 227.
Bronstein, P.M. and D.P. Crockett. 1976b. Exposure to the odor of food determines the eating preferences of rat pups. Behav. Biol. 18 387.

Bronstein, P.M., M.J. Levine and M. Marcus. 1975. A rat's first bite: The nongenetic cross-generational transfer of information. J. Comp. Physiol. Psychol. 89 295.

Buresova, O. and J. Bures. 1975. Functional decortication by spreading depression does not prevent forced extinction of conditioned taste aversion in rats. J. Comp. Physiol. Psychol. 88 47.

Burghardt, G.M. and E.H. Hess. 1966. Food imprinting in the snapping turtle, Chelhydra serpentina. Science 151 295.

Capretta, P.J. and L.L. Rawls, III. 1974. Establishment of a flavor preference in rats: Importance of nursing and weaning experience. J. Comp. Physiol. Psychol. 86 670.

Cheal, M. 1975. Social olfaction: A review of the ontogeny of olfactory influences on vertebrate behavior. Behav. Biol. 15 1.

Cheatle, M.D. and J.W. Rudy. 1979. Ontogeny of second-order odor-aversion conditioning in neonatal rats. J. Exp. Psychol.: Anim. Behav. Proc. 5 142.

Galef, B.G., Jr. 1979. Investigation of the functions of coprophagy in juvenile rats. J. Comp. Physiol. Psychol. 93 295.

Galef, B.G., Jr. and P.W. Henderson. 1972. Mother's milk: A determinant of the feeding preferences of weaning rat pups. J. Comp. Physiol. Psychol. 78 213.

Galef, B.G., Jr. and D.F. Sherry. 1973. Mother's milk: A medium for transmission of cues reflecting the flavor of mother's diet. J. Comp. Physiol. Psychol. 83 374.

Grote, F.W., Jr. and R.T. Brown. 1971. Rapid learning of passive avoidance by weanling rats: Conditioned taste aversion. Psychon. Sci. 25 163.

Kalat, J.W. and P. Rozin. 1973. "Learned safety" as a mechanism in long-term taste-aversion learning in rats. J. Comp. Physiol. Psychol. 83 198.

Leon, M. 1975. Dietary control of maternal pheromone in the lactating rat. Physiol. Behav. 14 311.

Leon, M., B.G. Galef, Jr. and J.H. Behse. 1977. Establishment of pheromonal bonds and dietary choice in young rats by odor pre-exposure. Physiol. Behav. 18 387.

Levine, M.J. and P.M. Bronstein. 1976. Replication report: I. Maternal rations affect the food preferences of weanling rats. Bull Psychonom. Soc. 8 230.

Moltz, H. 1979. On how the rat mother defends her young against enteric infections. Paper presented at the Symposium on Parental Behavior, at Rutgers University, Piscataway, New Jersey.

Porter, R.H. and F. Etscorn. 1974. Olfactory imprinting resulting from brief exposure in Acomys cahirinus. Nature 250 732.

Revusky, S.H. and E.W. Bedarf. 1967. Association of illness with prior ingestion of novel foods. Science 155 219.

Rudy, J.W. and M.D. Cheatle. 1977. Odor-aversion learning by neonatal rats. Science 198 845.

Smith, G.J. and N.E. Spear. 1978. Effects of the home environment on withholding behaviors and conditioning in infant and neo-natal rats. Science 202 327.

Thoman, E., A. Wetzel. and S. Levine. 1968. Learning in the neo-natal rat. Anim. Behav. 16 54.

EXPERIENCE AFFECTS BEHAVIORAL RESPONSES TO SEX ODORS

John Nyby[1] and Glayde Whitney[2]

[1]Dept. of Psychology, Chandler-Ullmann #17
Lehigh University, Bethlehem, PA 18015

[2]Dept. of Psychology, Florida State University
Tallahassee, FL 32306

INTRODUCTION

The term "pheromone" was originally coined to categorize specified chemical substances, excreted or secreted from one individual, which communicate information to another conspecific (Karlson and Butenandt, 1959). Much of the initial work investigating pheromonal communication utilized insects with the result that the term has been classically associated with relatively invariant behavioral or physiological responses elicited by quite specific chemicals.

However, some of the principles of chemical communication originally derived from insect work do not always appear to generalize to mammals (or to all insect cases). For example, in contrast to some insect behavior, the behavioral responses to mammalian pheromones often appear quite variable. This finding led one reviewer (Bronson, 1968) to suggest that the category of "releaser pheromones" based upon work with insects is not relevant for mammals and should be replaced by the term "signalling pheromones". A signalling pheromone was defined as one which provides information to the recipient, which may or may not lead to a change in the recipient's behavior.

There have been suggestions for some time that the plasticity of mammalian behavioral response to chemical signals may be due to experiential factors including learning (e.g., Bronson, 1974; Michael, Bonsall and Warner, 1975; Whitten, 1975). There seems, however, to have been a certain reluctance on the part of investi-

gators to intensively investigate the role of learning. Based
upon our own initial reluctance to pursue such a line of research
we suspect that there are a number of reasons for this neglect.
First of all the general background of many investigators does not
include much expertise in the general area of animal learning.
Thus, while a possible role for learning is generally acknowledged,
there seems to be the feeling that the study of such phenomena is
best left to someone else. However, there may be an even deeper
reason for neglect. Many of the persons examining olfactory
communication are either biologists or biologically-oriented
psychologists. As a general rule such persons have been strongly
influenced during their training by such seminal concepts as
"pheromones", "fixed-action patterns", "species-specific behaviors",
etc. All of these concepts entail a pervasive belief that those
animal behaviors most directly related to reproductive fitness and
most strongly shaped by natural selection are least modifiable by
the environment. Consequently, according to this line of reasoning,
any chemocommunication system that is highly modifiable by the environment or by learning is probably not a system of central
biological importance. Thus there may be a tendency for investigators to avoid studying chemocommunication systems in which learning
is obviously important. We will argue later, however, that learning variables can be very adaptive in modifying the responsiveness
of biologically important chemocommunication systems.

     Experiential factors are thought to potentially exert their
effects on the signal value of sex odors at two different periods
of ontogeny. For example, the signal value of sex odors may be
learned by infants early in development via an "imprinting" phenomenon. In addition, behavioral responsiveness to chemical signals
can be modified by adult experience. In this chapter we will
briefly review some of the research examining experiential determination and modification of the responsiveness to chemical signals
and then we will describe our own research with mice.

IMPRINTING AND SEX ODORS

     Both recordings of electrophysiological activity as well as
behavioral responses to odorants indicate that chemosensitivity
develops early in rodent postnatal life (Alberts, 1976). In
fact a growing body of literature has already documented the presence of maternally produced odors that are highly attractive to
preweaning rats (e.g., Leon and Moltz, 1971), spiny mice (e.g.,
Porter and Etscorn, 1974), hamsters (e.g., Gregory and Bishop,
1975), and housemice (e.g., Breen and Leshner, 1977). Since olfaction develops early and is extremely important for later
social behavior, considerable speculation has centered around the
possibility that macrosmatic animals may "learn" at an early age
the olfactory characteristics of their species in a fashion

analogous to visual imprinting described for birds (e.g., Marr and Gardner, 1965).

In general two different experimental strategies have been used to assess the possible role of early experience upon odor-mediated responses in adulthood. The first design involves cross-fostering newly born infants to lactating females of another species. The second approach involves exposing infants to unusual odors placed upon the parents, siblings, and/or in the air or cage shavings. In both approaches, responses to natural and artificial odors are later assessed in adulthood. Both approaches provide some support for the notion that adult preferences for certain odors can be altered by early experience with these odors.

Adult animals that had been crossfostered as neonates to dams of another species were generally found to have reduced preferences for conspecific odors and increased preferences for odors of the crossfostered species in housemice (Denenberg, Hudgens and Zarrow, 1964; Quadagno and Banks, 1970), pygmy mice (Quadagno and Banks, 1970), and southern grasshopper mice (McCarty and Southwick, 1977). Other research indicates that sexual preferences can be somewhat altered in guinea pigs (Beauchamp and Hess, 1973) and voles (McDonald and Forslund, 1978) as a result of early experience with nonconspecifics. Whether these latter two alterations were mediated by odors was unclear.

Placing unusual odors on the mother and siblings prior to weaning and assessing preferences for these odors in adulthood has been done with housemice (Mainardi, Marsan and Pasquali, 1965), rats (Marr and Gardner, 1965; Marr and Lilleston, 1969) and guinea pigs (Carter and Marr, 1970; Carter, 1972). While the existence of sex differences in adult preference appear (Mainardi et al., 1965) as well as variability in how long the preference lasts, these studies generally support an increased preference in adulthood for the unusual odor to which the animal had been exposed in infancy. In fact Carter (1972) reports that adult guinea pigs tended to show the highest levels of mating and feeding behavior under odor conditions similar to rearing.

Very recently Gilder and Slater (1978) have examined preference of adult mice for conspecific odors as a function of degree of kinship. When adult females were given a 4-choice preference, they preferred sawdust soiled by males of the same strain who were not their brothers over sawdust from their brothers, sawdust from males of another strain, or clean sawdust. Gilder and Slater suggest that such a preference pattern would favor mating with genetically similar males who are not brothers and thus lead to optimal levels of outbreeding.

Finally, while much of the evidence relating to olfactory imprinting comes from rodents, other macrosmatic species may also exhibit similar phenomena. For example, Muller-Schwarze and Muller-Schwarze (1971) raised a black-tailed deer fawn from birth with an artificial mother surrogate odorized with male pronghorn ischiadic scent. At 2.5 months of age, the fawn was observed to stay closest to a pronghorn female when also penned with a male and female black-tailed deer. Whether this preference extended into adulthood, however, was not reported.

ADULT EXPERIENCE AND SEX ODORS

While infant experience can certainly affect adult odor preferences in a variety of species, evidence also exists that adult preferences can be fine-tuned or even altered as a result of adult experience. For example, sexual experience by adult males leads to an increased preference for estrus over non-estrus female odors in rats (Carr, Loeb and Dissinger, 1965; Stern, 1970; Lydell and Doty, 1972; Landauer, Wiese and Carr, 1977), mice (Hayashi and Kimura, 1974), and dogs (Doty and Dunbar, 1974). Recent work also indicates that the male rhesus preference for female vaginal odors may be learned in adulthood (Goldfoot, Goy, Kravetz and Freeman, 1976). Additionally, adult sexual experience may interact with hormonal state to affect female preferences for gonadally intact vs. castrate male odors in rats (Carr et al., 1965) and for preputial odors in mice (Caroom and Bronson, 1971).

Although the attraction and preference of male hamsters for female vaginal secretion are not dependent upon previous sexual experience (Johnston, 1974), this normal preference is easily altered in adulthood via a conditioned taste aversion paradigm (Johnston and Zahorik, 1975). By pairing the consumption of female vaginal secretion with LiCl induced illness, male hamsters learned to avoid female vaginal secretion in subsequent preference tests. More recently Johnston, Zahorik, Immler and Zakon (1978) found that males that had been so conditioned also exhibited lower levels of male sex behavior. These authors argue that this conditioning procedure may have directly altered the signal value of the female odor.

Work described so far indicates that preferences for sex odors can be affected by both infant and adult experience. However, one of the drawbacks of much of the research described is that a simple attraction to or avoidance of an odor does not permit strong inferences concerning the signal value of the odor. A case in point concerns the recent finding that while the attraction to female hamster secretion can be largely accounted for by the dimethyl disulfide component of the secretion, this component does not seem to account for the ability of vaginal secretion to elicit copulatory behavior (Macrides, Johnson and Schneider, 1977). Rather than using preference testing methodology, a better approach for examining sex odor signal value would be to utilize a naturally occurring behavioral response that is directly related to sexual motivation and which will occur in response to appropriate odors alone.

MALE MOUSE ULTRASOUNDS AND SEXUAL AROUSAL

In our own attempts to examine whether the sex signalling value of female mouse odors are experientially acquired by male mice, we have used the naturally occurring behavior of ultrasonic emissions by male mice. Male mice emit 70 kHz vocalizations when courting and copulating with females. The ultrasounds are most prominent during initial investigatory behavior and temporally correlate with male behaviors that indicate a high degree of sexual motivation (Sales, 1972). The ultrasounds continue after the male begins mounting the female, decline across intromissions, and cease altogether by ejaculation. Whether subsequent ejaculatory sequences in a mating bout also contain male ultrasounds is not known. Females, however, seldom emit ultrasounds during male-female encounters (Whitney et al., 1973).

Several indirect lines of evidence suggest that male ultrasounds may serve a courtship function. In addition to the temporal correlation of ultrasounds with male sexual arousal, the motivation to produce ultrasounds is socially and hormonally modulated in a fashion very similar to various components of male sex behavior (Dizinno and Whitney, 1978; Nyby, Dizinno and Whitney, 1976, 1978; Nunez, Nyby and Whitney, 1978). Also supporting a functional role for ultrasounds, male mouse ultrasounds possess the genetic architecture of a trait directionally related to reproductive fitness (Coble, 1972; Whitney, Nyby, Coble and Dizinno, 1978; Nyby and Whitney, 1978).

It has been suggested that male "courtship" ultrasounds may have evolved as a result of sexual selection (Whitney, Stockton, and Tilson, 1971; Whitney et al., 1973). Because infant pup ultrasounds are thought to inhibit maternal aggression (Noirot, 1972; Sales, 1972) and because the adult male ultrasounds are physically similar to infant ultrasounds, the suggestion has been made that adult males, by mimicking infant ultrasounds, may reduce female aggression and withdrawal and thus enhance the probability of insemination (Whitney et al., 1973). Similar use of infantile behaviors by adult males during courtship behavior has been noted for many birds and mammals (Eibl-Eibesfelt, 1970).

MALE MOUSE ULTRASOUNDS: A BIOASSAY FOR FEMALE ODORS

Male mice will emit ultrasounds not only to females but also
to female odors. For example, bedding from female-occupied cages
elicits ultrasounds from males even when the males do not tactu-
ally contact the bedding (Whitney et al., 1974). Thus the effectiveness of female cues appears to reside in molecules that have
some degree of volatility. In unpublished work, olfactory bulbectomy was found to severely reduce the amount of ultrasound
emitted in response to females and altogether abolished the response to female bedding. While this manipulation is also consistent with chemosensory recognition of female-produced cues, the
results do not discriminate the importance of vomeronasal, trigeminal, nervus terminalis, septal organ, or olfactory chemosensory systems nor rule out central nonsensory effects of bulbectomy (Alberts, 1974; Graziadei, 1977).

FEMALE CUES THAT ELICIT ULTRASOUNDS

The female-produced chemosensory cues that elicit male ultrasounds appear to be located all over the female's body. Cotton
swabs containing female urinary, vaginal, or facial substances
were each effective elicitors of male ultrasound while corresponding male and control substances were relatively ineffective (Nyby,
Wysocki, Whitney and Dizinno, 1977). For reasons outlined elsewhere (Nyby and Whitney, 1978) we have recently used mainly urine
as the chemical stimulus of interest. Recent work (Nyby, Wysocki,
Whitney, Dizinno and Schneider, 1979) indicates that an ultrasound-eliciting substance in female urine is a water-soluble, heat-
resistant, relatively nonvolatile, organic substance whose production appears to be regulated primarily by pituitary factors
rather than ovarian hormones.

EARLY CUES AS TO THE IMPORTANCE OF LEARNING

In early work (e.g., Whitney et al., 1974), males seemed to
respond less and were much more variable in their ultrasonic response to female odors than to females. While a number of individual males in various experiments seemed to show "appropriate"
responding to urine, many males would not respond to urine at all,
while a few males emitted ultrasounds to "inappropriate" stimuli.

In this early research, the males who emitted ultrasounds to
female but not male urine were often males that had been used in

other behavioral research. Consequently, we began to come to the
subjective awareness that perhaps males had to have some association with females before they would emit ultrasounds to female
urine. On the basis of subjective feelings we decided to put a
group of male subjects through a systematic social experience
regime and then to test their ultrasonic responses to odors. Our
original motivation for giving males this social experience
regime was not because we were particularly interested in learning
but rather because we were interested in setting up a reliable behavioral assay for female odors.

Since the female estrous cycle is typically 4 or 5 days in the
mouse (Bronson, Dagg and Snell, 1968), we initially guessed that an
effective social experience regime might consist of a naive male
subject experiencing approximately 2 female cycles. Hence we
arbitrarily decided on an eight-day social experience regime.
This regime consisted of adult male subjects being given daily 3-
min exposures in their home cages to both male and female stimulus
mice. We also insured that male subjects experienced only victory
in their agonistic interactions with other males by presenting
stimulus males for the first 3 to 5 exposures using the dangling
method of Scott (1966). The reason for this method of male
stimulus presentation was that previous research (Nyby et al., 1976)
had indicated that social subordination greatly reduces ultrasound emissions from male mice. Female stimuli, however, were
simply placed inside the male subject's home cage. After several
days of this regime, the male subjects invariably attacked the
stimulus male while some mounting of female stimuli was also
observed. Sex behavior, however, seldom resulted in intromission
and never in ejaculation.

One of the first experiments for which we used "socially-
experienced" males involved testing male ultrasonic responsiveness
to female urine, male urine, and distilled water presented on cotton
swabs. The results are seen in Figure 1 and were by far the
clearest we had obtained that males could differentially recognize
the sex-signalling value of male and female urine.

It seemed subjectively clear that the excellent male responsiveness to female urine and the low responsiveness to male
urine as compared with our past unclear results were due to the
experiential regime that we put the males through in this experiment. The seemingly powerful effects of this social-experience
regime suggested that it might be appropriate to further examine
this phenomenon.

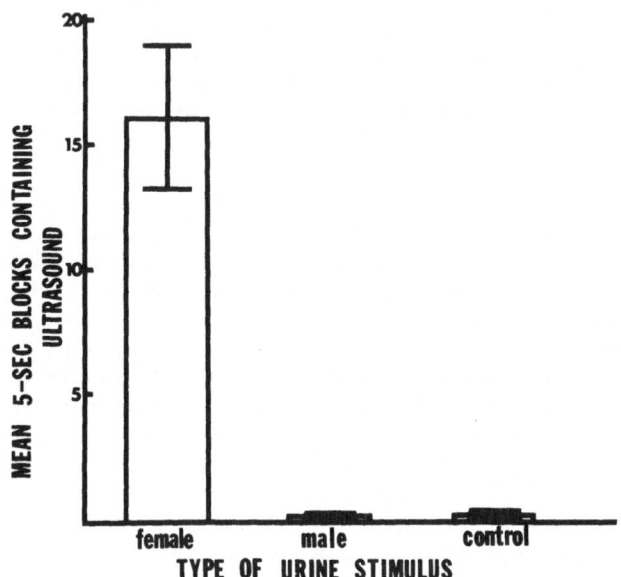

Figure 1.   Mean number of 5-sec blocks (+ SEM) containing 70-kHz
            ultrasonic vocalizations during a 3-min test emitted by
            DBA/2J male mice in response to female urine, male urine
            and water (control) on a cotton swab.   From Nyby,
            Wysocki, Whitney and Dizinno (1977).

EXPERIENCE DETERMINES FEMALE URINE SIGNAL VALUE

        In a set of experiments (Dizinno, Whitney and Nyby, 1978) we
attempted to explicitly test the role of social experience using
males from an inbred strain and a hybrid strain.   Subjects from
both of these 2 genotypes were randomly assigned to either a
"socially experienced" or a "socially inexperienced" condition.
The socially experienced males received the 8-day social experience regime already described while the socially inexperienced
males were treated similarly but received no exposures.   Thus the
socially inexperienced males had not physically contacted a
female mouse since weaning.

After the social-experience regime, the males, both experienced
and inexperienced, were each tested for their ultrasonic responsiveness to three different stimuli consisting of (1) male urine on a
cotton swab, (2) female urine on a cotton swab and (3) water on a
cotton swab. The swab tests occurred in 6-day sequences. Each
sequence consisted of presenting all three stimuli, each stimulus
presentation separated by 48 hours. Swab tests continued for a
subject until no ultrasounds were emitted to any stimulus for 2
consecutive sequences (12 days) or for a maximum of 15 sequences
(90 days). To determine whether animals would respond to females
after they had ceased responding to urine, all subjects were
given 2 post-tests with a female stimulus animal. The two post-
tests occurred 4 and 6 days after the last swab test.

Because the effects of experience were very similar for the 2
strains of males we graphically present here the combined results
from both groups of males. Ultrasounds were emitted only by the
socially experienced males during the social-experience regime.
In line with earlier findings, the ultrasounds were more evident
in the presence of a female than a male social-experience stimulus
animal.

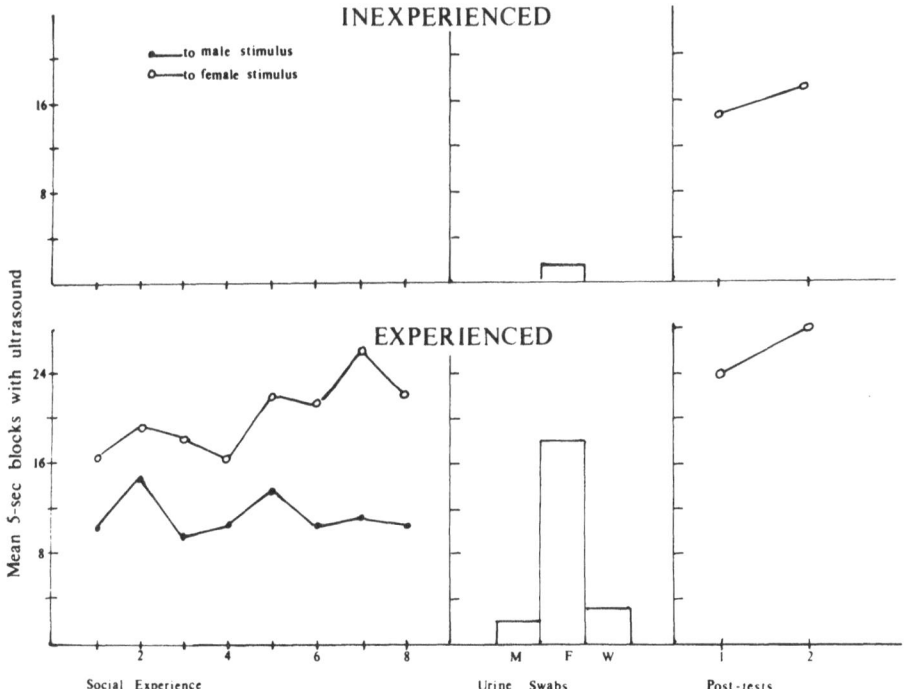

Figure 2. Mean number of 5-sec blocks containing 70-kHz ultrasonic
           vocalizations during a 3-min test by experienced and inex-
           perienced mice over the three phases of the experiment.
           M = male, F = female, W = water. Adapted from Dizinno,
           Whitney and Nyby (1978).

The results of the first 2 swab sequence tests are also summarized in Figure 2. Socially experienced males of both genotypes
emitted substantial ultrasound to female urine only, male urine
and water were relatively ineffective elicitors of ultrasounds.
However, more relevant to the "learning" hypothesis, socially
inexperienced males emitted little or no ultrasound to any stimuli.
Thus this experiment seemed to confirm our original impressions
that some degree of social experience in adulthood was necessary
for males to emit ultrasounds to female urine.

Also relevant to a "learning" hypothesis was the finding that
with repeated exposures to female urine, in the absence of the female herself, fewer and fewer of the males emitted ultrasounds in
response to the female urine. Of the 23 males that initially emitte
ultrasound in response to female urine on first exposure, only 6
were still emitting ultrasounds to female urine on the fifteenth
exposure. Thus male ultrasonic responsiveness to female urine
appeared to "extinguish" with repeated exposures.

After the "extinction phase" of the experiment, both socially
experienced and inexperienced males were found to emit ultrasounds in response to a stimulus female. However, of special
interest was the finding that of the 17 socially experienced males
who had "extinguished" their response to female urine, all 17
emitted ultrasounds upon first exposure to a female. This finding
indicated that both experienced and inexperienced males readily
discriminate females from female urine and in situations where
female urine is an ineffective cue for ultrasound elicitation,
the female herself is a potent cue. Thus the results of this
experiment seemed consistent with the notion that male mice were
learning the signal value of female urine via straightforward
Classical Conditioning. According to such an interpretation, female urine was a conditioned stimulus for adult mice while some
other aspect of the female was an unconditioned stimulus.

CONDITIONING VS. SENSITIZATION

However, in addition to having a chance to associate female
urinary odors with females, the socially experienced males of the
previous experiment also received more exposure to female urine.
Thus, the possibility remained that males "know" the signal value
of female urine but that repeated exposures (sensitization) are
necessary before males emit ultrasounds in response to urine.
Consequently, it was of interest to determine whether repeated
exposures to female urine, in the absence of the female, would
allow female urine to acquire ultrasound eliciting properties.

In conjunction with Frank Barbehenn at Lehigh University,
the ultrasonic responsiveness of 2 groups of male mice to male

and female urine was tested. At the beginning of the experiment
both groups of mice were sexually and socially naive. Both groups
were tested for their ultrasonic responsiveness to both female and
male urine each day for eight consecutive days. However, one of
the groups (socially experienced) was exposed to males and females
immediately following the presentation of a homotypical urine
swab, while the other group (socially inexperienced) was not.

The results of this experiment are seen in Figure 3. As in
the previous experiment, exposure to males and females caused
female urine to acquire ultrasound eliciting qualities. However,
the group that received no exposure to females or males emitted
virtually no ultrasound to either male or female urine. Thus it
appeared that exposure to female urine alone was not sufficient to
cause urine to acquire ultrasound-eliciting qualities; the female
urine must be paired with a female.

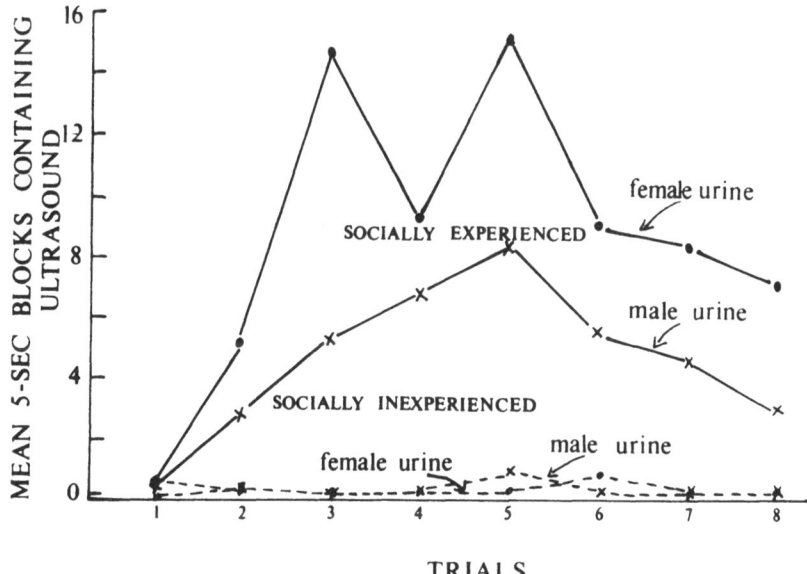

Figure 3. Mean number of 5-sec blocks containing 70-kHz ultra-
           sonic vocalizations during a 3-min test by socially
           experienced and socially inexperienced male mice in
           response to male and female urine.

This experiment had the additional advantage of allowing us
to examine the time course for this biologically relevent learning. As can be seen, two or three 3-min exposures to female
urine were sufficient for female urine to acquire full ultrasound-

eliciting potency. Thus this experiment seemed to confirm earlier observations that the learning occurs very quickly. This finding also suggested that males may be biologically predisposed to learn the signal value of female urine.

ACQUISITION AND EXTINCTION

In yet another experiment with Frank Barbehenn, we again attempted to map the speed with which the ultrasonic response to female urine is acquired. As in the previous experiment, male subjects were tested with female and male urine swabs each day. Immediately following measurement of the ultrasonic response to the swab, a social-experience stimulus animal was placed in the subject's home cage for 3 min. Half of the subjects had each urine swab immediately followed by a social-experience animal of the same sex while the remaining subjects had each urine swab immediately followed by an animal of the opposite sex. Since the effects of these differential sequential pairings of urine with stimulus animals had qualitatively similar effects, we describe here the combined results of the 2 groups.

As can be seen in Figure 4, the ultrasonic responsiveness to female urine again seemed to reach its peak after only two 3-min exposures to females. This finding again confirmed the relative ease with which males learn the signal value of female urine.

During the extinction phase of the experiment, the urinary swabs continued to be presented as before except that now the swab was not followed by exposure to a social-experience animal. As can be seen, the responsiveness to female urine declined across trials thus replicating our earlier findings (Dizinno et al., 1978).

CREATION OF AN ARTIFICIAL SEX PHEROMONE: ADULT EXPERIENCE VS. NEONATAL IMPRINTING

Because the salience of female urinary odor appeared to be learned, normally inappropriate odors, if suitably encountered, could conceivably become "artificial sex pheromones." In a recent experiment (Nyby et al., 1977) we examined whether males could be induced to emit ultrasounds to perfume which normally does not elicit ultrasounds and further whether the phenomenon would occur as a result of neonatal imprinting on the mother, adult male experience with the first adult female, or both. To test these possibilities, hybrid males were allowed contact with only two adult females from birth until testing. The first adult female con-

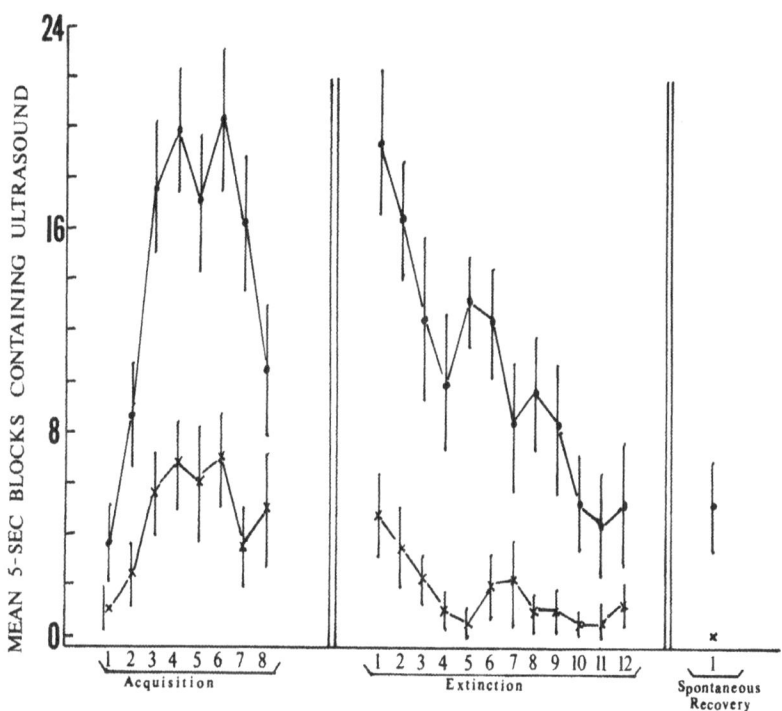

TRIALS AND PHASES

Figure 4.   Mean number of 5-sec blocks (+ SEM) containing 70-kHz
            ultrasonic vocalizations during a 3-min test by male
            mice in response to male (X—) and female (•—) urine.
            During acquisition, the 3-min urine tests were
            immediately followed by a 3-min exposure to a stimulus
            animal. During extinction and spontaneous recovery,
            urine presentations were not followed with a stimulus
            animal.

sisted of the male's mother with whom the male was housed from birth until weaning at 21 days of age. Later in early adulthood the subjects were exposed to the second adult female as well as an adult male in the 8-day social experience regime already described.

The male subjects were assigned to four different odor treatment groups: (1) P-P, perfume on the mother, perfume on the social experience female, (2) P-E, perfume on the mother, no perfume on the social experience female, (3) E-P, no perfume on the mother, perfume on the social experience female and (4) E-E, no perfume on either female. Females were odorized by whole-body spraying from atomizers containing either Wild Musk Spray (Coty) or ethanol (E) and allowed to dry for 15 min prior to their use as olfactory stimuli. This time course ensured that both the ethanol and perfume solvent had largely evaporated before the subjects encountered the females. The fathers and social experience males were not odorized.

As can be seen in Figure 5, males from all groups reliably responded to female urine (FU) but not the male urine (MU) or ethanol (E). However, only those males exposed to perfumed social experience females (Groups P-P and E-P) emitted much response to perfume. Males exposed prior to weaning to perfumed mothers and later exposed to non-perfumed, social experience females (Group P-E) did not emit ultrasounds to perfume. Thus the sex-signalling value of perfume appeared to acquire its ultrasound eliciting potency more as a result of adult experience than neonatal imprinting. However, since males exposed to perfumed females both before weaning and when adult (Group P-P) gave the greatest ultrasonic response to perfume, the adult learning appeared to be potentiated by the infant experience.

CONSTRAINTS ON CUES WHICH CAN ACQUIRE SEX SIGNALLING VALUE

Much current literature is concerned with "biological constraints" on learning (e.g., Hinde and Hinde, 1973). The fact that males can learn to emit ultrasounds to an artificial perfume as well as to naturally occurring biological odors suggests that the cues which can acquire ultrasound eliciting potency may not be tightly constrained. However, we should note that the perfume used may be similar to naturally occurring "musk-like" odors and secondly that 2 points on a possible continuum do not clearly indicate the degree of specificity which might be involved.

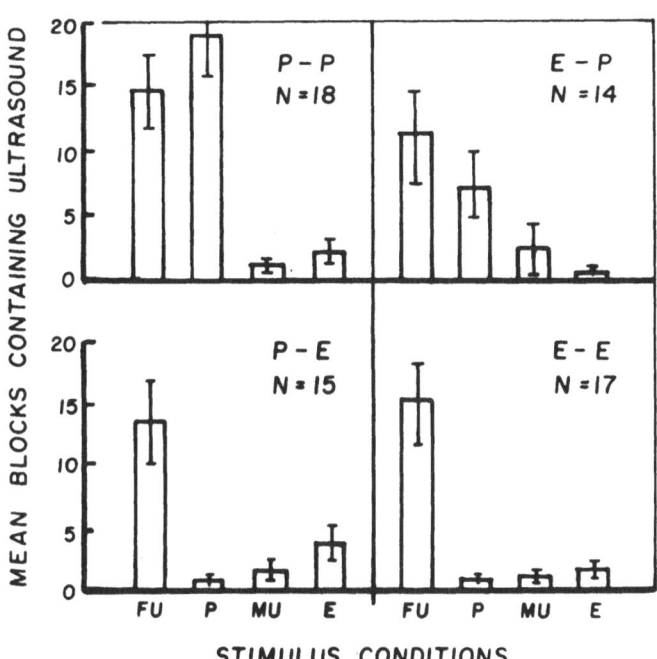

Figure 5. Mean number of 5-sec blocks ($\pm$ SEM) containing 70-kHz
ultrasonic vocalizations by four male groups in response
to four different stimuli. The four groups (N = number
of males per group) were defined by whether perfume (P)
or ethanol (E) was placed on the mother (first letter)
or on the female providing heterosexual exposure to the
adult male (second letter). The four stimuli were
female urine (FU), perfume (P), male urine (MU) and
ethanol (E). Each stimulus was presented for 3 min to
each subject. From Nyby, Whitney, Schmitz and Dizinno
(1978).

In a recently completed unpublished experiment both chemical substance and amount of experience were found to be important determinants of the cues to which adult male mice will emit ultrasounds. Independent groups of naive adult males were exposed to females which had been sprayed with either (a) Wild Musk Spray perfume, (b) coumarin or (c) ethanol (control group for wetting of females with solvent of the other 2 odorants). Each exposure consisted of 3 min in a cage with a female who had been previously sprayed with the respective odorant. Then after one exposure, three exposures and seven exposures, all males were tested in a counterbalanced fashion for ultrasonic responses to cotton swabs containing either (a) perfume, (b) coumarin, (c) male urine or (d) normal female urine. While both female urine and perfume acquired ultrasound eliciting potency, more responses occurred sooner to female urine than to perfume. However, after seven exposures, perfume was at least as potent as normal female urine. Response to coumarin over all trials was weak and inconsistent. It thus appears that at least one odorous substance, coumarin, does not easily acquire sex signalling potency.

CONCLUDING REMARKS

The research described here indicates that both the acquisition and maintenance of the male ultrasonic response to female urine are influenced by experiential factors. First, it appears that an adult male requires some experience with adult conspecifics before female urine acquires its ultrasound eliciting properties. Second, repeated presentations of female urine to a socially experienced male, in the absence of the female, cause the urine to cease eliciting ultrasounds. And thirdly, initially neutral perfume was found to acquire ultrasound eliciting properties if appropriately paired with a female. All of these results are consistent with the hypothesis that the salience of female urine is learned, and that the acquisition of this "pheromone" mediation of behavior may be an instance of classical conditioning.

The fact that the signal value of female urine is learned by adult male mice may be adaptive. For example, the male mouse acquisition of sex cue value of urinary substances may be similar to imprinting in some bird species in which sexual imprinting can occur at a substantially more advanced age than does filial imprinting (Immelman, 1975; 1978). Along these lines, we have speculated that a pubertal or dispersal age establishment of sex pheromone signal value among mice could be important in population dynamics, tending to reduce the amount of close inbreeding in natural populations (Whitney and Nyby, in press).

Behavioral plasticity may also be important in insuring that male mouse sex behavior occurs in nature at the appropriate place and time. The occurrence of sex behavior in mammals is typically very situation specific and often depends on the integration of information from a variety of sensory modalities. Thus, while an already learned sex odor may be readily recognized, the odor may or may not stimulate sexual arousal unless the animals are in the appropriate situation for sexual behavior to occur.

Whether the experiential component necessary for the mouse ultrasonic response to female urine is characteristic of mammalian signalling pheromone systems in general remains an open question. However, we suspect that learning may be centrally involved in the behavioral responses to sex odors by other species.

ACKNOWLEDGEMENTS

The preparation of this manuscript was supported by National Science Foundation Grant BNS 77-15265. We wish to thank Frank Barbehenn for allowing us to describe unpublished data.

REFERENCES

Alberts, J.R. 1974. Producing and interpreting experimental olfactory deficits. Physiol. Behav. 12: 657-670.

Alberts, J.R. 1976. Olfactory contributions to behavioral development in rodents. In Doty (Ed.) Mammalian Olfaction, Reproductive Processes and Behavior". pp. 67-91, Academic Press, New York, San Francisco, London.

Breen, M.F. and A.I. Leshner. 1977. Maternal pheromone: a demonstration of its existence in the mouse. Physiol. Behav. 18: 527-529.

Bronson, F.H. 1968. Pheromonal influences on mammalian reproduction. In M. Diamond (Ed.) Reproduction and Sexual Behavior". pp. 341-361, Indiana University Press, Bloomington.

Bronson, F.H. 1974. Pheromonal influences on reproductive activities in rodents. In M. Birch (Ed.) Pheromones. pp. 345-365, North Holland, Amsterdam, London.

Bronson, F.H., C.P. Dagg and G.D. Snell. 1968. Reproduction. In Green (Ed.) Biology of the Laboratory Mouse. Dover, New York.

Beauchamp, G.K. and E. Hess. 1973. Abnormal early rearing and sexual responsiveness in male guinea pigs. J. Comp. Physiol. Psychol. 85: 383-396.

Caroom, D. and F.H. Bronson. 1971. Responsiveness of female mice to preputial attractant: effects of sexual experience and ovarian hormones. Physiol. Behav. 1: 659-662.

Carr, W.J., L.S. Loeb and M.L. Dissinger. 1965. Responses of rats to sex odors. J. Comp. Physiol. Psychol. 59: 370-377.

Carter, C.S. 1972. Effects of olfactory experience on the behavior of the guinea-pig (Cavia porcellus). Anim. Behav. 20: 54-60.

Carter, C.S. and J.N. Marr. 1970. Olfactory imprinting and age variables in the guinea pig (Cavia porcellus). Anim. Behav. 18: 238-244.

Coble, J. 1972. The relative contributions of genotype and environment to the production of ultrasounds by adult Mus musculus in male-female pairings. Unpublished doctoral dissertation, Florida State University.

Denenberg, V.H., G.A. Hudgens, and M.X. Zarrow. 1964. Mice reared with rats: modification by early experience with another species. Science 143: 380-381.

Dizinno, G. and G. Whitney. 1977. Androgen influence on male mouse ultrasounds during courtship. Horm. Behav. 8: 188-192.

Dizinno, G., G. Whitney and J. Nyby. 1978. Ultrasonic vocalizations by male mice (Mus musculus) to female sex pheromone: experiential determinants. Behav. Biol. 22: 104-113.

Doty, R.L. and I. Dunbar. 1974. Attraction of beagles to conspecific urine, vaginal and anal sac secretion odors. Physiol. Behav. 12: 825-833.

Eibl-Eibesfeldt, I. 1970. Ethology, the Biology of Behavior. Holt, Rinehart and Winston, New York.

Gilder, P.M. and P.J.B. Slater. 1978. Interest of mice in conspecific male odours is influenced by degree of kinship. Nature 274: 364-365.

Goldfoot, D.A., R.W. Goy, M.A. Kravetz, S.K. Freeman. 1976. Reply to Michael, Bonsall and Zumpe; Reply to Keverne. Horm. Behav. 7: 373-383.

Graziadei, P.P.C. 1977. Functional anatomy of the mammalian chemoreceptor system. In Muller-Schwarze and Mozell (Eds.) Chemical Signals in Vertebrates. pp. 435-454, Plenum Press, New York.

Gregory, E. and H. Bishop. 1975. Development of olfactory-guided behavior in the golden hamster. Physiol. Behav. 15: 373-376.

Hayashi, S. and T. Kimura. 1974. A sex attractant emitted by female mice. Physiol. Behav. 13: 563.

Hinde, R.A. and J. Stevenson-Hinde. 1973. Constraints on learning: Limitations and Predispositions. Academic Press, London and N.Y.

Immelmann, K. 1975. Ecological significance of imprinting and early learning. Ann. Rev. Ecol. and Systematics 6: 15-37.

Immelmann, K. 1978. Genetical constraints on early imprinting: a perspective from sexual imprinting in birds (and personal communication with G.W.) Presentation at NATO Study Institute: Theoretical Advances in Behavior Genetics, Alberta, Canada.

Johnston, R.E. 1974. Sexual attraction function of golden hamster vaginal secretion. Behavioral Biology 12: 111-117.

Johnston, R.E. and D.M. Zahorik. 1975. Taste aversions to sexual attractants. Science 189: 893-894.

Johnston, R.E., D.M. Zahorik, K. Immler and H. Zakon. 1978. Alterations of male sexual behavior by learned aversions to hamster vaginal secretions. J. Comp. Physiol. Psychol. 92: 85-93.

Karlson, P. and A. Butenandt. 1959. Pheromones (ecto hormones) in insects. Ann. Rev. Entomol. 4: 39-58.

Landauer, M.R., R.E. Wiese Jr. and W.J. Carr. 1977. Responses of sexually experienced and naive male rats to cues from receptive and nonreceptive females. Anim. Learn. Behav. 5: 398-402.

Leon, M. and H. Moltz. 1971. Maternal pheromone: discrimination by preweaning albino rats. Physiol. Behav. 7: 265-267.

Lydell, K. and R.L. Doty. 1972. Male rat odor preferences for female urine as a function of sexual experience, urine age, and urine source. Horm. Behav. 3: 205-212.

Macrides, F., P.A. Johnson, S.P. Schneider. 1977. Responses of the male golden hamster to vaginal secretion and dimethyl disulfide: attraction versus sexual behavior. Behav. Biol. 20: 377-386.

Mainardi, D., M. Marsan and A. Pasquali. 1965. Causation of sexual preferences of the house mouse. The behaviour of mice reared by parents whose odour was artificially altered. Atti. Soc. Ital. Sci. Nat. 104: 325-338.

Marr, J.N. and L.E. Gardner, Jr. 1965. Early olfactory experience and later social behavior in the rat. Preference, sexual responsiveness and care of young. J. Genet. Psychol. 107: 167-174.

Marr, J.N. and L.G. Lilliston. 1969. Social attachment in rats by odor and age. Behaviour 33: 277-282.

Michael, R., R. Bonsall, and P. Warner. 1975. Primate sexual pheromones. In Olfaction and Taste, V. (Ed. D.A. Denton and J.P. Coghlan). pp. 417-424, Academic Press, N.Y.

McCarty, R. and C.H. Southwick. 1977. Cross-species fostering: effects on the olfactory preference of Onychomys torridus and Peromyscus leucopus. Behav. Biol. 19: 255-260.

McDonald, D.L. and L.G. Forslund. 1978. The development of social preferences in the voles Microtus montanus and Microtus canicaudus: effects of crossfostering. Behav. Biol. 22: 497-508.

Müller-Schwarze, D. and C. Müller-Schwarze. 1971. Olfactory imprinting in a precocial mammal. Nature 229: 55-56.

Noirot, E. 1972. Ultrasounds and maternal behavior in small rodents. Dev. Psychobiol. 5(4): 371-387.

Nunez, A.A., J. Nyby and G. Whitney. 1978. The effects of testosterone, estradiol, and dihydrotestosterone on male mouse (Mus musculus) ultrasonic vocalizations. Horm. Behav. 11: 264-272.

Nyby, J., G.A. Dizinno and G. Whitney. 1976. Social status and ultrasonic vocalizations of male mice. Behav. Biol. 18: 285-289.

Nyby, J., G.A. Dizinno and G. Whitney. 1977. Sexual dimorphism
    in ultrasonic vocalizations of mice (Mus musculus): gonadal
    hormone regulation. J. Comp. Physiol. Psychol. 91: 1424-1431.
Nyby, J., C.J. Wysocki, G. Whitney and G. Dizinno. 1977. Phero-
    monal regulation of male mouse ultrasonic courtship (Mus
    musculus). Anim. Behav. 25: 333-341.
Nyby, J., G. Whitney, S. Schmitz, and G. Dizinno. 1978. Post-
    pubertal experience establishes signal value of mammalian
    sex odor. Behav. Biol. 22: 545-552.
Nyby, J. and G. Whitney. 1978. Ultrasonic communication by
    adult rodents. Neurosci. Biobehav. Rev. 2: 1-14.
Nyby, J., C.J. Wysocki, G. Whitney, G. Dizinno, and J. Schneider.
    1979. Elicitation of male mouse ultrasonic vocalizations.
    I. Urinary cues. J. Comp. Physiol. Psychol. In press.
Porter, R.H., and F. Etscorn. 1974. Olfactory imprinting re-
    sulting from brief exposure in Acomys cahirinius. Nature
    250: 732-733.
Quadagno, D.M. and E.M. Banks. 1970. The effect of reciprocal
    cross fostering on the behaviour of two species of rodents,
    Mus musculus and Baiomys taylori ater. Anim. Behav. 18:
    379-390.
Sales, G.D. 1972. Ultrasound and mating behaviour in rodents
    with some observations on other behavioral situations. J.
    Zool. 168: 144-164.
Scott, J.P. 1966. Agonistic behavior of mice and rats, A review.
    Am. Zool. 6: 683-701.
Stern, J.J. 1970. Responses of male rats to sex odors. Physiol.
    Behav. 5: 519-524.
Whitney, G., Stockton, M.D. and Tilson, E.F. 1971. Possible
    social function of ultrasounds produced by adult mice (Mus
    musculus). Amer. Zool. 11: 72 (abstract).
Whitney, G , J.R. Coble, M.D. Stockton, and E.F. Tilson. 1973.
    Ultrasonic emissions: Do they facilitate courtship of mice?
    J. Comp. Physiol. Psychol. 84: 445-452.
Whitney, G., M. Alpern, G. Dizinno, G. Horowitz. 1974. Female
    odors evoke ultrasounds from male mice. Animal Learning and
    Behavior 2: 13-18.
Whitney, G. 1976. Genetic considerations in studies of the
    evolution of the nervous system and behavior. In R.B.
    Masterson, W. Hodos and H. Jerison (Eds.) Evolution, Brain
    and Behavior, Persistent Problems. Lawrence Erlbaum Associates,
    Hillsdale.
Whitney, G., J. Nyby, J.R. Coble, G.A. Dizinno. 1978. Genetic
    influences on 70 kHz ultrasound production of mice (Mus
    musculus). Behav. Gen. 8: 574 (abstract).
Whitney, G. and J. Nyby. 1979. Cues that elicit ultrasounds.
    Amer. Zool., in press.
Whitten, W. 1975. Responses to pheromones by mammals. Olfaction
    and Taste V (Ed. D.A. Denton and J.P. Coghlan). pp. 389-395,
    Academic Press, N.Y.

DEVELOPMENT OF OLFACTORY ATTRACTION BY YOUNG NORWAY RATS

Michael Leon

Department of Psychology, McMaster University
Hamilton, Ontario, Canada L8S 4K1

EMISSION, SYNTHESIS AND ATTRACTION TO MATERNAL ODOR

Young rats orient toward maternal olfactory cues as early as 2-3 days postpartum (Altman, et al., 1973). The orientation to these cues becomes more and more reliable, until by days 12-14, they not only orient, but they move toward the odor from a distance (Altman et al., 1973; Leon and Moltz, 1972). The pups continue to be attracted to the mother during the period that the young are capable of leaving the nest site but must reunite with the mother for periodic nursing bouts (Rosenblatt and Lehrman, 1963). Termination of the approach response to maternal odor occurs at about 27 days postpartum, which is the time of weaning (Leon and Moltz, 1972).

For their part, mothers emit odors in increased concentrations precisely during the period during which the young normally approach them (Leon and Moltz, 1972). Maternal odors facilitate mother-young reunions by marking the nest material (Gregory and Pfaff, 1971), the pups (Leon, 1974), and the mother (Leon, 1974). The most salient maternal odor is synthesized in the caecum by the action of microorganisms on the partially digested food during the formation of caecotrophe (Leon, 1974). Non-lactating rats normally re-ingest most of their caecotrophe as it is defecated (Lutton and Chevallier, 1973), but as a result of a prolactin-induced rise in food intake, lactating dams increase their caecotrophe defecation without a concomitant increase in the amount of caecotrophe that they re-ingest (Leon, 1974). Mothers thereby emit a great deal more caecotrophe to the environment, and are consequently more attractive than non-lacting rats (Leon, 1974). Maternal caecal odor is "maternal" only in the sense that dams are more likely to

emit the odor-bearing caecotrophe to the environment. A summary
diagram of the synthesis and emission of the maternal odor is
presented in Figure 1.

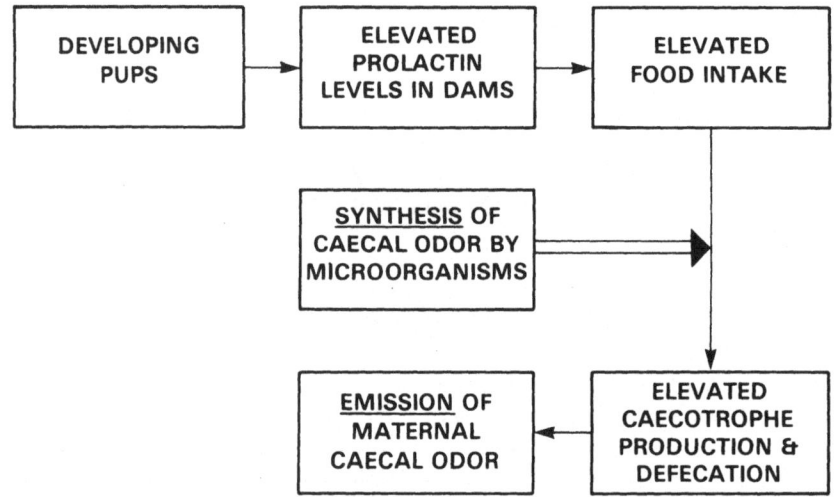

Figure 1. A model for maternal caecal odor emission and synthe-
           sis. Developing pups stimulate high levels of circula-
           ting prolactin in their mothers, which induces the dams
           to increase their food intake. When the food enters
           the caecum, a portion of it is acted upon by caecal
           microorganisms to produce the odor. The odor is emitted
           when the dam defecates the caecotrophe without reingest-
           ing that material.

DEVELOPMENT OF THE APPROACH RESPONSE

     It appears that young rats must acquire their attraction to
the odor of their mother by means of experience. When mothers
were placed on a diet with sucrose as its only constituent carbo-
hydrate, the dams were unable to maintain caecal microbial growth
under these conditions and did not synthesize a maternal caecal
odor. Their pups were attracted neither to mothers fed the sucrose
diet, nor to normally fed colony dams that were highly attractive
to colony reared young (Leon, 1974). On the other hand, when
sucrose diet dams raised their pups in a room with the odor of
normal colony mothers, the sucrose diet reared pups became
attracted to the odor of colony mothers (Leon, 1975). These data

suggest that the experience of pups with maternal odors determines
their subsequent responsiveness to such odors, but what is the importance of an ability to develop an attraction to different odors
for young rats? The answer apparently lies in the fact that caecal
microorganisms produce different odors on different dietary substrates provided by mother rats, who are omnivores. Pups raised
with mothers eating either of two very different diets showed clear
preferences for the maternal odor with which they had previous
experience. The young were not attracted to the odor, with which
they had no experience, that was produced by mothers eating the
other diet (Leon, 1975). The rat pup, therefore, has the ability
to develop specific olfactory attachments to different maternal
odors.

While this experiment suggested a function for the acquired
attraction, it did not address the mechanism by which the young
developed their olfactory affinities. One possibility is that
approach behavior is classically conditioned. Young rats uncon-
ditionally approach the warmth and nourishment of their mothers
(Gustafsson, 1948). After repeated pairings of odor and mother
the maternal odor may function as a conditional stimulus that
eventually elicits approach behavior. Young rats, however, need
not experience the odor in their mother's presence to develop an
affinity to it. Leon et al. (1977) allowed pups to be reared by
sucrose diet dams (that did not have a caecal odor), and exposed
them, while isolated from their mother and siblings, to the odor
of a normal colony dam for 3 hr/day. Control pups from the same
litters simply had fresh air blown over them while in isolation.
Pups that had been pre-exposed to maternal odor in the absence
of concurrent maternal care approached the odor of colony mothers
with as great a probability as pups actually reared by colony dams
(Figure 2). Control pups were not attracted to that odor. Exposure to an arbitrarily selected non-maternal odor (peppermint)
was also sufficient to induce pups to approach that odor in a
test situation. Simple exposure of young rats to odors is, therefore, sufficient to establish an affinity to such cues in the
absence of maternal reinforcers.

While simple exposure allows pups to acquire an affinity to an
odor, the nature of the mediating mechanism is not clear. That is,
how does exposure promote the acquisition of a special affinity to
odors by pups? One possibility is that continuing exposure to the
odor of the mother imprints onto the nervous system a special
sensitivity to that particular odor. Døving and Pinching (1973)
have provided some support for that notion. They reported that
exposure of young rats to a single refined odor for as little as
2 weeks produced morphological changes (akin to transneuronal
degeneration) in the mitral cells, which are the second-order
neurons in the primary olfactory system. The pattern of these
mitral changes within the bulb were specific to the odor to which
the young had been exposed, with different odors altering different

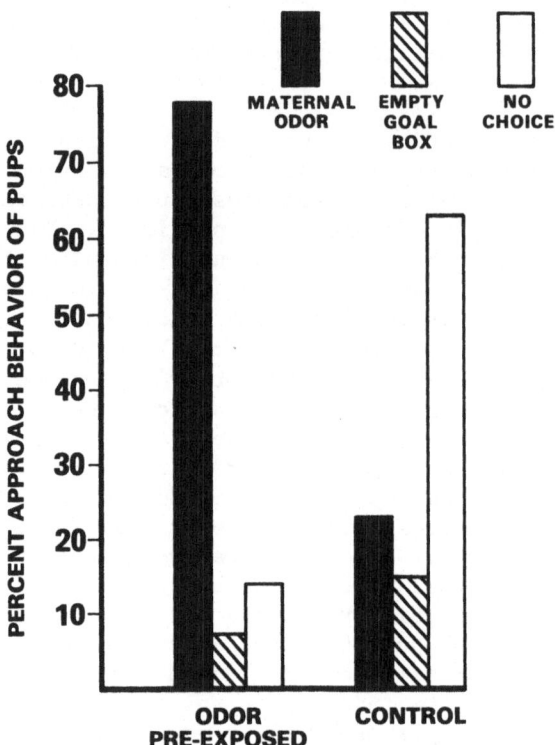

Figure 2. The percent of pups pre-exposed while in isolation
either to maternal odor or to fresh air that approach
the maternal odor in a test situation.

areas within the mitral cell layer (Pinching and Døving, 1974).
These topographically specific changes in the bulb after continuing
exposure of young rats to a single chemical apparently mediate a
differential sensitivity to the pre-exposed pure odor (Laing and
Panhuber, 1978). Since young rats are continuously exposed to the
odor of their mother for the first two weeks postpartum (albeit
in low concentrations) and since the young subsequently develop a
special affinity for that odor, it seemed possible that different
maternal odors may normally be imprinted onto particular portions
of the bulb. A special sensitivity to maternal odor mediated by
those bulb areas would account for the development of an affinity
to such cues. If topographically specific olfactory imprinting of
maternal odors normally occurs at the level of the mitral cells,
then one would expect that pups raised with mothers on one of two
distinct diets would have differentially altered mitral cell layers.

Ten Wistar mothers were given the sucrose based diet (Teklad
test diet #TD69446) and ten were given Purina Rat Chow ad lib in
mid pregnancy. Recall that the sucrose diet was previously shown
to suppress the maternal caecal odor, while the Purina diet allowed
dams to be highly attractive to pups. To insure that there was no
common caecal odor present in the cages of the two groups, the
dams were given clean cages with new wood chip bedding on the day
before parturition and the groups were placed in separate rooms
with separate exhaust systems. Litters were reduced to 8 in
number on the day of birth (day 0) and the mothers and young were
left undisturbed, except for replenishment of food and water, until
day 28 postpartum. At that time, one pup was taken at random from
each litter and anesthetized with Nembutal. The weanling was then
respirated with 95% oxygen and perfused with buffered saline and
a formaldehyde-glutaraldehyde fixative. After the head was immersed
in fixative for a day, the olfactory bulbs were removed and subsequently embedded in paraffin. Coronal sections of 3 μm were taken
through the entire bulb and every tenth section stained with
methylene blue and Azure II.

We found no difference in any observable morphological comparison between the mitral cells (shown in Figure 3) in any single
animal or between the mitral cells in the two groups. Although
there may be some form of olfactory neural imprinting, it is not
observable on a gross anatomical level in the mitral cells. It
seems likely that the reported mitral cell changes reported by
others were due to restriction of olfactory experience to a single
pure odor (see Laing and Panhuber, 1978); the more complex odors
of a mother rat, even without the caecal odor, are sufficient to
support normal mitral cell growth.

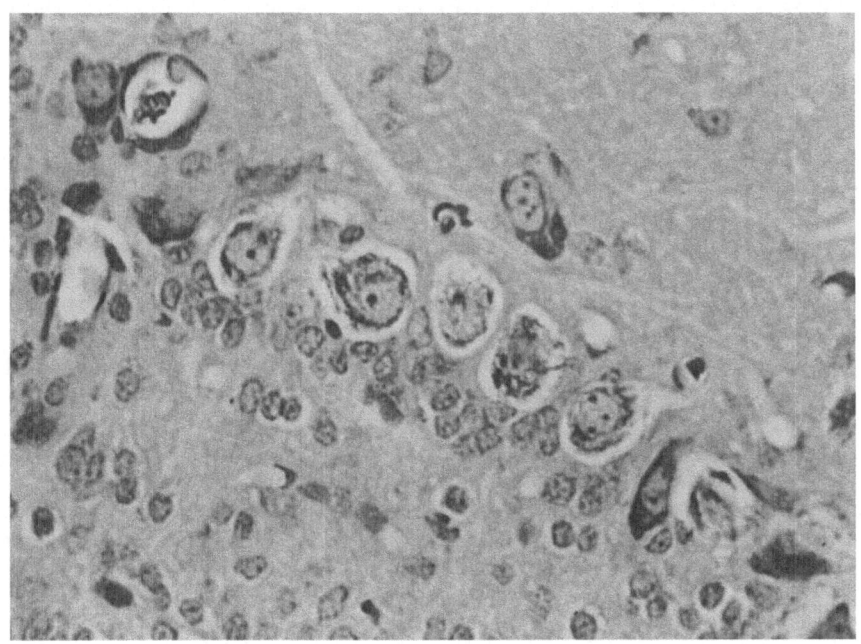

Figure 3. Mitral cells from pups reared with the odors of dams
           eating diet P (Purina Chow).

        While there are no obvious morphological changes at the mitral
cell level, young rats may have other specialized neural mechanisms
that cause olfactory information to be processed in such a way as
to produce a subsequent affinity to their mother's odor. In fact,
there are some differences between the adult and the infant olfactory system that might account for the special sensory coding
that results in the development of olfactory attraction in young
rats. I will describe the adult olfactory system in terms of
structure and function, and then discuss the differences between
the adult olfactory system and the system of the young rat. Finally,
I will suggest how the immature system might be specially adapted
to allow the establishment of odor attraction in rat pups.

STRUCTURE AND FUNCTION OF THE ADULT PRIMARY OLFACTORY SYSTEM

        The primary olfactory sensation and perception system may be
divided into three functional sections: (1) the receptors,

(2) the olfactory bulb with its central projections, and (3) the centrifugal projections back to the olfactory bulb, which together form a feedback loop system that mediates and modulates olfactory perception with respect to both the external chemical stimulation and the internal state of the organism.

Packed into turbinates located in the posterior nasal cavity are the primary olfactory receptors, which are slender ciliated neurons, the fine axons of which terminate on the second order neurons. The more prominent of these second order output neurons are the mitral cells. They receive their convergent sensory input from the olfactory receptor neurons in a dense synaptic formation known as a glomerulus. Smaller tufted cells may also function as output neurons in the bulb. The mitral axons collect and emerge from the bulb as the lateral olfactory tract on the surface of the olfactory peduncle. The lateral olfactory tract serves as the primary olfactory afferent, and synapses in the anterior olfactory nucleus, the olfactory tubercle, and cortical amygdala, and then extends throughout the piriform cortex (Heimer, 1968, 1972; Scalia and Winans, 1975; Valverde, 1965; White, 1965; see Figure 4).

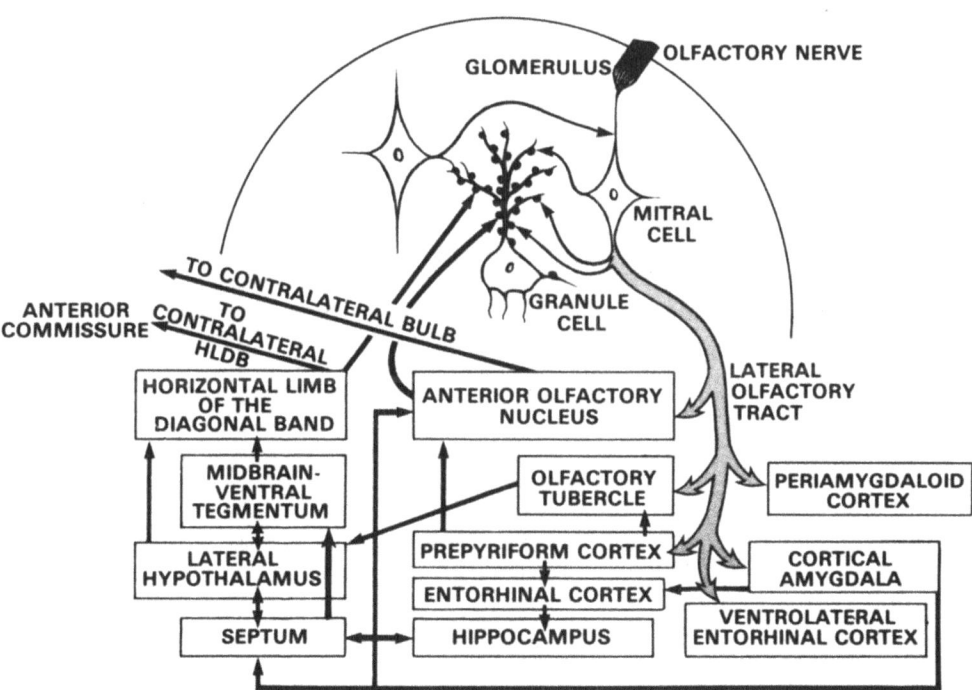

Figure 4.   A schematic representation of olfactory output and input
            systems to the mitral cells, with the mediation of such
            input by the granule cells.

The granule cells are the most numerous of the cell types within the olfactory bulb and their function will become central to our discussion. These cells synapse with the dendrites of the mitral cells, the axon collaterals of mitral cells, and the axons of the contralateral tufted cells, mediating the network of negative feedback within the olfactory bulb on these second order neurons (Shepherd, 1972). Five levels of the inhibitory neural feedback loop have been identified in the olfactory system and they are depicted in Figure 4. First, Rall and Shepherd (1968) suggested that excitation of mitral dendrites can excite a granule cell, which then inhibits the subsequent firing of that same neuron at the dendritic level. This type of mitral cell inhibition may serve to code high intensity olfactory stimuli by preventing an individual neuron from firing repeatedly. Strong stimuli may thereby be efficiently encoded by means of a patterned distribution of neural activity. Mitral cell response decrement may also be mediated by granule cell action of input either from within or from outside the bulb. Secondly, mitral cells can inhibit other mitral cells when axon collaterals which synapse onto granule cells inhibit neighboring mitral cell firing (Rall, Shepherd, Reese and Brightman, 1966). This lateral inhibition of the firing of other mitral cells probably works as it does in other sensory systems to sharpen the coding of incoming stimuli and thereby enhance the ability of the system to discriminate between olfactory cues. Lateral inhibition also appears to be served at the glomerular level by the short axon horizontal cells that connect glomeruli (Pinching and Powell, 1971a, b; Yamamoto, Yamamoto, and Iwama, 1963).

The granule cells also mediate much of the centrifugal input that acts to influence bulb activity, since the centrifugal fibers synapse on granule cells which then affect mitral firing (Heimer, 1968; Shepherd, 1972). One source of these efferents, forming the third source of inhibition, is from the anterior olfactory nucleus, which sends fibers to granule cells that synapse on mitral cells in both ipsilateral and contralateral bulbs (Price and Powell, 1970a, b). This nucleus also mediates a fourth centrifugal input from the piriform cortex (Heimer, 1968) which proceeds to the granule cells of the olfactory bulb. The fifth source of modulation by granule cells on mitral cell activity arises from the horizontal limb of the diagonal band (HLDB; Price and Powell, 1970a, b; Shute and Lewis, 1967), the fibers of which enter the bulb via the lateral olfactory tract. The HLDB receives input from both midbrain (ventral tegmentum) and hypothalamus, particularly its lateral aspect, which itself is the recipient of strong afferent connections from piriform cortex, amygdala, and olfactory tubercule (Powell, Cowan and Raisman, 1965). Direct input to the HLDB may emanate from the amygdala, thalamus and frontal cortex. This system also has access to the midbrain-hypothalamic interactions as well as having primary olfactory afferent inputs to the hypothalamus (Pfaff and Pfaffman, 1969; Scott and Chapin, 1975; Scott and Leonard, 1971; Scott and Pfaffman, 1967). The afferent connections of the HLDB appear to allow information about the internal state of the organism to modulate the encoding of olfactory stimuli by the mitral cells (Pager, 1974a, b; Pager, Giachetti, Holley and LeMagnen, 1972; Pager and Royet, 1976).

The point of this very simplified description of the primary olfactory system is to provide a basis for the understanding of the mechanisms by which the adult system can sense odors. The mature system can gate high intensity signals by reciprocal inhibition, and can have fine olfactory discrimination ability by means of lateral inhibition of other mitral cells. Moreover, centrifugal inhibition allows what might be considered to be a modulation of mitral activity by the internal state of the organism.

STRUCTURE AND FUNCTION OF THE DEVELOPING PRIMARY OLFACTORY SYSTEM

The rat olfactory system is structurally and, therefore, functionally immature at birth, with a portion of its development continuing postnatally. I will first discuss the morphological development of the primary olfactory system in young rats and then describe the possible functional significance of these changes.

The neonatal olfactory bulb has at its interior and occupying a large portion of the entire structure, the olfactory ventricle, the walls of which are composed of ependymal tissue with a large subependymal surround (Altman and Das, 1966). The olfactory receptor axons and the glomeruli are present in the rat neonate, as are the mitral cells, which are more closely packed than they are in the adult (Altman and Das, 1966; Singh and Nathaniel, 1977).

Emanating from the bulb is the lateral olfactory tract, which is visible as early as the first postnatal day, and is composed of unmyelinated axons with a number of mature as well as immature synaptic sites in the piriform cortex (Westrum, 1975). Myelination of the lateral olfactory tract can be seen by days 10-12, continuing through day 21. Day 1-2 pups have the adult laminar termination pattern in the cortex, although the proportion of fibers coming directly from the olfactory bulb is greater than in the adult (Moxley and Price, 1975). By day 14 postpartum, the piriform cortex resembles, in many respects, that of the adult (Westrum, 1975).

The afferent olfactory system, then, appears to be functional, although not fully mature, at birth. The receptors synapse on mitral cells allowing the system to receive, code and project olfactory information to the partially mature piriform cortex. The

organism would, therefore, be expected to be immediately responsive
to odors, although perhaps not at very low levels of stimulation,
since few active synapses are present.

The most striking difference between the neonate and the adult
bulb is the lack of granule cells to synapse with the mitral cells
(Altman and Das, 1966). The granule cell population of the bulb
forms postnatally by the proliferation, differentiation, and migra-
tion of cells from the subependymal to the mitral cell layer.
There is a steady growth of mature granule cells until day 12
postpartum, when growth is substantially stabilized, although some
granule cell proliferation continues until 22 days after birth.
Changes in the electrophysiological output of the olfactory cells
correspond closely with the growth of the granule cell population.
The initial spontaneous activity of the structure is low, but develops rapidly until about day 12, when the rate of maturation
plateaus, reaching adult patterns by day 24 (Salas et al., 1969).

Without the granule cells that mediate mitral inhibition,
the pups should not be capable of fine discriminations between
odors that would require lateral inhibition in the first two weeks
postpartum. In fact, the ability of olfactory neurons to discriminate between odors is initially quite poor. There is, for
example, no differential neural response to maternal odor, food
odor or fresh air on day 6. By day 12, however, a clear discrimination is evident, differentiating further by day 21 (Salas et al.,
1970). Neither should the mitral cells be capable of self inhibition and should continue to fire in response to strong stimuli
during that time. Habituation to odors may also be slow to appear
before granule cell growth stabilizes. In addition to the maturation of inhibitory function with granule cell growth, it seems
likely that the afferent system becomes more sensitive to olfactory
stimuli with the growth of central synaptic connections, increasing
olfactory acuity as the pups grow.

A NEUROBEHAVIORAL MODEL FOR THE ESTABLISHMENT OF OLFACTORY
AFFINITIES

The acquisition of olfactory preferences in young rats may be
dependent on the particular capabilities of the developing nervous
system during the first two weeks of life, when the affinity to
maternal odors develops. The olfactory system of growing rats
initially should be highly responsive, but slow to decrement its
response to salient maternal odors. Indeed, young rats appear to
have a nervous system that continues to respond to maternal odors
soon after birth, probably without much response decrement. This
aspect of the young olfactory system may allow a pup to continually
process the salient olfactory stimulus to which it is exposed by
its mother. Such stimulation may allow the rats to become

sufficiently exposed to the maternal odor to become familiar with
it. Since lateral inhibition should not yet be present, pups
should be unable to make subtle distinctions between maternal odors.
Such a mechanism is consistent with the failure of the pups to alter
their approach behavior to the maternal odor, despite subtle changes
in the emitted odor (verified by gas chromatography) probably due
to changes in diet and the accumulation of these odors in their
nest (Leon, unpublished observations).

After the olfactory affinity is established, the inhibitory
modulation of the olfactory system matures, allowing the olfactory
output system to modulate incoming stimuli, and perhaps habituate
and discriminate more and more readily between odors. New olfactory stimuli may no longer have the chance of constant,
familiarizing access to the central nervous system any longer. In
fact, while exposure of young rats from days 1-19 to an arbitrary
odor (peppermint) induced the pups to approach that odor (Leon et
al., 1977), similar exposure of adult rats did not induce them to
approach it (Leon, unpublished observations). Drickamer (1972)
also found that young Peromyscus bairdi preferred to eat food that
was scented with an odor with which they had been pre-exposed, but
odor pre-exposure did not alter the behavior of adults. It should
be noted here that the notion of mere exposure increasing the
preference for a stimulus is not a new one (see Hill, 1978 for a
review), but to this point, neither the differential sensitivity
of neonates to olfactory familiarization, nor the special neural
basis underlying such a process have been emphasized.

There is neural evidence that suggests a special responsiveness of the young rat to olfactory stimuli. Pager (1974a) added
a neutral odor to the diet of adult rats. After a month, the
adults displayed no differential mitral response to that odor,
whether they were hungry or satiated. On the other hand, when
rat pups were raised in the presence of the odor in their food,
they showed a significant increase in mitral firing when presented with the odor, but only when they were hungry. These data
indicate both that the central nervous system of young rats is
more susceptible than adults to becoming familiarized with a
specific cue and that olfactory coding is modulated by the internal state of the organism. Pager (1974b) further showed that
fibers from the HLDB mediated the central modulation of mitral
activity.

The point is that the output cells of the immature bulb
should respond to maternal odors with a neural response that is
probably unique in the lifetime of the rat. This unique type of
stimulation may underlie the mechanism by which odors attain
their special status for the young animals. The immaturity of
bulb may be an adaptation to allow the establishment of olfactory

affinities. Rather than considering an immature system to be merely an incomplete version of the adult system, it may be more profitable to regard the young system as highly evolved to meet the special problems that the organism encounters early in life. For example, an interesting comparison to the Norway rat may exist in the precocial spiny mouse, Acomys cahirinus. These rodents become attracted to maternal and artificial odors after mere exposure for as little as a single day after birth (Porter and Etscorn, 1974; Porter and Doane, 1976). If these pups acquire their attraction to odors by means of a neural process similar to that proposed here, then one would expect an immature olfactory system in these otherwise mature individuals.

During the first two weeks postpartum, pups orient toward, and eventually begin to approach the familiar maternal odor, but why do they begin to approach it? More specifically, is there a particular internal state that develops in the pups that induces them to monitor and to approach the maternal odor? There is some evidence that suggests that a maturation of phobic responses to novel stimuli induces the young to approach familiar odors and also precludes the acquisition of affinities to novel odors. For example, when placed in an unfamiliar environment away from mothers and siblings, young rats demonstrate a striking increase in locomotor activity beginning at an age when they start to approach the maternal odor (Bronstein and Dworkin, 1974; Campbell et al., 1969; Campbell and Mabry, 1972; Randall and Campbell, 1976; Feigley et al., 1972; Goodrick, 1975; Hofer, 1973a, b, 1975; Leon and Moltz, 1972; Moorcroft, 1971; Moorcroft et al., 1971; Tobach et al., 1967). On the other hand, the presence of the mother with her familiar odor subdues the young (Hofer, 1975; 1976; Randall and Campbell, 1976). Pups made anosmic either by surgical disruption of the olfactory system or by chemical destruction of the olfactory epithelium respond as though they were in a novel environment, when they are actually in the presence of their mothers and littermates (Alberts 1976; Hofer, 1975, 1976; Singh and Tobach, 1975; Singh et al., 1977). This striking phenomenon is independent of the thermal or nutritive loss engendered by the anosmic pups, for when both parameters are maintained, the enhanced locomotor activity remained (Hofer, 1973a, b; Randall and Campbell, 1976).

The young rat, therefore, responds to a novel environment devoid of familiar odors by moving, presumably to escape the situation, or/and increase the probability of encountering the mother, siblings, nest, or their residual cues. Familiar odors seem to decrease the novelty and thereby decrease the fearfulness of a new surround. Perhaps rat pups begin to approach the maternal odor when their inherent phobic responses to novel environments mature, because the odor is the most familiar, and therefore least aversive, stimulus in their surroundings. Support for this notion comes from an experiment in which pups were pre-exposed to the olfactory test area in the absence of maternal cues. The pups subsequently took much longer to approach the maternal odor in the test area than pups pre-exposed to a neutral area (Jans, Galef and Leon, unpublished observations). A similar conceptualization of the mechanism involved in the ontogeny of the visually guided approach behavior of ducklings to their mother, with similar supporting data, has previously been put forward (Moltz, 1963).

What might be the neural basis for the maturation of phobic responses to novelty in the environment? One might well expect the input mediating phobic responses to be predominately from the limbic system. These limbic influences would have to mature postnatally, particularly during the time of the pheromonal bond, to play their specific role in the development of pup approach behavior to maternal odor. Indeed, the hippocampus matures post-natally (Altman and Das, 1965, 1966; Bliss et al., 1974), and similarities (including increased locomotion in unfamiliar areas) have been drawn between the behavior of a juvenile rat with an immature hippocampus, an adult with that structure removed, and an adult whose hippocampal development has been prevented by postnatal irradiation (Altman et al., 1973; Bayer, Brunner, Hine and Altman, 1973). The hippocampus might monitor the internal state of the animal and modulate both behavioral responsivity in new environments and the neural coding of odors in a manner similar to the motivational modulation of mitral responsiveness to food odors that probably originates in the hypothalamus. Perhaps it is only with the maturation of the limbic capabilities to integrate information about the internal state of the organism with the external stimuli impinging upon it that there is a decline in the approach response to familiar stimuli. The adult, while still neophobic, may be able to deal with new stimuli in a more sophisticated and less rigid manner than approaching only a few salient, familiar stimuli.

ACKNOWLEDGEMENTS

I thank M. Daly and A. Fleming for their comments and S. Putns for her assistance. This research was supported by NSERC grant #A8578.

REFERENCES

Alberts, J.R. 1976. Olfactory contributions to behavioral de-velopment in rodents. In Mammalian Reproductive Processes and Behavior (R. Doty, Ed.). pp. 67-94, Academic Press, New York.

Altman, J. and Das, G.D. 1965. Autoradiographic and histological
    evidence of postnatal hippocampal neurogenesis in rats. J.
    Comp. Neurol. 124: 319–336.

Altman, J. and Das, G.D. 1966. Autoradiographic and histological
    studies of postnatal neurogenesis. I. A longitudinal investiga-
    tion of the kinetics, migration and transformation of cells
    incorporating tritiated thymidine in neonate rats, with special
    reference to postnatal neurogenesis in some brain regions. J.
    Comp. Neurol. 126: 337–390.

Altman, J., Brunner, R.L., Bulut, F.G. and Sudarshan, K. 1973.
    The development of behavior in normal and brain-damaged infant
    rats, studied with homing (nest-seeking) as motivation. In
    Drugs and The Developing Brain (A. Vernadakis and N. Weiner,
    Eds.). pp. 321–348. Plenum Press, New York.

Altman, J., Brunner, R.L. and Bayer, S.A. 1973. The hippocampus
    and behavioral maturation. Behav. Biol. 8: 557–596.

Bayer, S.A., Brunner, R.L., Hine, R., Altman, J. 1973. Behavioural
    effects of interference with postnatal acquisition of hippo-
    campal granule cells. Nature 203: 222–224.

Bliss, T.U.P., Chung, S.H. and Stirling, R.V. 1974. Structural
    and functional aspects of the mossy fibre system in the hippo-
    campus of the post-natal rat. J. Physiol. 239: 92–94.

Bronstein, P.M. and Dworkin, T. 1974. Replication: The persis-
    tent locomotion of immature rats. Bull. Psychon. Soc. 4:
    124–126.

Campbell, B.A., Lytle, L.D. and Fibiger, H.C. 1969. Ontogeny of
    adrenergic arousal and cholinergic inhibitory mechanisms in
    the rat. Science 166: 637–638.

Campbell, B.A. and Mabry, P.D. 1972. Ontogeny of behavioral
    arousal: A comparative study. J. Comp. Physiol. Psychol.
    81: 371–379.

Døving, K.B. and Pinching, A.J. 1973. Selective degeneration of
    neurons in the olfactory bulb following prolonged odor
    exposure. Brain Res. 52: 115–129.

Drickamer, L.C. 1972. Experience and selection behavior in the
    food habits of Peromyscus: use of olfaction. Behaviour 41:
    269–287.

Goodrick, C.L. 1975. Adaptation to novel environments by the rat:
    Effects of age, stimulus intensity, group testing and tempera-
    ture. Dev. Psychobiol. 8: 287–296.

Gregory, E.H. and Pfaff, D.W. 1971. Development of olfactory
    guided behavior in infant rats. Physiol. Behav. 6: 573–576.

Gustafsson, P. 1948. Germ-free rearing in the rat. Acta Path.
    Microbiol. Scand. Suppl. 72–75: 1–130.

Heimer, L. 1968. Symaptic distribution of centripital and
    centrifugal nerve fibers in the olfactory system of the rat.
    An experimental anatomical study.   J. Anat. 103: 413–432.

Heimer, L. 1972. The olfactory connections of the diencephalon in
    the rat. An experimental light- and electron-microscope study

with special emphasis on the problem of terminal degeneration. Brain Behav. Evol. 6: 1-6.

Hill, W. 1978. Effects of mere exposure on preferences in non-human mammals. Psych. Bull. 85: 1177-1198.

Hofer, M.A. 1973a. Maternal separation affects infant rat behavior. Behav. Biol. 9: 629-633.

Hofer, M.A. 1973b. The role of nutrition in the physiological and behavioral effects of early maternal separation on infant rats. Psychosom. Med. 35: 350-359.

Hofer, M.A. 1975. Studies on how early maternal separation produces behavioral change in young rats. Psychosom. Med. 37: 245-264.

Hofer, M.A. 1976. Olfactory denervation: Its biological and behavioral effects in infant rats. J. Comp. Physiol. Psychol. 90: 829-838.

Laing, D.G. and Panhuber, H. 1978. Neural and behavioral changes in rats following continuous exposure to an odor. J. Comp. Physiol. 124: 259-265.

Leon, M. 1974. Maternal pheromone. Physiol. Behav. 13: 441-453.

Leon, M. 1975. Dietary control of maternal pheromone in the lactating rat. Physiol. Behav. 14: 311-319.

Leon, M. 1978. Filial responses to olfactory cues in the laboratory rat. In: Advances in the Study of Behavior, Vol. VIII (J. Rosenblatt, Ed.). pp. 117-153, Academic Press, New York.

Leon, M. and Moltz, H. 1972. The development of the pheromonal bond in the albino rat. Physiol. Behav. 8: 638-686.

Leon, M., Galef, B.G., Jr. and Behse, J. 1977. Establishment of pheromonal bonds and diet choice in young rats by odor preexposure. Physiol. Behav. 18: 387-391.

Lutton, C. and Chevallier, F. 1973. Coprophagie chez le rat blanc. J. Physiol. Paris. 66: 219-228.

Moltz, H. 1963. Imprinting: An epigenetic approach. Psych. Rev. 70: 123-138.

Moorcroft, W.H. 1971. Ontogeny of forebrain inhibition of behavioral arousal in the rat. Brain Res. 35: 513-522.

Moorcroft, W.H., Lytle, L.D. and Campbell, B.A. 1971. Ontogeny of starvation-induced behavioral arousal in the rat. J. Comp. Physiol. Psychol. 75: 67-69.

Moxley, G.F. and Price, J.L. 1975. Aspects of the development of afferent projections to the olfactory cortex. Neurosciences Society Mettings, New York. p. 346.

Pager, J., Giachetti, I., Holley, A. and LeMagnen, J. 1972. A selective control of olfactory bubl electrical activity in relation to food deprivation and satiety in rats. Physiol. Behav. 9: 573-579.

Pager, J. 1974a. A selective modulation of the olfactory bulb electrical activity in relation to the learning of palatability in hungry and satiated rats. Physiol. Behav. 12: 189-195.

Pager, J. 1974b. A selective modulation of olfactory input suppressed by lesions of the anterior limb of the anterior commissure. Physiol. Behav. 13: 523-526.

Pager, J. and Royet, J. 1976. Some effects of conditioned aversion on food intake and olfactory bulb electrical activity in relation to food deprivation and satiety in rats. Physiol. Behav. 9: 573-579.

Pfaff, D.W. and Pfaffmann, C. 1969. Olfactory and hormonal influences on the basal forebrain of the male rat. Brain Res. 15: 137-156.

Pinching, A.J. and Powell, T.P.S. 1971a. The neuropil of the glomeruli of the olfactory bulb. J. Cell Sci. 9: 347-377.

Pinching, A.J. and Powell, T.P.S. 1971b. The neuropil of the periglomerular region of the olfactory bulb. J. Cell Sci., 9: 379-409.

Pinching, A.J. and Døving, K.B. 1974. Selective degeneration in the rat olfactory bulb following exposure to different odors. Brain Res. 82: 195-204.

Porter, R.H. and Etscorn, F. 1974. Olfactory imprinting resulting from brief exposure in Acomys cahirinus. Nature 250: 732-733.

Porter, R.H. and Doane, H. 1976. Maternal pheromone in the spiny mouse (Acomys cahirinus). Physiol. Behav. 16: 75-78.

Powell, T.P.S., Cowan, W.M. and Raisman, G. 1965. The central olfactory connections. J. Anat. 99: 791-813.

Price, J.L. and Powell, T.P.S. 1970a. An experimental study of the origin and the course of the centrifugal fibers to the olfactory bulb in the rat. J. Anat. 107: 215-237.

Price, J.L. and Powell, T.P.S. 1970b. The afferent connections of the nucleus of the horizontal limb of the diagonal band. J. Anat. 107: 239-256.

Rall, W., Shepherd, G.M., Reese, T.S. and Brightman, M.W. 1966. Dendrodendritic synaptic pathway for inhibition in the olfactory bulb. Exp. Neurol. 14: 44-56.

Rall, W. and Shepherd, G.M. 1968. Theoretical reconstruction of field potentials and dendrodendritic synaptic interactions in the olfactory bulb. J. Neurophysiol. 31: 884-915.

Randall, P.K. and Campbell, B.A. 1976. Ontogeny of behavioral arousal in rats: Effect of maternal and sibling presence. J. Comp. Physiol. Psychol. 90: 453-459.

Rosenblatt, J.S. and Lehrman, D.S. 1963. Maternal behavior in the laboratory rat. In: Maternal Behavior in Mammals (H.C. Reingold, Ed.). pp. 8-57, Wiley, New York.

Salas, M., Guzman-Flores, C. and Schapiro, S. 1969. An ontogenetic study of olfactory bulb electrical activity in the rat. Physiol. Behav. 4: 699-703.

Salas, M., Schapiro, S. and Guzman-Flores, C. 1970. Development of olfactory bulb discrimination between maternal and food odors. Physiol. Behav. 5: 1261-1264.

Scalia, F. and Winans, S.S. 1975. The differential projections of the olfactory bulb and accessory olfactory bulb in mammals. J. Comp. Neurol. 161: 31-56.

Schapiro, S. and Salas, M. 1970. Behavioral response of infant rats to maternal odor. Physiol. Behav. 5: 815-817.

Scott, J.W. and Pfaffman, C. 1967. Olfactory input to the hypothalamus: Electrophysiological evidence. Science 158: 1592-1594.

Scott, J.W. and Leonard, C.W. 1976. The olfactory connections of the lateral hypothalamus in the rat, mouse and hamster. J. Comp. Neurol. 141: 331-334.

Scott, J.W. and Chapin, B.R. 1975. Origin of olfactory projections to lateral hypothalamus and nuclei gemini of the rat. Brain Res. 88: 64-68.

Shepherd, G.M. 1972. Synaptic organization of the mammalian olfactory bulb. Physiol. Res. 52: 864-917.

Shute, C.C.D. and Lewis, P.R. 1967. The ascending cholinergic reticular system: neocortical, olfactory and subcortical projections. Brain 90: 497-539.

Singh, D.N.P. and Nathaniel, E.J.H. 1977. Postnatal development of mitral cell perikaryon in the olfactory bulb of the rat. A light and ultrastructural study. Anat. Rec. 189: 413-432.

Singh, P.J. and Tobach, E. 1975. Olfactory bulbectomy and nursing behavior in rat pups. Dev. Psychobiol. 8: 151-164.

Singh, P.J., Tucker, M. and Hofer, M. 1977. Effects of nasal $ZnSO_4$ irrigation and olfactory bulbectomy on rat pups. Physiol. Behav. 17: 373-382.

Tobach, E., Rouger, Y. and Schneirla, T.C. 1967. Development of olfactory function in the rat pup. Am. Zool. 7: 792.

Valverde, F. 1965. Studies on the Piriform Lobe. Cambridge, Mass. Harvard University Press.

Westrum, L.E. 1975. Electron microscopy of synaptic structure in olfactory cortex of early postnatal rats. J. Neurocytol. 4: 713-732.

White, L.E., Jr. 1965. Olfactory bulb projections of the rat. Anat. Rec. 152: 465-480.

Yamamoto, C., Yamamoto, T. and Iwama, K. 1963. The inhibitory system in the olfactory bulb studied by intra-cellular recording. J. Neurophysiol. 26: 403-415.

ODOR AVERSION LEARNING BY NEONATAL RATS: ONTOGENY OF OSMIC

MEMORY

Jerry W. Rudy and Martin D. Cheatle

Psychology Department
Princeton University
Princeton, NJ 08540

INTRODUCTION

"The importance of remote signs (signals) of objects can be
easily recognized in the movement reactions of the animal. By
means of distant and even accidental characteristics of objects
the animal seeks his food, avoids his enemies, etc. (Pavlov, 1927,
p. 52)". The obvious adaptive significance of signal directed
behavior raised a fundamental question for Pavlov and several
generations of researchers: How does one stimulus event acquire
the capacity to act as a signal for another stimulus event?

The research we describe is also concerned with this question.
We are especially interested, however, in discovering ontogenetic
influences on the processes mediating the acquisition of signal
properties by olfactory stimulation. We hasten to add that our
decision to employ olfactory stimulation as the potential signalling event was not motivated by any special interest in chemical
senses. Rather, this choice was dictated by contraints imposed
upon us by the animal we chose to investigate. Because of our
interest in the ontogenetic determinants of signal learning, we
have employed an altricial animal, the laboratory rat, as the
subject. At birth the rat's central nervous system (CNS) is quite
underdeveloped. It matures rapidly, however, such that the CNS
of the 4-5 week old is quite comparable to that of the adult
(Campbell and Coulter, 1976). Development of the rat's CNS is
especially rapid during the first two weeks after birth (Campbell
and Coulter, 1976).

For this reason, we thought that our chances of detecting
important ontogenetic influences on signal learning would be

increased if we could develop a conditioning preparation that could
be employed with the newborn rat. It is well known, however, that
the rat's visual and auditory senses do not become functional until
the rat is about two weeks old, and this fact eliminates visual and
auditory stimuli as potential signals for the newborn rat.

In contrast, the rat's olfactory system is functional at birth.
For example, olfactory cues enable the newborn pup to locate its
food source, the dame's nipple (Teicher and Blass, 1976, 1977). We
thus thought it possible to develop a Pavlovian conditioning procedure that would modify the neonate's reaction to olfactory stimulus and permit us to determine if the processes involved change as
the rat makes the transition from infancy to adulthood.

The Pavlovian-like methodology we developed for this purpose
is quite simple (Rudy and Cheatle, 1977). Neonatal rats are exposed to a paired presentation of a novel odor (McCormick's lemon
extract), the conditioned stimulus (CS), and an injection of an
illness-inducing drug, lithium chloride (LiCl), the unconditioned
stimulus (UCS). Several days later the pup is tested for a conditioned aversive reaction to the odor. A single pup is placed in
the center of Plexiglas chamber that has a wire mesh floor beneath
which are two equal-sized containers, one holding pine shavings
scented with the odor CS, the other holding shavings scented with
another odor (McCormick's garlic juice). The pup is free to move
and we record the percentage of test time (150 sec) it spends over
shavings scented with the odor CS. The odor CS is then inferred to
have acquired aversive signalling properties if the pups that
received the odor-LiCl pairing spend less time over it than do pups
in appropriate control conditions.

In the remainder of this paper we will describe some of the
more notable results of this research program. Special attention
will be given to experiments that have revealed ontogenetic influences on odor aversion learning. Because this work has been
described previously (Rudy and Cheatle, 1977, 1979; Cheatle and
Rudy, 1978, 1979), we will omit a detailed description of our
methods of procedure and ask the interested reader to consult
these papers for this information.

ODOR AVERSION LEARNING BY THE INFANT RAT

The initial goal of our research was to determine the earliest
age at which the rat can learn to react to an odor as a signal for
an aversive event. Apparently, it can do so when only 2 days old
(Rudy and Cheatle, 1977). For example, in one experiment, 2-day-
old pups were sequentially exposed to a lemon scent for 5-min,
injected with 3 meg of a 0.15-mole solution of LiCl, and then
reexposed to the lemon scent for 30 min. When tested at 8 days of

age, these pups displayed a pronounced aversion to the lemon scent
($\bar{x}$ = 16 percent of test time). In comparison control pups, that
received either (a) a comparable 35-min exposure to the lemon scent
60 min prior to the LiCl injection or (b) the LiCl injection   60
min prior to the lemon scent, displayed a slight preference for
the lemon scent ($\bar{x}$ = 58 and 64 percent of test time, respectively).
We also have reported that the conditioned aversion to the lemon
scent can be extinguished if following the lemon scent-LiCl pairing
pups receive non-reinforced exposure to the lemon scent (Rudy and
Cheatle, 1979).

We also have compared the acquisition of odor aversions by
pups trained at ages ranging from 2-14 days old and found that
they display equivalent aversion to the lemon scent (Rudy and
Cheatle, 1979). This outcome might suggest that the mechanisms
mediating odor aversion learning by the rat are fully developed at
birth. Then it should be noted that the training parameters we
employed were calculated to maximize the opportunity for the pups
to learn the odor aversion. Specifically, the pups were exposed
to the odor prior to the LiCl injection and then immediately
reexposed to it for 30 min. Such a procedure should insure that
the pup contiguously experienced the odor and the reinforcing
consequence of the drug, and thereby should maximize the possibility
that the pup would associate the two events. Other training parameters, however, might be more successful in revealing an ontogenetic influence on odor aversion learning. One such potentially
interesting parameter is the interval separating odor exposure and
LiCl injection.

Long-Delay Odor Aversion Learning

It is well established that drugs such as LiCl are capable of
supporting conditioned aversions to novel flavors even when the
flavor and LiCl events are separated by several hours (c.f.
Revusky and Garcia, 1970). This phenomenon is often referred to
as long-delay learning. Adult rats also acquire aversion to odors
that precede LiCl injection by several hours (Taukulis, 1974).
Perhaps the newborn pup, although quite capable of acquiring the
odor aversion when the odor and illness are concurrently experienced, does not yet have the capability to acquire this aversion
when the two events are separated in time. Our research (Rudy and
Cheatle, 1979) indicates that this is the case.

Consider the results of the experiment presented in Figure 1.
As this figure indicates pups either 2, 4, 6 or 8 days old received
one of two types of conditioning treatments. Pups in the 0 Delay
treatment received training that insured that the odor and illness
were experienced concurrently. Pups in the 15' Delay treatment,

however, received a 5-min exposure to the lemon scent 15 min prior
to the LiCl injection.

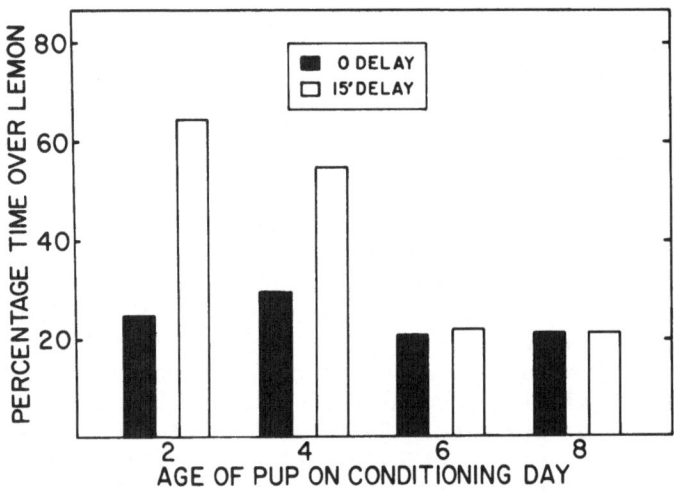

Figure 1.   Mean percentage of time spent over lemon-scented shavings
            by pups trained at either 2, 4, 6, or 8 days of age.
            (From Rudy and Cheatle, 1979. Copyright 1979 by
            Lawrence Erlbaum Associates. Reprinted by permission).

        It is clear from Figure 1 that, irrespective of training age,
all pups in the 0 Delay conditions displayed an aversion to the
lemon scent. In contrast, pups in the 15' Delay conditions only
displayed the aversion if they were 6 or 8 days old at the time of
training. The 2- and 4-day-old pups failed to acquire an aversion
to the lemon scent. Furthermore, we have reported that the 2-day-
old pups failed to acquire the odor aversion even if the 5-min exposure to the lemon scent preceded the LiCl injection by only about
15 sec, whereas 8-day-old pups acquired the aversion even if the
odor and LiCl events were separated by 90 min (Rudy and Cheatle,
1978, 1979).

        The results of the long-delay learning experiments argue
against the view that the mechanisms of odor aversion learning are
fully developed at birth. Instead, they suggest that these processes undergo considerable development during the first week
following birth. We have examined two other variations of the
odor aversion task that also have revealed an ontogenetic influence.
One set of data comes from an investigation of the so-called
stimulus-preexposure effect, another set from an investigation of
second-order odor-aversion conditioning. We will now describe
these experiments.

Ontogeny of the Stimulus-Preexposure Effect

One of the major variables that influence the behavioral product of a Pavlovian conditioning experience is the subject's famili-
arity with the stimulus serving as the CS (Lubow, 1973). Conditioning is much more rapid if the CS is novel (not previously experienced
by the subject) than if it is familiar (experienced by the organism
prior to the conditioning trial).

We have reported several experiments that have examined the
effect of stimulus preexposure on odor aversion conditioning (Rudy
and Cheatle, 1979). The results can be easily summarized. When
pups either 2- or 8-days old are given a 30-min preexposure to a
lemon scent and 60 min later receive a pairing of the lemon scent
and LiCl, they later fail to display an aversion to the lemon scent
in comparison to pups that only receive the pairing experience.
Thus, when the stimulus-preexposure treatment only precedes the
conditioning episode by 60 min, 2- and 8-day olds behave alike.
They react to the lemon scent during testing as if it had never
been paired with LiCl.

The more interesting results were obtained when we repeated
this basic experiment but separated the stimulus-preexposure treatment from the pairing operation by 4 hours. The right panel of
this Figure 2 displays the test results for pups trained at 8 days
of age. Note that the stimulus-preexposure treatment markedly
affected the aversion to the lemon scent. The pups that received
this treatment (Group L/L-UCS) displayed no aversion, whereas the
pups that only received the pairing operation (Group MS/L-UCS)
displayed a substantial aversion. In fact the pups in the stimulus-
preexposure condition did not differ from pups in nonconditioned
control groups (Groups L/L-Sham and MS/L-Sham).

The left panel of Figure 2 displays the test data for the
pups trained at 2 days of age. In contrast to the outcome observed
with the 8-day-old pup, the stimulus-preexposure treatment failed
to affect the aversion to the lemon scent by the 2-day old pup.
Group L/L-UCS did not differ from Group MS/L-UCS and both groups
displayed marked aversions to the lemon scent in comparison to
the controls (Group L/L-Sham and Group MS/L-Sham).

In summary, the results of these experiments were clear.
Stimulus preexposure prevented acquisition of an odor aversion by
both 2- and 8-day old pups when a relatively short interval (60
min) separated the preexposure and conditioning treatments. But
when a relatively long interval (4 hours) separated these two
experiences only the 8-day old pups failed to acquire the aversion.
Two day olds conditioned to the odor as if it were novel.

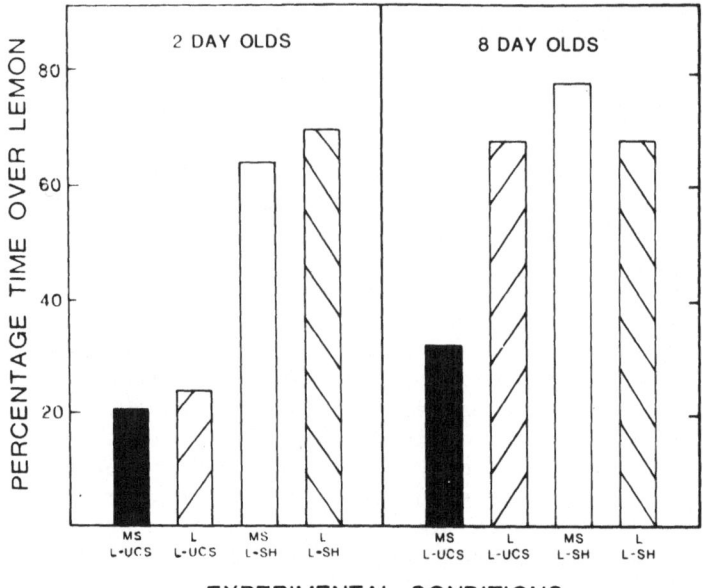

Figure 2. Mean percentage of time spent over the lemon scented
           shavings by pups trained at 2 or 8 days of age.
           (From Rudy and Cheatle, 1979. Copyright 1979 by
           Lawrence Erlbaum Associates. Reprinted by permission).

Second-Order Conditioned Odor Aversions

The Pavlovian first-order conditioning operation--exposing
the neonatal rat to an odor (CS)-LiCl (UCS) pairing--clearly
modifies that pup's behavioral reaction to the odor. The odor
CS is avoided as if it were a signal for danger. Pavlov, of course,
also reported that a stimulus that had acquired signalling properties by virtue of a first-order conditioning was also capable of
reinforcing the acquisition of signal properties by a second stimulus event. He termed this phenomenon second-order conditioning.

We might abstractly represent the experimental operations for
producing second-order conditioning as a two-phase training procedure, S_1-UCS; S_2-S_1, where S_1-UCS represents the Phase 1, first-
order conditioning operation of pairing a neutral stimulus, S_1,
with the UCS, and S_2-S_1 represents the Phase 2, second-order operation of pairing a second neutral stimulus, S_2, with the now-condi-
tioned stimulus, S_1.

The experiments just described clearly indicate that an olfactory stimulus can acquire first-order signalling properties

when paired with a LiCl UCS. Can such a signal also reinforce the acquisition of a second-order conditioned reaction?

Our initial attempts to address this question employed rats 7 days old at the time of the Phase 1, S_1-UCS pairing. These pups were given a 5-min exposure to the lemon scent (S_1), injected with LiCl (UCS), and reexposed to the odor for 30 min. The Phase 2, S_2-S_1 pairing was administered when the pups were 8 days old. It consisted of a 5-min exposure to a novel odor, (S_2) an orange scent (McCormick's orange extract), following immediately by a 30-min exposure to S_1.

The results of one experiment (Cheatle and Rudy, 1978, Experiment 1) that examined this treatment are presented in Figure 3. Two of the conditions represented in this figure, Group P/P/MS and Group P/P/E, each received the two-phase S_1-UCS; S_2-S_1 treatment. The only difference between these two groups was that, prior to the tests for aversion to S_2 and S_1, the pups in Group P/P/E received 5 non-reinforced exposures to S_1 that were calculated to extinguish its aversive properties. Groups labled B/P/MS and B/P/E were controls that were treated exactly like Groups P/P/MS and P/P/E, respectively, except that in Phase 1 they received the LiCl injection 60 min prior to experiencing the lemon scent (S_1).

As can be seen in Figure 3, Group P/P/MS displayed a first-order aversion the lemon scent (S_1) in comparison to the groups that received the backward pairing of S_1 and the UCS. It is equally clear that the extinction operation administered to Group P/P/E eliminated the aversion to S_1. Of principal interest, however, is that the pups in both Group P/P/MS and P/P/E displayed strong aversions to the orange scent (S_2) in comparison to the pups in the control conditions (Groups B/P/MS and B/P/E).

The results of this experiment thus provide a demonstration of second-order odor-aversion conditioning by neonatal rats. Apparently, not only does an odor (S_1) paired with LiCl acquire aversive signalling properties, it also acquires the ability to reinforce aversive signalling properties to another odor (S_2).

This particular experiment is also instructive on another matter. When the pup experienced the Phase 2, S_2-S_1 training event, there were several potential associations it could acquire that could mediate a conditioned aversion to S_2 (c.f. Rizley and Rescorla, 1972). For example, it could have associated representations of the two odors, orange scent (S_2) and lemon scent (S_1). The conditioned reaction to S_2, on this account, would reflect the outcome of a chain of associations whereby, during testing, S_2 activated a representation of S_1 which in turn evoked the conditioned reaction. Alternatively, the pup could have associated S_2 with a conditioned aversive reaction evoked by S_1

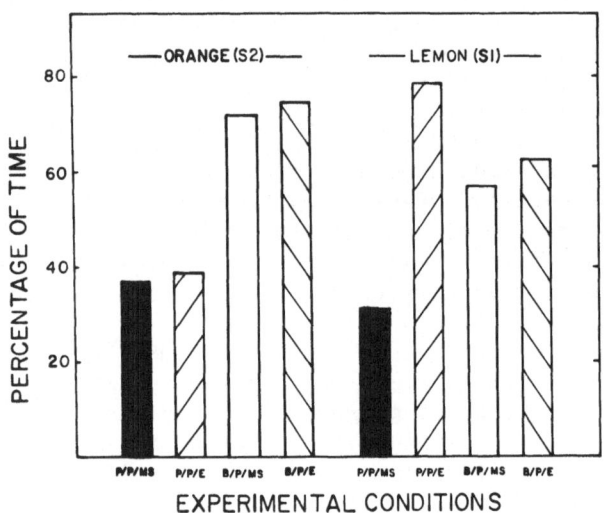

Figure 3.   Mean percentage of time spent over the second-order
stimulus, the orange scent (S$_2$) and the first-order
stimulus, the lemon scent (S$_1$). (From Cheatle and
Rudy, 1978. Copyright 1979 by the American Psychological Association. Reprinted by permission).

on the occasion of the S_2-S_1 pairing, and S_2 would then directly activate the aversive reaction during testing.

The experiment just described argues against the view that the second-order reaction depended upon the pup associating the two odors. If this association was critical to the second-order reaction (displayed by Group P/P/MS) then extinguishing the conditioned reaction to S_1 (of subjects in Group P/P/E) should have eliminated the second-order aversion to S_2, but it did not. To produce second-order conditioning it would appear that S_1 need only possess conditioned properties at the time of the S_2-S_1 pairing. Once established, the second-order conditioned reaction to S_2 appears to be independent of the first-order conditioned reaction to S_1. Such data make it likely that the conditioned response to S_2 was mediated by pup associating S_2 with a conditioned reaction evoked by S_1 during Phase-2 training (see also, Rizley and Rescorla, 1972; Holland and Rescorla, 1975).

Ontogeny of Second-Order Conditioning

The preceding experiment makes the point that neonatal pups can acquire a second-order odor aversion. But it should be appreciated that pups in this experiment were 8 days old when they experienced the S_2-S_1 pairing. Recall that we observed appreciable differences between 8- and 2-day old pups in both the long-delay learning and stimulus-preexposure experiments. Perhaps an ontogenetic investigation of second-order odor-aversion learning might also reveal differences between young and old pups.

Figure 4 prsents the results of an experiment (Cheatle and Rudy, 1979, Experiment 1) designed to address this issue. Pups in groups labeled P/P received the two-phase, S_1-UCS; S_2-S_1 training sequence. Groups labeled B/P and P/U were control conditions included to insure ourselves that any aversion to S_2 displayed by the P/P conditions was indeed a conditioned reaction (see Cheatle and Rudy, 1979 for more detail). The pups were either 2, 4, 6 or 8 days old on the day of the Phase 2 S_2-S_1 pairing.

It is clear from the top panel of Figure 4 that the age of the pup on the day of the Phase 1, S_1-UCS pairing did not influence acquisition of a first-order aversion. This confirms our previous findings with this treatment. In contrast, the bottom panel of Figure 4 reveals that the age of the pup on the day of the Phase 2, S_2-S_1 pairing had a major influence. Older pups (6-8 day olds) acquired the aversion to S_2 but younger pups (2-4 day olds) apparently did not, because they did not differ from their appropriate controls (Group B/P and P/U).

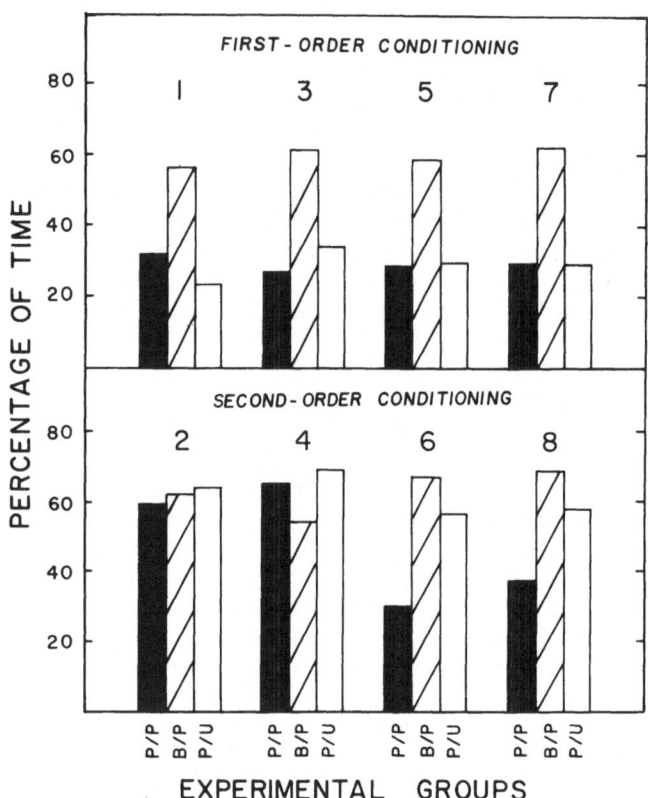

Figure 4.   Mean percentage of time spent over the lemon scent,
            S_1, (top panel) and the orange scent, S_2, (bottom
            panel).   The numbers 1, 3, 5, and 7 designate the age
            at which the pups received Phase-1 training, and the
            numbers 2, 4, 6 and 8 designate the age at which the
            pups received Phase-2 training. (From Cheatle and
            Rudy, 1979. Copyright 1979 by the American Psychologi-
            cal Association. Reprinted by permission).

Acquisition of a second-order aversion with these training procedures thus was strongly dependent upon the age of the pup. It is especially interesting that the age at which differences emerged corresponded exactly to the age at which we previously had found differenced among neonates in the long-delay learning task (see Figure 1). In both cases, 2-4 day-old pups behaved alike in failing to display odor aversions, and 6-8 day-old pups behaved alike in displaying aversion to the scent. This may be happenstance, or it may indicate that the effect of age in both cases was the result of an ontogenetic influence on a common process, a matter about which more will be said.

For the moment, however, it must be appreciated that the age-related differences observed in second-order conditioning might reflect the possibility that the younger pups did not acquire a sufficient first-order aversion to S_1 to allow it to reinforce a second-order aversion to S_2. This could be true even though the pups at all ages displayed an aversion to S_1 (see Figure 4).

To evaluate this possibility, we conducted another experiment (Cheatle and Rudy, 1979, Experiment 2). All pups were exposed to the Phase 1, S_1-UCS pairing when only 3 days old. Following this

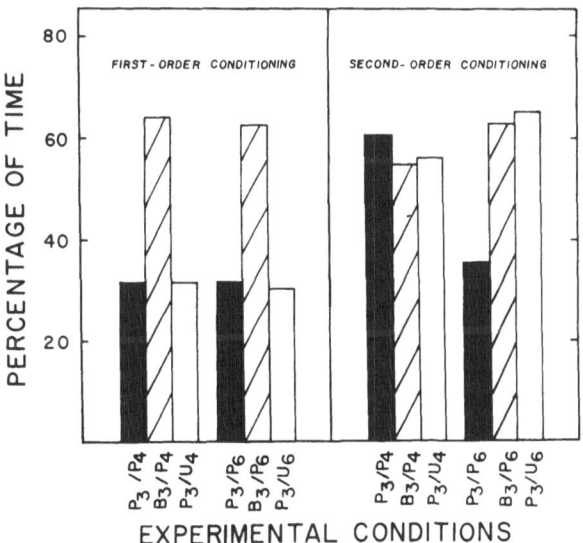

Figure 5.   Mean percentage of time spent over the lemon scent, S_1, (left panel) and orange scent, S_2, (right panel). The subscripts designate the age of the pup at the time of Phase-1 and Phase-2 training. (From Cheatle and Rudy, 1979. Copyright 1979 by the American Psychological Association. Reprinted by permission).

training, however, some pups were exposed to the Phase 2, S_2-S_1 pairing when 4 days old (Group P3/P4) but others did not receive the S_2-S_1 pairing until they were 6 days old (Group P3/P6). We reasoned that if the 4-day old pups failed to acquire the second-order aversion because S_1 was not capable of reinforcing an aversion to S_2 then, regardless of the age of the pup at the time of the S_2-S_1 pairing (4 versus 6 days old), the pups should fail to acquire a second-order aversion to S_2. On the other hand, if the pup's age at the time of the S_2-S_1 pairing was the critical variable influencing second-order conditioning to S_2, then the pups in Group P3/P6 should acquire an aversion to S_2 but the pups in Group P3/P4 should not.

The results of this experiment were clear. As can be seen in right panel of Figure 5, the pups that received the Phase-2 pairing when 6 days old (Group P3/P6) acquired an aversion to the orange scent (S_2), whereas the pups that received the Phase 2 pairing when only 4 days old (Group P3/P4) did not. This outcome indicates that the first-order aversion to S_1 acquired by 3-day old pups was indeed sufficient to reinforce a second-order conditioned reaction to S_2. It further indicates that in order for the training parameters of Phase 2 to result in second-order conditioning, the pups must be 6 days old when they experience the S_2-S_1 pairing.

The age of the pup is clearly a determinant of whether or not the Phase 2, S_2-S_1 pairing produces a second-order aversion to S_2. One might interpret the failure of the 2-4 day-old rat to acquire a second-order aversion to mean that it does not yet have the capability to do so. Then the results of the final experiment we describe are instructive. They reveal that even the 4-day old pup can acquire a second-order aversion to S_2 if the parameters of Phase-2 training are modified so that S_2 and the reinforcing stimulus, S_1, are experienced concurrently.

The results of this experiment (Cheatle and Rudy, 1979, Experiment 3) are presented in Figure 6. The 4-day old subjects in the group labeled P/P received the standard sequential presentation of S_2 and S_1 that was employed in the previous second-order conditioning experiments. And, as was the case in those experiments, they displayed no evidence of a second-order aversion to S_2. The 4-day old subjects in the group labeled P/MP, however, received a modified S_2-S_1 pairing. They experienced a 5-min exposure to S_2 followed immediately by a 30-min exposure to both S_2 and S_1. Note that pups in Group P/MP displayed a pronounced second-order aversion to S_2. Presenting S_2 concurrently with S_1 thus compensated for the deficiency that prevented the 4-day old pup from acquiring a second-order aversion to S_2 when exposed to the standard, sequential S_2-S_1, Phase-2 treatment.

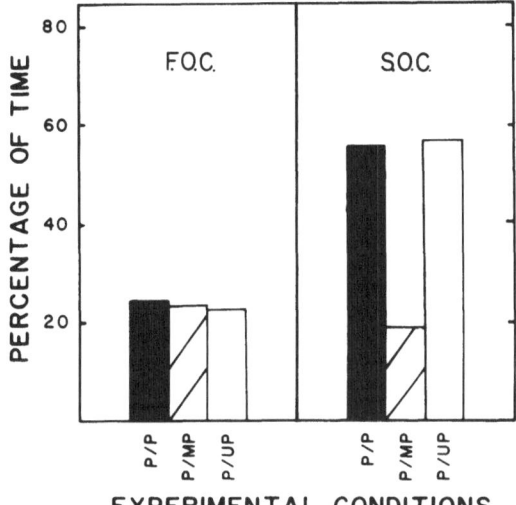

Figure 6.   Mean percentage of time spent over the lemon scent,
            S_1 (left panel) and orange scent, S_2 (right panel).
            F.O.C., first-order conditioning; S.O.C. second-order
            conditioning. (From Cheatle and Rudy, 1979. Copy-
            right 1979 by the American Psychological Association.
            Reprinted by permission).

CONCLUSION

     Although we noted at the outset of this paper that we had no
special interest in the chemical senses, we cannot help but be
impressed with the degree to which the behavior of the neonatal
rat is controlled by olfactory stimulation. After birth, the
pup is guided to its first meal by olfactory stimulation provided
by deposits of the mother's saliva and/or fluid from the pup's
amniotic sac on the nipple, and suckling behavior remains under
the control of olfactory cues for at least two weeks (Blass et al.,
in press; Teicher and Blass, 1976, 1977). Moreover, the enormous
behavioral activity emitted by the isolated 15-16 day old rat
(Campbell, Lytle and Fibiger, 1969) is greatly attenuated by the
presence of odors emanating from its nest shavings (Campbell and
Raskin, 1978). Most recently, the presence of nest odors also
have been found to improve the 16-day old rat's performance in
a spatial alternation task and its learning of a passive avoidance
behavior (Smith and Spear, 1978).

Learned Determinants of Olfactory Control

     The research described in this paper also documents the con-
trol olfactory stimulation exerts on the behavior of the neonatal
rat. It has especially emphasized the contribution of _learning_
to the determination of the neonate's behavior reaction to an
olfactory stimulus. The researcher, of course, does not directly
observe learning. He infers that learning contributes to the determination of behavior when he observes that an organism's experience with a stimulus on one occasion influences its behavioral
reaction to that stimulus on some other occasion.

     It is in this sense that our experiments clearly have revealed
that learning processes can contribute to control of the neonatal
rat's behavior by olfactory stimulation. Neonates exposed to a
novel odor in conjunction with a LiCl injection subsequently react
differently to that odor than neonates that have not had this
experience, they avoid it.

     What is perhaps most remarkable, is that rats only 2 days old
at the time of the odor-LiCl pairing later evidence such learning.
Prior to this work, no clear evidence of learning had been observed
in rats less than 6 days old. Our choice to employ olfactory
stimulation in this research obviously was a fortunate one.

     To be sure, the signal value of an olfactory stimulus can be
modified by learning even in the newborn rat. This does not mean,
however, that the processes mediating this learning are fully
developed at birth. To the contrary, three classes of experiments
suggest that these processes undergo rapid development during the
first week following birth.

     First, consider the results of the long-delay learning experi-
ments. They revealed that the 2-4 day old pup acquired the odor
aversion so long as the odor and reinforcing consequences of LiCl
were experienced concurrently. When the odor and LiCl experiences
were separated by 15 min, however, pups trained at these ages
failed to evidence the odor aversion. In fact, 2-day olds failed
to acquire the aversion even when the odor and LiCl experiences
were separated by only 15 sec. In contrast, pups trained when
6-8 days acquired the odor aversion both when the odor and LiCl
experiences were concurrent and when they were separated by 15
min. Furthermore, pups trained when 8 days old acquired an odor
aversion even when the odor and LiCl events were separated by 90
min (Rudy and Cheatle, 1978, 1979).

     Second, consider the results of the stimulus-preexposure
experiments. Both 2- and 8-day old pups failed to acquire an
odor aversion when they were preexposed to the odor 60 min prior
to the odor-LiCl pairing. If, however, the interval separating

odor preexposure from the odor-LiCl pairing was 4 hours, the 2-day old pups acquired an aversion to the odor but the 8 day old failed to do so.

Finally, consider the results of the second-order conditioning experiments. When the Phase-2 second-order conditioning experience consisted of a 5-min exposure to one odor, S_2, and a 30-min exposure to a previously conditioned odor, S_1, pups only 2-4 days old failed to acquire the second-order aversions to S_2. Pups given this experience when 6-8 days old, however, displayed a second-order aversion to S_2.

Ontogeny of Osmic Memory

The processes mediating odor aversion learning clearly are not fully developed at birth. We think it might be useful to suppose that the ontogenetic influences observed in these experiments are mediated by age-related differences in a common process. We also think that the developmental changes of functional significance are likely in the olfactory system. Specifically, the results of these experiments may reflect the ontogeny of osmic memory processes. For instance, an olfactory stimulus may activate a transient memory representation or trace that decays more rapidly for 2-4 day old pups than for 6-8 day olds. Alternatively, one might assume that the memory trace activated in young pups by olfactory stimulation is more vulnerable to retroactive interference than is the memory trace activated in the older pup. Regardless of the source of forgetting, however, the proposition is that older pups better retain a memory of an olfactory experience than do younger pups.

Support for the assertion that a common ontogenetic effect underlies the observed age-related differences is offered by noting that the age at which differences emerged in each of the three classes of experiments was quite similar. In each class there were substantial differences between 2- and 8-day old pups. Moreover, where comparable age ranges were sampled, as was the case in the long-delay learning and second-order conditioning experiments, the correspondent effect of the age variable was striking. In both cases, 2- and 4-day olds behaved alike but differed from 6- and 8-day olds. The fact that the age at which differences in the pup's reaction to the training procedures was quite similar, of course, does not necessarily imply that they were mediated by a difference in a common process. But it does make the hypothesis worthy of serious consideration.

Support for the proposition that what changes as the pup ages is its ability to retain a memory representation of an olfactory stimulus experience is offered by a consideration of the variables

that interacted with the pups age to determine the results of these
three classes of experiments. In both the long-delay learning
and stimulus-preexposure experiments, the age variable interacted
with the delay or <u>retention interval</u> separating exposure to the
odor and some subsequent experimental treatment.

For example, in the long-delay learning experiments, 2-6 day
old pups and 4-8 day old pups acquired comparable odor aversions
when the odor and LiCl episodes were experienced concurrently
but only the 4-8 day olds acquired the aversion when the odor
and LiCl event were separated by 15 min. In the stimulus-preexposure experiments, 2- and 8-day old pups equally failed to acquire
the odor aversion when odor preexposure preceded the odor-LiCl
pairing by only 60 min, but only the 8-day old failed to acquire
the aversion when that interval was 4 hours.

A similar case can be made that the influence of age observed
in the second-order conditioning experiment was dependent upon
the retention interval separating exposure to S_2 and exposure to
S_1. Specifically, in comparison to the 6-8 day pup, the 4-day old
pup failed to acquire the aversion to S_2 when S_2 and S_1 were
experienced sequentially. The difference between 4- and 6-8 day
olds, however, was ameliorated when the 4-day old pups experienced
S_2 and S_1 concurrently.

This entire set of findings might be summarized by saying
that young pups (2-4 day olds) were influenced more by variation
in the retention interval between odor exposure and the subsequent
training experience than were older pups (6-8 day olds). And,
this influence of age on the effect of retention interval might
profitably be supposed to reflect ontogenetic variation in the
neonates' osmic memory processes.

When, for example, the odor and reinforcing events (LiCl or
S_1) are not concurrently experienced (as was the case in the long-
delay learning and second-order conditioning experiments), it is
reasonable to assume that the pup must be able to retain a memory
representation of the odor experience if it is to associate it
with the reinforcer. The difference between the young and old
pups obtained in these two tasks then can be explained as being
due to the older pup's superior retention of the odor experience.

The stimulus-preexposure effect also can be reasonably assumed
to depend upon osmic memory processes. If prior exposure to the
odor is to interfere with subsequent conditioning to that odor,
then the pup, in some sense, must be able to retain a memory representation of the initial odor experience during the interval
separating this experience and the odor-LiCl pairing. The fact
that odor exposure 4 hours prior to the odor-LiCl pairing prevented

the 8-day old but not the 2-day old from acquiring the odor aversion then can be attributed to better retention of the initial odor experience by the older pups.

We think that it is reasonable to interpret these experiments as reflecting the ontogeny of osmic memory processes. We recognize, however, that this interpretation is not necessarily compelled by the data. A potential problem with these experiments as they relate to the osmic memory hypothesis is that the behavioral dependent variable was not influenced just by the pups' past experience with olfactory stimulation. It was determined by the pups' experience with two events, the odor and the LiCl injection. We do not think that it is likely that these particular results were influenced by ontogenetic changes in the processes mediating the reinforcing consequences of the LiCl. But we presently cannot completely rule out this possibility.

ACKNOWLEDGEMENT

This research was supported in part by National Science Foundation Grant GB-38322 to J.W. Rudy.

REFERENCES

Blass, E.M., Teicher, M.H., Cramer, C.P., Bruno, J.P. and Hall, W.G. In press. Olfactory, thermal, and tactile controls of suckling in preauditory and previsual rats. J. Comp. Physiol. Psychol.

Campbell, B.A. and Coulter, X. 1976. The ontogenesis of learning and memory In Neural mechanisms of learning and memory (M.R. Rosenzweig and E.L. Bennett, Eds.), pp. 209-235, MIT Press, Cambridge.

Campbell, B.A., Lytle, L.D. and Fibiger, H.C. 1969. Ontogeny of adrenergic arousal and cholinergic inhibitory mechanisms in the rat. Science 166: 637.

Campbell, B.A. and Raskin, L.A. 1978. Ontogeny of behavioral arousal: The role of environmental stimuli. J. Comp. Physiol. Psychol. 92: 176.

Cheatle, M.D. and Rudy, J.W. 1978. Analysis of second-order odor-aversion conditioning in neonatal rats: Implications for Kamin's blocking effect. J. Exp. Psychol: Animal Behav. Processes 4: 237.

Cheatle, M.D. and Rudy, J.W. 1979. Ontogeny of second-order odor-aversion conditioning in neonatal rats. J. Exp. Psychol: Animal Behav. Processes 5: 142.

Holland, P.C. and Rescorla, R.A. 1975. Second-order conditioning with food unconditioned stimulus. J. Comp. Physiol. Psychol. 88: 459.

Lubow, R.E. 1973. Latent inhibition. Psychol. Bull. 79: 398.

Pavlov, I.P. 1927. Conditioned Reflexes. Oxford Press, London.

Revusky, S.H. and Garcia, J. 1970. Learned associations over long delays. In Psychology of Learning and Motivation: Advances in Research and Theory (G.H. Bower and J.T. Spence, Eds.), pp. 1–84, Academic Press, New York.

Rizley, R.C. and Rescorla, R.A. 1972. Associations in second-order conditioning and sensory preconditioning. J. Comp. Physiol. Psychol. 74: 151.

Rudy, J.W. and Cheatle, M.D. 1977. Odor-aversion learning by neonatal rats. Science 198: 845.

Rudy, J.W. and Cheatle, M.D. 1978. A role for conditioned stimulus duration in toxiphobia conditioning. J. Exp. Psychol: Animal Behav. Processes 4: 399.

Rudy, J.W. and Cheatle, M.D. 1979. Ontogeny of associative learning: Acquisition of odor aversions by neonated rats, In The Ontogeny of Learning and Memory (N.E. Spear and B.A. Campbell, Eds.), pp. 157–188, Erlbaum, Hillsdale, N.J.

Smith, G.J. and Spear, N.E. 1978. Effects of the home environment on withholding behaviors and conditioning in infant and neo-natal rats. Science 202: 327.

Teicher, M.T. and Blass, E.M. 1976. Suckling in newborn rats: Eliminated by nipple lavage, reinstated by pup saliva. Science 193: 422.

Teicher, M.H. and Blass, E.M. 1977. The role of olfaction and amniotic fluid in the first suckling response of newborn albino rats. Science 198: 635.

Taukulis, H. 1974. Odor aversions produced over long CS–US delays. Behav. Bio. 10: 505.

THE INFLUENCE OF PHEROMONES ON PUBERTY IN RODENTS

John G. Vandenbergh

Department of Zoology
North Carolina State University
Raleigh, NC 27650

This discussion will focus on evidence indicating that the
onset of puberty can be influenced by both acceleratory and inhibitory pheromonal stimuli in a variety of species. It is important to understand the factors controlling puberty because it, along
with birth and death, are the three most critical events in the
life history of an organism. At a fundamental level, the attainment of puberty permits an organism to contribute its genes to the
gene-pool of the species, and at a more immediate level, the
attainment of puberty results in the addition of another individual
to the breeding population of a species. Twenty-five years ago
Cole (1954) described the important role of the age of puberty in
population dynamics. A slight reduction in the mean age of attain-
ing puberty can markedly enhance the reproductive potential of a
population by reducing generation time. Pheromones may be one of
several factors playing an important role in regulating puberty.

The work "pheromone", coined twenty years ago by Karlson and
Lüscher (1959), has come under critical scrutiny recently in print
(Beauchamp et al., 1976) and at a symposium held in 1978 at the
Eastern Regional Animal Behavior Meeting. The critical comments
are directed primarily at the designation of mammalian signalling
chemicals as pheromones and less so at chemical signals which prime
physiological functions. In my opinion, the word "pheromone" is
useful, especially with regard to priming substances utilized by
various mammals to coordinate reproductive and perhaps other
physiological functions.

Puberty in the Female

Puberty in an individual female house mouse as measured by
first vaginal cornification, can occur between 24 and 75 days of
age in my laboratory, and the mean age at puberty following various
treatments has varied from 28.0 to 57.1 days (Vandenbergh, 1967;
Vandenbergh et al., 1972). There was no evidence of differences
in reproductive success even though the age at which breeding began
varied so widely. The social environment is the primary factor influencing the age at which a female mouse attains puberty. Rearing
female mice in the presence of a male results in the early onset
of puberty (Vandenbergh, 1967), and rearing females in the presence
of other females delays the onset of puberty (Vandenbergh et al.,
1972; Drickamer, 1974).

It became evident that this effect of the social environment
on puberty was due largely to chemical communication when puberty
was found to be accelerated in juvenile females exposed to bedding
material soiled by adult males (Vandenbergh, 1969). The point was
more conclusively made when Cowley and Wise (1972) and Colby and
Vandenbergh (1974) showed that puberty in female mice could also be
markedly accelerated by exposure of females to the urine of adult
males. Chemical stimuli account for most but not all of the effect
of the male on female puberty. Contact stimuli appear to contribute the remaining portion of the effect. Drickamer (1974) was
able to demonstrate the contributory role of contact behaviors by
measuring puberty in immature females after exposure to a neonatally
androgenized adult female. In rats, such females display masculine
copulatory behaviors as adults even in the absence of concurrent
androgens (Sachs et al., 1973), and it was found that female mice
responded similarly after neonatal androgen exposure. As shown in
Figure 1, Drickamer found that the presence of male bedding alone
produced an intermediate age of puberty onset in contrast to the
early puberty resulting from male presence and the late puberty
found when females are raised in isolation. The presence of a
female showing male-like behavior, i.e., the neonatally androgenized
female, resulted in the same intermediate level of stimulation as
the male bedding alone. But, when both male-soiled bedding and a
neonatally androgenized female were present, the effect was equivalent to the effect induced by the presence of an adult male.
Drickamer concluded from this that both contact stimuli resulting
from masculine behavior patterns and a chemical stimulus from the
male were responsible for the male's acceleratory effect on females.
Bronson and Maruniak (1975), using a somewhat different experimental
approach, reached the same conclusion.

Once it became clear that male urine contained a pheromone that
accelerated the onset of puberty in females, preliminary steps were

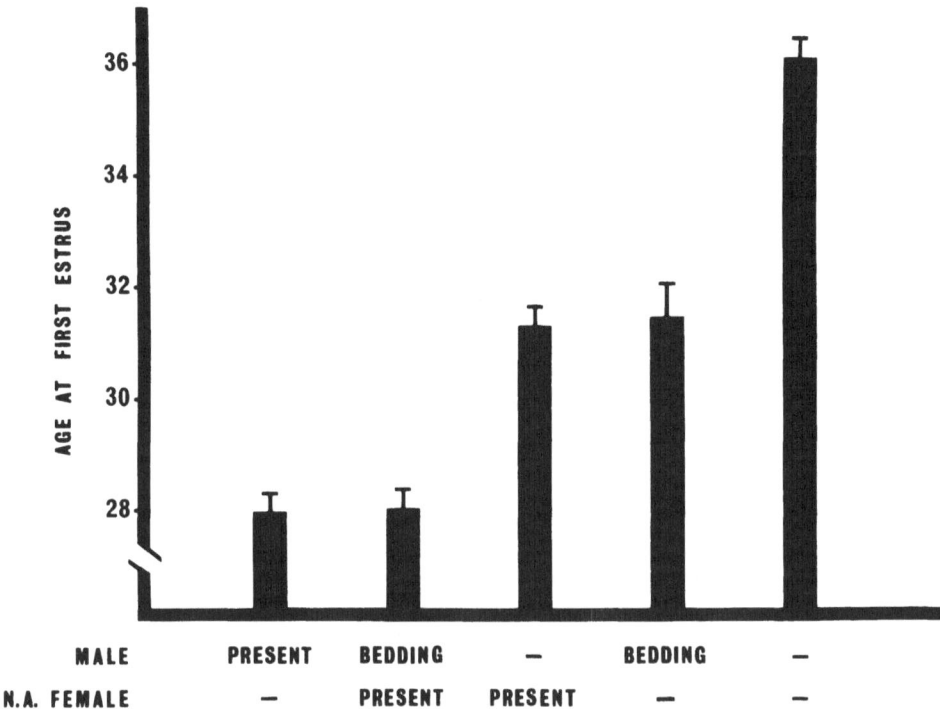

Figure 1.   The age at first estrus, in days, of female mice exposed
            to an adult male, bedding material soiled by a male, or
            an adult female which had received a masculinizing dose
            of testosterone propionate at 1 day of age (N.A. Female).
            Exposure to these treatments was continuous from wean-
            ing at 21 days of age. These data were drawn from
            Drickamer, 1974.

taken to characterize the substance or substances involved. To
accomplish this task, a more rapid and convenient bioassay had to be
devised to replace the repeated vaginal lavage we had been using
to detect first estrus. The bioassay we developed is based on the
uterine response to putative pheromones placed on the nares and
oronasal groove of juvenile females (Vandenbergh et al., 1975).
The uteri of females exposed to male urine in this manner daily from
28 to 35 days of age are significantly heavier than females exposed to water or to female urine. The difference is usually two
to three times the weight of controls.

It was found that the active ingredient was androgen dependent,
heat labile, and was associated with the protein component of male

mouse urine (Vandenbergh et al., 1975). Further chemical studies
using column chromatography to separate urinary fractions revealed
that a fraction with a peak molecular weight of 860 Daltons and
showing positive reactions for peptides contained the active pheromone (Vandenbergh et al., 1976). Tests for the presence of steroids
in this fraction were negative. Subsequent work in my laboratory
and elsewhere (Whitten, pers. comm.) has confirmed these results.
It is still not clear that the pheromone is a peptide or mixture of
peptides. The active material may be bound to a peptide and thus
detected only in association with peptides. The evidence strongly
suggests that the active ingredient is not a metabolite of testosterone but rather a secondary product produced by androgen sensitive tissue. Female mice injected with testosterone propionate
over a two-week period void urine capable of accelerating pubertal
development in other females with a potency equalling that of intact
adult males (Lombardi, 1976).

Although a great deal remains to be done, work with the puberty-
acceleration system indicates that the adult male mouse produces a
substance which is excreted in the urine and which has a specific,
long-lasting effect on the maturing female reproductive system.
This specificity of effect and apparent chemical specificity
permits application of the term "pheromone" as first described by
Karlson and Lüscher (1959).

Work with puberty inhibition due to female stimulation has not
progressed as rapidly as work on puberty acceleration. The inhibitory effect of females on other females was noted by Vandenbergh
et al. (1972) when they found that grouped females attained puberty
later than isolated females. Drickamer (1974) has continued this
line of research and has shown that bedding material soiled by a
group of females was equally effective as the presence of females
in suppressing the onset of puberty. Interestingly, bedding
material capable of suppressing puberty is only produced by females
living in groups. The bedding from isolated females, even when
pooled to control for the number of producers, is ineffective. This
finding suggests that the production of an inhibitory pheromone is
influenced by the social environment of the female. Recently,
McIntosh and Drickamer (1977) have shown that bladder urine collected
from females contains the active suppressant of puberty. Their data
also suggest that deactivation of the inhibitory pheromone may
occur in the urethra since, in groups of caged females, bladder
urine is inactive. Further work has shown that ovarian secretions
are not responsible for the maturation-delaying effect (Drickamer
et al., 1978).

The onset of puberty in females has been found to be subject
to social stimuli in several species in addition to the house
mouse (Table 1), and in a few cases urine has been demonstrated to
be the bearer of the message. Teague and Bradley (1978) have shown

Table 1. Mammals in which female puberty has been shown to be
          accelerated by male stimuli.

Species	Days Advanced	Uterine Weight	Male Presence	Urine	References
Mus musculus	20	x2	x	x	Vandenbergh, 1967 and Vandenbergh et al., 1975
Rattus norvegicus	9	-	x		Vandenbergh, 1976
Peromyscus maniculatus	-	x2		x	Teague and Bradley, 1978
Microtus ochrogaster	14	-	x		Hasler and Nalbandov, 1974
Dicrostonyx groenlandicus	14	-	x		Hasler and Banks, 1975
Sus scrofa	27	-	x		Brooks and Cole, 1970

that urine from males accelerates female sexual development in the
prairie deermouse. Other studies have shown that the presence of
an adult male accelerates female sexual development in the Norway
rat, prairie vole, collared lemming, and domestic pig (see Table 1
for references). The stimulatory effect of the male on female
sexual maturation may not be limited to mammals as recent work
with the parasitic helminth, Schistosoma mansoni, reveals.

The reproductive system of female S. mansoni fails to develop
when these helminths are maintained in single-sex culture (Erasmus,
1973). Complete reproductive development is dependent upon residence by the female in the male's gynecophoral canal (Shaw, 1977).
Such a close (and to those of us used to dealing with vertebrates,
residence within the body of the opposite sex is very close)
physical relationship suggests that either tactile contact or a
chemical stimulus accelerates sexual maturation. Recently, Shaw
et al. (1977) have shown that extracts of males induce the complete
development of the female's reproductive system. Similar extracts
of female worms have no effect.

Puberty in the Male

Evidence that sexual development in the male as well as the female rodent is subject to modification by the social environment has also been accumulating, and some recent evidence strongly suggests a role for pheromones in this effect as well. Early work by Steinach (1936) showed that unmated male rats displayed reproductive atrophy when denied access to females, and he suggested that female odor may be necessary for maintenance of the male's reproductive system. Folman and Drori (1966), on the other hand, demonstrated that reproductive atrophy occurred in all-male groups of rats even when presented with female odors, indicating that heterosexual activity rather than odor alone may be essential for reproductive maintenance. More recently, Fox (1968), working with mice, showed that the presence of an adult female slightly accelerated sexual development in the male. Vandenbergh (1971) confirmed this finding and also found that the presence of an adult male inhibited sexual development in immature males.

An intensive series of studies of the role of male stimuli in controlling male sexual maturation is underway in the laboratory of J.M. Whitsett using the prairie deermouse, Peromyscus maniculatus, and the early results reveal an important role for chemical communication in this system. Bediz and Whitsett (1979) showed that prairie deermice reared in all-male groups had lighter testes and seminal vesicles than isolate-reared controls at 5, 9, and 16 weeks of age (Figure 2). Although the reproductive organs were suppressed, there was only an effect on body growth in one of five experiments. Additional experimentation revealed that the presence of females had no significant effect on male maturation. Thus, male prairie deermice have a specific inhibitory effect on the reproductive development of cohabiting males. This series of experiments set the stage for an investigation of the male stimulus or stimuli producing the inhibitory effect.

Lawton and Whitsett (in prep.) tested whether bedding material soiled by adult male prairie deermice could inhibit sexual development in immature males. Their results revealed a significant inhibitory effect of soiled bedding on testicular and seminal vesicle growth. The seminal vesicle weights, following exposure to soiled bedding, were 37 percent lower than those following exposure to clean bedding (Figure 3). These investigators then showed that the male urine contained in the soiled bedding was responsible for this effect by demonstrating that adult male urine applied directly to the nares of immature males induced a similar inhibitory effect (Figure 4). Urine from females had no effect. To determine whether the effects seen in these experiments were dependent on an intact

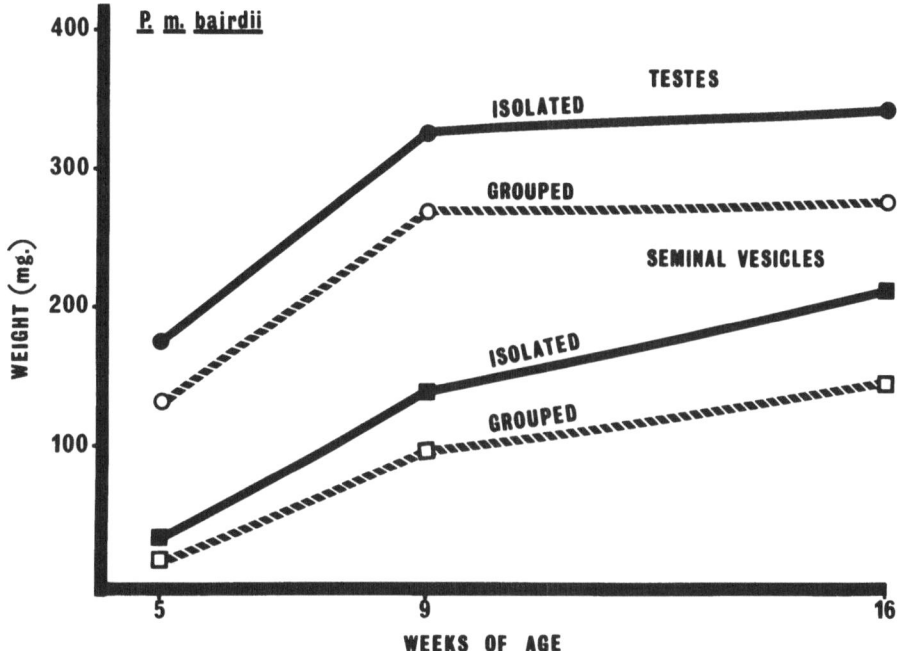

Figure 2. Testicular and seminal vesicle weight of male prairie
          deermice reared in groups of 5 or 9 to 11 or in isola-
          tion. These data were drawn from Bediz and Whitsett,
          1979.

olfactory system, Lawton and Whitsett exposed both bulbectomized
and sham-operated immature males to water or male urine. Following removal of the olfactory bulbs, the bulbectomized males failed
to show reproductive inhibition when exposed to male urine
(Figure 5). The major conclusions emerging from this series of
experiments are that urine from male prairie deermice contains a
suppressant of sexual maturation and that this suppressant acts
via the olfactory system in the immature male.

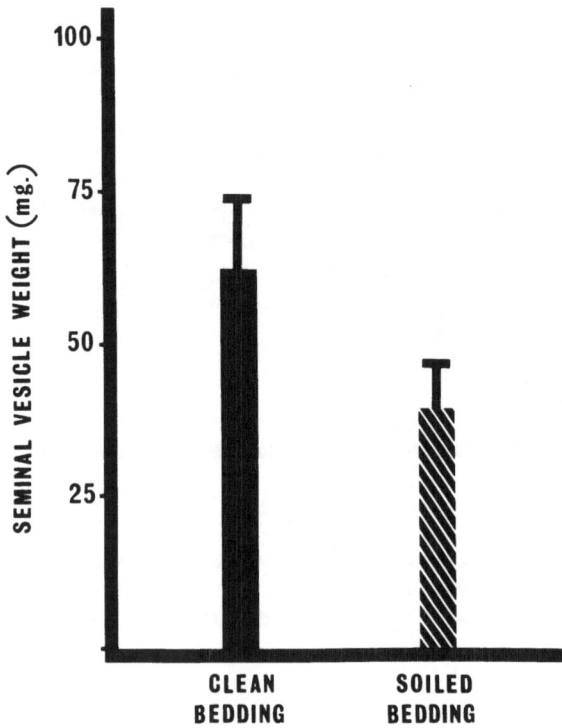

Figure 3.   The effect of bedding soiled by mature males on the development of seminal vesicles in juvenile prairie deermice.   These data were drawn from Lawton and Whitsett,
in preparation.

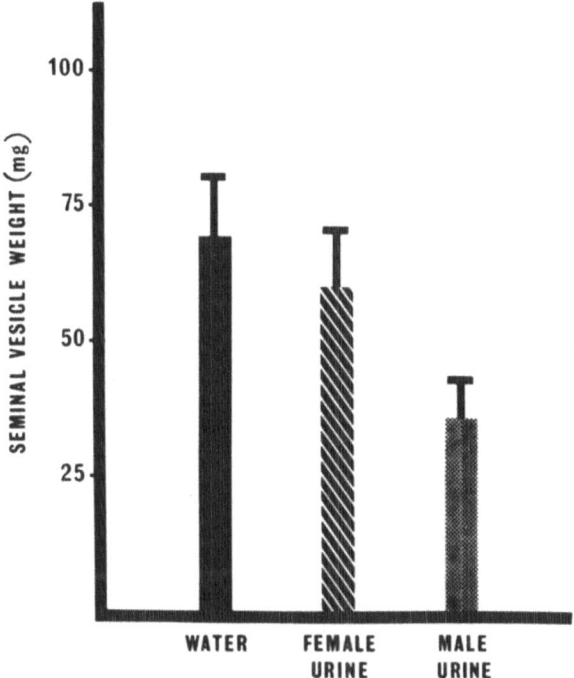

Figure 4. The effect of urine from adult males and females on the
            development of seminal vesicles in juvenile prairie deer-
            mice. These data were drawn from Lawton and Whitsett,
            in preparation.

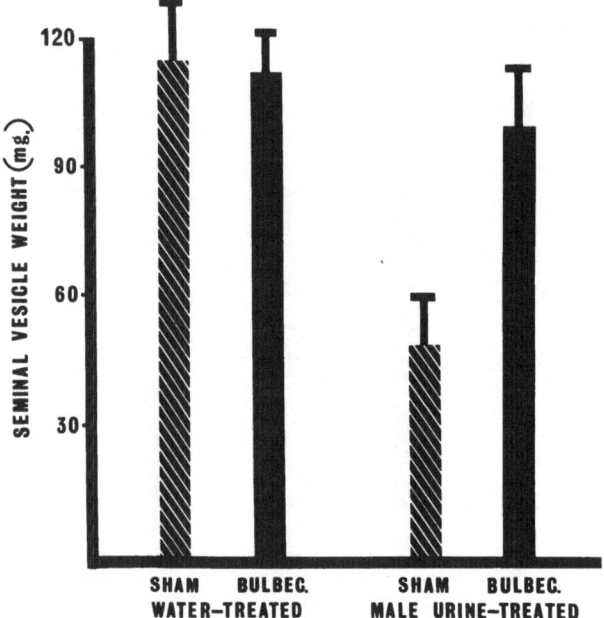

Figure 5.   The effect of olfactory bulbectomy on the development
of seminal vesicles in juvenile prairie deermice following exposure to male urine or water.   These data were
drawn from Lawton and Whitsett, in preparation.

SUMMARY

In summary, the onset of puberty in a variety of species is influenced by signals received from the social environment. These signals can be either stimulatory or inhibitory, and the intensity or even presence of the signal may be influenced by the environment of the sender. The result of such an interactive scheme is the coordination of reproduction within a population. It is quite possible that signals modifying sexual development can play an important role in maintaining populations appropriate to particular environments. I have focused on chemicals as signals, i.e., "pheromones", but it is clear that pheromones are only one of the many modes of communication in animals. In songbirds, for example, the quality of the male song has recently been shown to modify sexual development of females (Kroodsma, 1976). It is likely that future studies will show that stimuli from the same or opposite sex may modulate puberty in many species, and in some of these the stimulus will prove to be a pheromone.

REFERENCES

Beauchamp, G.K., Doty, R.L., Moulton, D.G. and Mugford, R.A. 1976. The pheromone concept in mammalian chemical communication: a critique, in "Mammalian Olfaction, Reproductive Processes, and Behavior" (R.L. Doty, ed.), pp. 143-160, Academic Press, New York.

Bediz, G.M. and Whitsett, J.M. 1979. Social inhibition of sexual maturation in male prairie deermice, J. Comp. and Physiol. Psych. (in press).

Bronson, F.H. and Maruniak, J.A. 1975. Male-induced puberty in female mice: Evidence for a synergistic action of social cues, Biol. of Reprod. 13:94.

Brooks, P.H. and Cole, D.J.A. 1970. The effect of the presence of a boar on the attainment of puberty in gilts, J. Reprod. and Fertil. 23:435.

Colby, D.R. and Vandenbergh, J.G. 1974. Regulatory effects of urinary pheromones on puberty in the mouse, Biol. Reprod. 11: 268.

Cole, L.C. 1954. The population consequences of life history phenomena, Quart. Rev. Biol. 29:103.

Cowley, J.J. and Wise, D.R. 1972. Some effects of mouse urine on neonatal growth and reproduction, Anim. Behav. 20:499.

Drickamer, L.C. 1974. Contact stimulation, androgenized females and accelerated sexual maturation in female mice, Behav. Biol. 12:101.

Erasmus, D.A. 1973. A comparative study of the reproductive system of mature, immature and unisexual female Schistosoma mansoni, Parasitol. 67:165.

Folman, Y. and Drori, D. 1966. Effects of social isolation and
   of female odours on the reproductive system, kidneys and
   adrenals of unmated male rats, J. Reprod. Fertil. 11:43.

Fox, K.A. 1968. Effects of prepubertal habitation conditions on
   the reproductive physiology of the male house mouse, J. Reprod.
   Fertil. 17:75.

Hasler, J.F. and Banks, E.M. 1975. The influence of mature males
   on sexual maturation in female collared lemmings (Dicrostonyx
   groenlandicus), J. Reprod. Fertil. 42:583.

Hasler, M.J. and Nalbandov, A.V. 1974. The effect of weanling
   and adult males on sexual maturation in female voles (Microtus
   ochrogaster), Gen. and Comp. Endocrinol. 23:237.

Karlson, P. and Lüscher, M. 1959. "Pheromones": A new term for
   a class of biologically active substances, Nature (London)
   183:55.

Kroodsma, D.E. 1976. Reproductive development in a female songbird:
   Differential stimulation by quality of male song, Science 192:
   574.

Lawton, A.D. and Whitsett, J.M. Pheromonal influences on male
   maturation in prairie deermice. (in prep.).

Lombardi, J.R., Vandenbergh, J.G. and Whitsett, J.M. 1976. Androgen control of the sexual maturation pheromone in house mouse
   urine, Biol. Reprod. 15:179.

McIntosh, T.K. and Drickamer, L.C. 1977. Excreted urine, bladder
   urine, and the display of sexual maturation in female house
   mice, Anim. Behav. 25:999.

Sachs, B.D., Poliak, E.L., Kreiger, M.S. and Barfield, R.J. 1973.
   Sexual behavior: Normal male patterning in androgenized female
   rats, Science 181:770.

Shaw, J.R. 1977. Schistosoma mansoni: Pairing in vitro and development of females from single-sex infections, Exper.
   Parasitol. 41:54.

Shaw, J.R., Marshall, I. and Erasmus, D.A. 1977. Schistosoma
   mansoni: In vitro stimulation of vitelline cell development
   by extracts of male worms, Exper. Parasitol. 42:14.

Steinach, E. 1936. Zur Geschichte des männlichen Sexualhormons
   und seiner Wirkung am Säugetier und beim Menschen, Wien.
   Klin. Wschr. 49:161.

Teague, L.G. and Bradley, E.L. 1978. The existence of a puberty
   accelerating pheromone in the urine of the male prairie deermouse (Peromyscus maniculatus bairdii), Biol. Reprod. 19:314.

Vandenbergh, J.G. 1967. Effect of the presence of a male on the
   sexual maturation of female mice, Endocrinology 81:345.

Vandenbergh, J.G. 1969. Male odor accelerates female sexual maturation in mice, Endocrinology 84:658.

Vandenbergh, J.G. 1971. The influence of the social environment
   on sexual maturation in male mice, J. Reprod. Fertil. 24:383.

Vandenbergh, J.G., Drickamer, L.C. and Colby, D.R. 1972. Social
   and dietary factors in the sexual maturation of female mice,
   J. of Reprod. and Fertil. 28:397.

Vandenbergh, J.G. Finlayson, J.S., Dobrogosz, W,J,, Dills, S,S, and Kost, T.A. 1976. Chromatographic separation of puberty accelerating pheromone from male mouse urine, Biol. Reprod. 15: 260.

Vandenbergh, J.G., Whitsett, J.M. and Lombardi, J.R. 1975. Partial isolation of a pheromone accelerating puberty in female mice, J. Reprod. Fertil. 43:515.

THE MODULATION OF REPRODUCTION BY PRIMING PHEROMONES IN HOUSE

MICE: SPECULATIONS ON ADAPTIVE FUNCTION

F.H. Bronson and Arthur Coquelin

The Institute of Reproductive Biology
Department of Zoology, University of Texas
Austin, TX 78712

INTRODUCTION

It is well established now that some of the reproductive processes of some mammals are regulated in part by chemical cues (priming pheromones) emanating from conspecifics. It is important to note that most of the research supporting this conclusion has utilized only a single species, the house mouse. Furthermore, most of the relevant experiments have involved long-domesticated stocks of mice and all of the research in this area has been conducted in a laboratory setting. Thus, while we have developed a solid appreciation of the laboratory biology of mouse primers, we have little knowledge about the occurrence of similar phenomena in other mammals, and we can only speculate at this time about the functional utility of priming pheromones to any wild population.

The objective of this paper is to conceptualize the adaptive functions served by priming pheromones in wild populations of house mice. Such speculations have been made in the past but, typically, these have been superficial attempts based on a poor appreciation of the field biology of this species (e.g., Whitten and Bronson, 1970). The present review will emphasize the relevant ecological literature in an attempt to provide a more firm foundation for speculations about adaptive functions. Specifically, first we intend to review two of the most obvious ecological strategies of Mus, those of opportunism and colonization. These two strategies have dominated the known evolutionary history of house mice and, in part, they find their physiological bases in the unique flexibility of the environmental regulation of this species' reproduction.

Thus these particular strategies seem those most likely to provide insight into the problem of adaptive functions for the pheromonal cueing of reproduction. Next, we will present a detailed considera- tion of the social structuring of natural populations. Con- ceptualizing a functional role for any pheromonal action must start by questioning whether or not the social conditions prerequisite for the action ever occur in a natural population. Finally, after a brief review of our laboratory-derived knowledge of mouse primers, we intend to develop two arguments. First, one and possibly two of the phenomena previously linked to primer activity in the labora- tory actually may be artifacts of the laboratory environment. The precise social conditions necessary to produce these particular phenomena seem unlikely to occur in any wild population and, further- more, it is difficult or impossible to visualize their selective advantages. Secondly, the remaining priming phenomena can best be viewed as a single cueing system whose adaptive value seems to reside in its support of this species' opportunism and, in particular, in support of its colonization tactic.

OPPORTUNISM AND COLONIZATION

The house mouse is a small, nocturnal rodent that can exist either commensally with man or in total independence of man's activities. Commensal populations occur almost globally in man's dwellings, outbuildings or grain storage facilities. The non- commensal type of population, often found in grasslands and culti- vated areas, will be referred to here as a "feral" population (Berry, 1977). The ecological opportunism of this species has been noted frequently (e.g., Berry, 1970; Anderson, 1970). That is, the house mouse has demonstrated an outstanding capacity for exploiting an unstable environment in which favorable habitats appear and disappear rapidly. Some man-made habitats capable of supporting house mice are stable over long periods of time; many are not. Grain, for example, often is stored only for a period of months, thereby creating a highly favorable but transient environment. Seasonal patterns in cultivation create short-lived areas for exploitation by feral mice. Aperiodic variation in rainfall like- wise influences the potential for particular areas to support feral populations. The house mouse flourishes in such unstable environ- ments, commensal or feral, its populations springing up rapidly and then dying out after having seeded still more populations in other areas.

The mice that form the progenitors of either temporary or more permanent populations are of two types: dispersing young or adults that are shifting their home ranges. As documented later in this paper, home range shifting by adults may be either common or rare, depending upon the population under consideration. The dispersal of young, on the other hand, seems to be a constant in all house

mouse populations and results from three factors: (a) a high
fertility rate for the species; (b) intra-population aggression;
and (c) a saturated carrying capacity (see Lidicker, 1975). Indeed,
Anderson (1970) estimates that 80 percent of the annual production
of a typical Mus population is expended in this manner. All dispersing young mice are potential colonizers. Most are lost. Some,
however, find exploitable and unoccupied habitats and start a new
population. For the purpose of this paper then, colonization will
be conceptualized as that component of the house mouse's opportunism
that involves dispersing young.

The house mouse has a history of successful colonization that
is paralleled by only one or two other mammals. The modern genetic
complex known as the house mouse is thought to have originated on
the dry steppes of southeastern Russia (Schwarz and Schwarz, 1943).
This species early adopted a commensal existence with man and,
therefore, has been transported widely during man's travels. Today
commensal populations may be found in all of the major land masses
of the temperate and tropical zones, parts of tropical Africa are
an exception (Berry, 1970; Berry et al., 1978). Commensal populations of house mice may be found as far north as Iceland and the
Lofoten Islands (65°N and 69°N, respectively; Berry and Jakobson,
1975). Feral populations occur almost as widely as do commensal
populations, ranging from the tropics up to 62°N and to 54°S (the
Faeroes and Macquarie Island, respectively; Berry et al., 1978).

Dispersal is a common feature of most rodent populations. Two
types of adaptations have complimented this process in Mus, thereby
allowing the near-global spread of this species from Russia. First,
house mice are metabolically-adapted to dry climates (Haines and
Schmidt-Nielsen, 1967). Thus they have been able to exploit some
of the most arid environments known: man's abodes, his outbuildings
and his transports. Second, Mus exhibits a unique flexibility in
the environmental regulation of its reproduction; they can breed
under a truly remarkably broad set of environmental conditions
(Bronson, 1979). These two adaptations, when combined with a high
rate of dispersal, form the basis for this species' unique success
as a colonizer. Finally, it should be noted that major colonization
by Mus has continued in recent times; the Hawaiian Islands weren't
colonized until about 200 years ago (Wheeler and Selander, 1972)
and the Galapagos Islands weren't colonized until after World War
II (Eibl-Eibesfeldt, 1950).

The Social Organization of House Mouse Populations

Our knowledge about the social structures exhibited by house
mice is derived from two sources: direct observation of experimentally confined mice and, secondly, either trapping data or blood

enzyme analyses from natural populations. The natural populations
that have been studied by ecologists or geneticists typically fall
into one of two categories: relatively stable, high density,
commensal populations in barns or chicken coops versus relatively
unstable, low density, feral populations in grasslands or cultivated areas. A third type of mouse population, one occurring
seasonally in stacked grain in Britain, also has been studied intensively. This type of population will be considered here simply
as a variant of the low density, feral type since its founders
typically come from feral habitats.

A constant feature of all mouse populations, regardless of
density or stability, is the propensity for males to be aggressive
toward other males. This behavior apparently forms the basis for
all social structuring in this species. The precise nature of the
social organization is influenced additionally by the distribution
of the food supply and the physical complexity of the environmental
substrate (see Stueck and Barrett, 1978). Social structures in
natural populations will be considered separately for the two types
of populations that have been investigated most commonly.

Stable, High Density, Commensal Populations

When mice inhabit man-made structures where food is plentiful
and stable, and where the physical substrate is reasonably complex,
the result is a population that may endure at high density for many
months to a few years. This type of population occurs commonly
throughout the tropical and temperate zones wherever man has created
the appropriate conditions. A host of laboratory studies in complex
cages, rooms, or outdoor enclosures, and a few studies of natural
populations, have documented the existence of social substructures
in these high density, commensal populations (e.g., the following
represent a list of over 50 references pertaining to such populations: Eibl-Eibesfeldt, 1950; Davis, 1958; Petrusewicz and
Andrezjewski, 1962; Crowcroft and Rowe, 1963; Christian et al.,
1965; Anderson and Hill, 1965; Crowcroft, 1966; Reimer and Petras,
1967; Selander, 1970; Adamczyk and Walkowa, 1971; Harrington, 1976;
and Lidicker, 1976).

Typical of the social organization of these dense and stable
commensal populations is a division of the living area into
territories, each of which is defended by a single male. Territorial boundaries are reinforced by the routine use of urinary
cues (e.g., Harrington, 1976). Each territorial division usually
includes several breeding females (typically less than 10), some of
their offspring, and a few subordinate males, all in turn dominated
by the one territorial male. Indirect evidence (freedom of movement and reproductive success) suggests a loose hierarchical
ranking among the adult females in a territory, but the aggressive

reinforcement of such ranking is not overt (Lloyd and Christian, 1969). Females nevertheless, particularly if pregnant, may aid the dominant male in territorial defense (Crowcroft, 1966). Home range movements are severely limited under these conditions, typically totaling only a few square meters (e.g., Young et al., 1950).

Gene flow between territories in this type of population is restricted because of aggressive defense; hence, each insular division of the mouse's living space has been conceptualized as a deme as well as a territory (Anderson, 1970; Selander, 1970). Given chronological stability of the ecological milieu, several generations may be born under these conditions, and densities may approach one mouse per square foot. The number of territories apparently remains more or less constant, regardless of density (Lidicker, 1976). Thus a high rate of dispersal by young mice always is a characteristic of these populations. Dispersing females sometimes successfully enter a dense and well-established territory; mortality of dispersing young males and adult subordinate males, however, is high.

Low Density, Feral Populations

House mice often occur as feral populations that are character- ized by quite low density, e.g., one mouse per 100-1,000 square meters (e.g., Justice, 1962; Breakey, 1963; Caldwell, 1964; Lidicker, 1966; DeLong, 1967; Berry, 1968; Myers, 1974). The carrying capacity of most grasslands, old-fields, and cultivated areas is much lower than it is for most situations offering commensal possibilities, hence the low density expectation. Home ranges in such situations often have been reported up to 400 square meters (e.g., DeLong, 1967). In a thorough, 12-day tracking study, Justice (1961) reported maximum linear limits of two home ranges in a rye field to be 25 and 30 meters, certainly in each case encompassing several hundred square meters.

Unfortunately, we know little about the type, or even the existence of social structures in low density, feral populations. The available evidence, however, argues against the existence in feral populations of the socially-rigid and temporally-stable, demic organizations observed in barns and chicken coops. For example, one of the most obvious features of feral populations of Mus is their spatial instability. Seasonal or aperiodic changes in habitat quality, such as those related to cultivation practices or climatic patterns, result in considerable home range shifting (e.g., Justice, 1962; Newsome, 1969; Berry and Jakobson, 1974; and Newsome and Corbett, 1975). Indeed, Caldwell (1964) considered feral Mus to be seminomadic and even questioned whether or not the home range concept has relevance to feral populations of Mus. A

second characteristic of feral populations is their high turnover
rate (relative to commensal populations). Mortality may run as
high as 30 percent per month (Lidicker, 1966), and overwintering
losses may run as high as 90 percent (Berry, 1968). Additionally,
these populations typically (but not always) have seasonal reproduction and dispersal. All of these factors summate to yield
marked spatial and temporal instability which, in turn, argues
against the existence of the stable, deme-territory substructures
described earlier.

Even given temporal stability of the resident adult component
of a feral population, rigid territoriality intuitively seems impossible to maintain at the low densities often found in feral
habitats. The constant patrolling and harrassment necessary both
for the maintenance of dominance within a territory and for the
defense of the territory itself seems incompatible with the size
of the mouse and a home range measured in thousands of square
meters. The use of enduring olfactory cues for boundary marking
would not necessarily modify this conclusion since,to be effective,
such communication probably requires prior conditioning (subordination) and considerable reinforcement. Attempts to establish more
directly the existence of territoriality in feral populations have
led to confusion. Over-dispersion of centers of activity, and/or
non-overlap of home ranges, both could indicate territoriality in
a homogeneous habitat but conflict exists in the conclusions derived from past analyses of these parameters (c.f., Anderson, 1970;
Lidicker, 1966; and Myers, 1974). Similarly, a failure to observe
transmission of introduced alleles, and/or a non-random distribution
of particular forms of serum enzymes, also would indicate territorial
behavior in a grassland population; again, however, the existing
data are in conflict (c.f., Anderson et al., 1964; Myers, 1974;
and Berry and Jakobson, 1975).

As noted previously, seasonal populations occurring in stacked
grain (corn ricks) in Britain have been studied intensively. Here,
grain is harvested in an unshucked form and then stacked in the late
summer, remaining in this condition until the spring. Feral mice
that live during the remainder of the year in fields and hedgerows
rapidly find and invade these stacks. Breeding often occurs through
the winter, and spring densities sometimes exceed 15 mice per cubic
meter (Southern and Laurie, 1946; Southwick, 1958; Rowe et al.,
1963, 1964). Loosely stacked (unshucked) grain could offer the
structural complexity necessary for the development of many small
territories. Whether or not this type of social organization
actually occurs, however, is unknown. The larger than expected
home ranges hinted at in such stacks (Southern and Laurie, 1946)
argues against the capacity to rigidly defend territories in this
structurally complex situation. The high rates of wounding that
have been found in such populations probably indicate unstable
social conditions (e.g., Rowe et al., 1964). Likewise, the move-

ment out of these ricks by both adult males and females (Rowe et al., 1963) suggests more social chaos than that encountered in barns where the dispersers are usually young. Thus the available data again argue against stable territoriality in this kind of population.

Generalities about the Social Organization of Mouse Populations

The available data argue that mouse populations vary tremendously in the precise mode of their social organization. The rigid, stable and slightly cooperative deme-territory configuration, so elegantly described by so many different workers, seems to be the normal result only of an ecological milieu characterized by a plentiful, stable food supply and a physical substrate that is intermediate in complexity. The common assumption that this social mode is universal among mice, however, seems unwarranted. Certainly, at another extreme, one can visualize an absolutely homogenous, feral substrate with a low carrying capacity, where territoriality would be impossible to maintain because of the necessity for such a small animal to maintain such a large home range. Relative dominance relationships between nearest male neighbors seems a much more reasonable possibility in such a situation (c.f. Davis, 1958). Heterogeneity of such a substrate would promote territoriality and a demic character to the population but temporal instability would promote social instability. Given the rapid appearance and disappearance of habitats suitable for mice, it is probable also that a relatively high degree of social instability is characteristic of all feral populations. Thus social plasticity, but always within the framework of inter-male aggression, must be the basic rule for the social organization of house mouse populations.

PHEROMONAL CUEING OF REPRODUCTION IN HOUSE MICE

The laboratory biology of mouse priming pheromones has been reviewed extensively (e.g., Vandenbergh, 1973, 1975; Whitten and Champlin, 1973; and Bronson, 1976a, 1979). The present endeavor, therefore, will consider this topic only briefly and often without specific reference to many of the original papers. By way of background, both sexes of mice mark their home substrate with urine and, hence with urinary pheromones, while locomoting. Under certain conditions, urinary cues elicit or inhibit gonadotropin secretion in other mice, thus changing their reproductive potential. Starting with the observations of Dr. W.K. Whitten in the mid-50's, several specific actions of mouse primers have been identified. Partly because of the sequence in which these actions were discovered, and partly because different individuals were involved in their discoveries, it has become routine to think in terms of as many as six distinct priming phenomena. Several of these phenomena obviously are related and the present review will con-

sider only two categories of primer action, those acting in females
and those acting in males.

The Regulation of Ovulation by Priming Pheromones

The **best-investigated action of priming pheromones in**
mice involves the temporal regulation of ovulation. Urinary cues
emanating from male mice accelerate this process; urinary cues from
female mice decelerate it. As shown in a series of studies by
Vandenbergh (e.g., 1967), these actions are particularly dramatic
in regard to the pubertal ovulation. Indeed, the presence of an
adult male is absolutely crucial for the proper organization of
puberty in this species (Stiff et al., 1974). Puberty can be
induced in young females solely by exposure to male urine (Colby
and Vandenbergh, 1974). Extremely small amounts of urine will
evoke uterine growth in young females; as little as 10 μl per day,
delivered as an aerosol, has been used successfully in our laboratory. Rapid induction of the pubertal ovulation, however, requires
a synergism of the male's urinary cues and a tactile cue (Drickamer,
1974a; Bronson and Maruniak, 1975). The precise nature of the
tactile cue is not known but it is inactive in its own right. The
young female's hormonal responses to the male's cues include a
sequential release of LH and then estradiol, followed by a secondary
suppression of follicle-stimulating hormone (FSH) and an elevation
in serum prolactin (both occur in response to the elevated estrogen
titer) and, finally, an elevation of serum progesterone and the
preovulatory release of LH, usually on the third or fourth day of
male-exposure (Bronson and Desjardins, 1974, and Bronson and
Maruniak, 1976).

Puberty in young females also is modulated by urinary cues
emanating from other females (Vandenbergh et al., 1972; Drickamer,
1974b). Urine from either adult or immature females is effective
in this regard but, regardless of the age of the urine donors, they
must have been housed in groups; tactile stimulation is necessary
for pheromonal potency of female urine (Drickamer, 1974b). Importantly, female urinary cues not only delay the pubertal ovulation
in other females, they almost totally block any acceleratory action
of a male for a considerable period during development. The hormonal actions by which such inhibition is effected have not been
studied in immature females.

Ovulation in the adult female is regulated by urinary cues
emanating from both sexes in a manner quite similar to that seen
in pubertal females. As initially demonstrated by Whitten (1956,
1959), female mice exhibit a shorter ovulatory cycle when housed
with males, a longer cycle when housed in isolation (and in the
absence of male odor), and the longest cycle of all when they are
housed in groups in an environment free of male odor (see also

Bronson, 1968). As little as 100 μl of male urine per day, given
as an aerosol for two days, will synchronize the cycles of groups
of eight wild house mouse females (Bronson, 1971). The urinary
basis for the mutual suppression of ovulation exhibited by grouped
adult females has not been verified but soiled bedding has been
found to be effective in this regard (Champlin, 1971). Experiments
with ovariectomized adults bearing estrogen implants again suggest
an exclusive action of the male's urinary factors on LH in the
adult female; grouping of these adult females suppresses their
serum LH while elevating serum prolactin, FSH titers being unre-
sponsive to either male or female cues (Bronson, 1976b).

The Action of Female Urinary Cues on Male Mice

Male mice respond to females or to female urine by secreting
LH and testosterone (Macrides et al., 1975; Maruniak and Bronson,
1976; Batty, 1978). Volatility of the relevant urinary cues, over
at least a few centimeters, has been documented. Of particular
interest is the fact that the gonadal state of the female is
unimportant in determining the potency of her urine. Urine samples
obtained from ovariectomized females, from pregnant females, or
from normally cycling females either just before or just after
ovulation, all yield a similar release of LH in test males; urine
from female hamsters or from intact or castrated male mice, however, is ineffective in this regard (Maruniak and Bronson, op. cit.).
The male's responsiveness to female urine develops between 24 and
36 days of age and habituates completely and rapidly (1-3 hours)
to repeated exposures to the same urine sample (Maruniak et al.,
1978).

The reproductive function of a transient elevation in serum
concentrations of LH, and hence testosterone, has proven both
puzzling and difficult to investigate. Exposing a female to male
odor yields an obvious and discrete event, ovulation. Spermatogenesis, on the other hand, proceeds at a constant rate in adult
male mice and apparently they are always ready to copulate under
laboratory conditions. Therefore no obvious, discrete event could
result from the experimental exposure of a male to a female cue of
any kind. To complicate this situation further, male mice like
many other male mammals frequently experience dramatic, acute
episodes of LH release and testosterone secretion in the laboratory (Bartke et al., 1973). Based upon our present knowledge
about male sexual behavior and spermatogenesis, one would suspect
little if any effect of a transiently elevated LH titer on latency
to mate or on ejaculatory potency in a normally active male. Indeed, considerable work in our laboratory now has verified this
probably already obvious fact. One also might suspect a meaningful interaction between the male's LH responses to female urine

and his rate of sexual development (see Fox, 1968). Such probably
is not the case. The response develops too late to be of great
benefit in this regard and the habituation characteristic of this
response prevents prolonged action (Maruniak et al., 1978).

One experiment concerning the male's response to female urine
has revealed a robust effect. Isolated adult males were implanted
with capsules containing testosterone at a dosage designed to elevate their average blood testosterone levels twofold. Urine from
these males proved markedly more effective in inducing uterine
growth in young females than was urine collected from control males
bearing blank capsules. Thus our "standard laboratory male" (i.e.,
one living in isolation or one chronically dwelling with a female
in a small cage) does not provide very potent urine. Regarding a
function for the male's LH response to female urine then, one can
speculate along the following lines. It is conceivable that blood
levels of LH and testosterone, and hence urinary potency, could be
somewhat suppressed even in dominant males in natural populations.
This condition could arise because of an absence of females, because of habituation of the females present, because of threshold
environmental conditions, or because of stress (as this term is
defined in the laboratory). If one or more of the conditions described above actually exist in the wild, male mice would need to
be stimulated by female urine in order to elevate their own urinary
potency to, in turn, stimulate the female. Obviously this is a
highly speculative argument, but a large number of experiments now
have searched for potential functions of the male's LH response to
female urine and the only positive result has been that involving
an increase in his urinary potency. In the absence of better data,
the premise that the male participates in a urinary feedback loop
with the female to time her ovulation will be incorporated in this
paper.

Laboratory Artifacts

It is important to recognize the possibility that some of the
phenomena previously linked to the primer pheromonal system in mice
indeed may be laboratory artifacts. This is not to say that the
mechanisms underlying the phenomena in question are not real; it
does imply that a mechanism could evolve to serve one adaptive
function in the wild yet find expression in other ways in the
artificial confines of the animal room. An excellent candidate for
such a possibility is the mutual suppression of ovulation that is
observed among groups of adult females, whether or not this suppression takes the form of pseudopregnancy or anestrus (c.f. Whitten,
1959; Ryan and Schwartz, 1977). The possible occurrence of dense,
all-female social structures, with no mature male living anywhere
in the vicinity, seems a totally unreasonable expectation in any

natural population of house mice, commensal or feral. Conceptualizing an adaptive function for the suppression of ovulation in the adult likewise is difficult. The only conceivable benefit would be a synchrony of births among groups of females when simultaneously exposed to a male, and hence, a periodic saturation of a predator-prey system. There is absolutely no empirical evidence that such synchrony ever occurs in a natural population of mice. Thus, all things considered, it seems best at this time to categorize the mutual suppression of ovulation among adult females as an interesting and historically important laboratory artifact. The presence of the suppressive mechanism in adults probably represents a hold-over from prepubertal development where the total antagonism of the male's ovulation-promoting cues has more obvious importance.

Another possible candidate for laboratory artifact status is the pregnancy block which follows exposure to a strange male (Bruce, 1959). While an extremely interesting phenomenon, this action likewise may represent past selection for a core mechanism that has real utility in the non-pregnant animal but which might be observed in the pregnant female only under laboratory conditions. Bruce's work with the pregnancy block represents an important milestone in the field of mammalian pheromone biology. As shown many times now, recently inseminated females often fail to experience implantation if their stud male is removed and replaced with a strange male. The magnitude of the effect is increased if the strange and stud males differ markedly in their genetics. The relevant cue is urinary and dependent upon androgen. Olfactory bulbectomy protects the female. The most obvious physiological correlate of the blocked pregnancy is a failure of prolactin support for the corpus luteum and, hence, a failure of progestational dominance, a failure to implant, and a return to the ovulatory cycle (reviewed by Bruce, 1966).

Three types of evidence argue against a meaningful role for the pregnancy blocking phenomenon in natural populations: (a) a reasonable argument can be made that the pregnancy blocking mechanism actually is the same as that used by males to induce ovulation in non-pregnant females and, therefore, the block need not be thought of as a physiological entity in its own right; (b) it is difficult to visualize the occurrence in a natural population of the precise social conditions necessary to produce a block; and (c) the adaptive functions that have been suggested so far for the blocking phenomenon are not convincing.

Regarding the first argument, that the pregnancy blocking mechanism actually may be the same as that used by males to induce ovulation in non-pregnant females, two observations repeatedly have been taken as evidence that the blocking mechanism is a distinct entity: the fact that the block arises because of a discrimination between two males and that the primary action of the second male is

on prolactin rather than on LH. Modern hormone assays have not been
applied to this phenomenon but the necessity for a discrimination
between two males simply is not true. For example, some stocks
of mice when grouped together at estrus become pseudopregnant in
the absence of any male (Ryan and Schwartz, 1977). Pseudopregnancy
in this case is the hormonal equivalent of the early stages of
pregnancy. Exposing such pseudopregnant females to a male results
in a nonsignificant depression in serum prolactin, a dramatic increase in serum progesterone, and the induction of ovulation, all
on the same time course as that occurring in the pregnancy block
itself (Ryan and Schwartz, personal communication). Thus, a discrimination between two males is not a necessary prerequisite for
a **block**. Furthermore, it has been shown repeatedly that urine
pooled from groups of males can induce a block in recently inseminated females, thus there is no need for individuality of the blocking males. Noting that (a) preimplantation blocks can be induced
with exogenous gonadotropin (Hoppe and Whitten, 1972), (b) the
post-ovulatory ovary has a different steroidogenic substrate than
does the preovulatory ovary, and (c) the male's response to female
urine habituates rapidly (Maruniak et al., 1978), one then can
postulate another basis for this block. Pairing a male and a female should result in an immediate surge in the excretion of
pheromonally-rich urine on the male's part, but his urine should
rapidly become less potent. Removing this relatively ineffective
stud and replacing him with a new male could simply reinitiate the
whole process, evoking an LH response on the part of the female,
blocking her pregnancy and returning her to a preovulatory condition (as argued previously by Macrides et al., 1975). This
possibility actually fits the available data as well or better than
does the possibility that the strange male block is endocrinologically and behaviorally distinct from a male's induction of ovulation in a non-pregnant female.

The second argument noted above is that it is difficult to
visualize the occurrence in the field of the social conditions
necessary for the pregnancy **block**. Replacing the stud by a strange,
vigorous male, might or might not occur in a natural population
depending upon the specific population under consideration. The
rigid deme-territory structure that has proven typical of many
high density commensal populations would defy effective entrance
by a strange male. Indeed, under more controlled but similar
experimental conditions, replacement of the dominant male, when
necessary, is often by one of his own offspring (Reimer and
Petras, 1967); there would be neither unfamiliarity between the
"strange" male and the female nor much genetic difference between
the two males in this case. Furthermore, the close, physical
presence of other adult females is a constant feature of the deme-
territory system and it is well established that such a condition
protects the inseminated female from the blocking action of a
strange male (Bruce, 1963). The loosely structured, somewhat

chaotic, low density populations typical of many feral conditions, on the other hand, might provide the necessary social opportunities, depending upon a host of unknown factors. Unfortunately we have no hard information for or against this possibility.

Finally, the adaptive functions that have been suggested for the blocking phenomenon simply are not convincing. These include its possible action as a density-dependent limiting factor for population growth (e.g., Chipman and Fox, 1966) and its possible utility for the prevention of inbreeding in small demes (Bruce, 1966). A third possibility, namely that it is more efficient to lose one's young while early in pregnancy rather than have them cannibalized by a strange male after birth, also should be mentioned. In the first case, it should be noted again that the precise social conditions that protect the female from a pregnancy block are those most likely encountered at high density. Both the first and second possibilities require acceptance of the principle of group selection, a questionable action in light of modern evolutionary thought. The third possibility assumes that cannibalization by strange males actually occurs in field populations but we have absolutely no data suggesting that this is true. Indeed, in the total absence of field data showing that any of these possibilities could occur, one really needs to be comfortable at this point with the question of how a blocked pregnancy could be advantageous to an individual male, an individual female, and to their offspring. In the case of the male, it is obvious how selection would favor the capacity to cancel another male's insemination with one's own. Of what advantage, however, is the loss of pregnancy to a female? Until this question can be handled both empirically and conceptually it seems parsimonious to view the blocking phenomenon as possibly representing an important mechanism (the induction of ovulation by male urinary factor) that shows itself as a blocked pregnancy only in a carefully controlled laboratory setting. Obviously, in view of all of the imponderables, this is a highly questionable conclusion. Nevertheless, because of the uncertainties associated with it, the strange male pregnancy block has been omitted from the general synthesis to be developed in this paper.

CONCLUSIONS

It seems reasonable at this time to conceptualize the actions of priming pheromones in house mice within the framework of a single cueing system designed strictly for the temporal regulation of ovulation, pubertal or otherwise. Figure 1 presents a schematic model of such a view. The major components of Figure 1 are mutual stimulation between adult males and adult or prepubertal females and, secondly, female-female antagonism of the males' action only among prepubertal females. Male urinary pheromones, with or

without synergizing tactile stimulation, release LH in recipient
females thereby yielding ovulation via the pathways shown. This
action is totally overridden in prepubertal females by urinary
cues emanating from other females. The pathways suggested in
Figure 1 for this inhibitory action are generalized from studies
of adults. The action of female urinary cues on males is assumed
to be an "insurance" action, i.e., its utility is to increase
pheromonal synthesis in the male, if such is needed, thereby
allowing a mutual feedback loop whose objective still is the
stimulation of ovulation in the female.

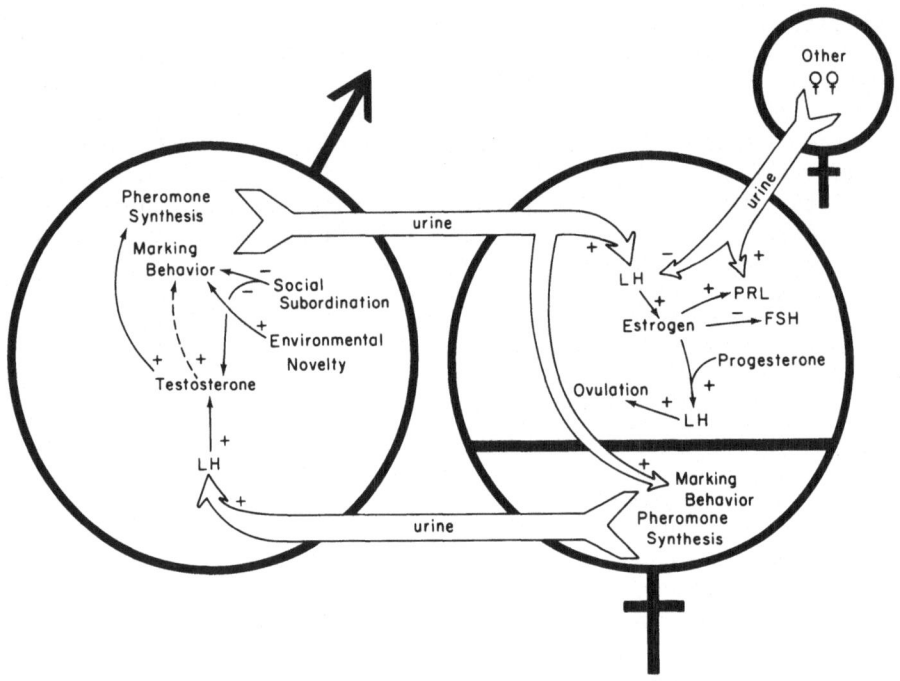

Figure 1.

Figure 1 also indicates the known regulation of pheromone
synthesis and pheromone dissemination via marking behavior (reviewed by Bronson, 1976a). Thus the separation of female hormonal
responses from the female cueing aspects of this system by a
horizontal line indicates the gonadal independence of these processes in the female. On the other hand, testosterone support is
necessary in the male for both pheromone synthesis and marking
behavior. The dashed arrow denoting the support of marking
behavior by testosterone merely indicates the uncertainty about
day-to-day modulation of this behavior by this hormone. The solid

arrow indicating testosterone support for pheromone synthesis in the male, in contrast, implies momentary modulation via agonistic encounters and female stimulation. There can be little doubt, for example, that if subordination of a male mouse is sufficiently severe, it will be accompanied by depressed testosterone secretion, and pheromonal potency then will decrease also (e.g., Lombardi and Vandenbergh, 1977). Thus the possibility that subordinate males are incapable of participating in this cueing system certainly should be noted, if on no other basis than that they simply do not mark their home substrate with urine (Desjardins et al., 1973). Finally, it has been established that past sexual experience is not important in the male's pheromone synthesis (Eisen, 1973) nor is this factor important in the male's LH responsiveness (Maruniak and Bronson, 1976). Likewise there is no evidence for a modulating role for sexual experience in the female's side of this cueing system. Therefore we are viewing a cueing system that either is under rigid genetic control or one that is established by very early social experience (probably the former, see Kirchhof-Glazier, 1979).

Speculations about the functional utility of the mouse's pheromonal cueing system will be made first within the framework of its day-to-day operation in the field and then in regard to this species' colonization strategy. In the first case, several striking differences in the manner in which the two sexes regulate their own cueing and responding functions may indicate the way this system acts in natural populations (Table 1). For example, the two sexes differ completely in the dependence of both marking and urine potency on gonadal steroids. This could be a critical feature allowing mutual stimulation even in a situation wherein both sexes are gonadally-quiescent. Thus as noted previously, a reproductively quiescent female (e.g., a dispersing female) could invoke the mutual stimulation feedback loop (i.e., stimulate a quiescent male), thereby inducing her own ovulation. In general, then, mice would never be faced with a situation wherein neither sex could stimulate and/or respond to the other because both the stimulus and the response were dependent upon fully active gonads in both sexes.

The other sexual differences shown in Table 1 seem designed to conform to other overriding physiological or ecological needs rather than being of direct benefit themselves. Both the fact that marking behavior habituates rapidly in males while that of females doesn't habituate at all, and the fact that males mark in response to novelty while females mark in response to a male, correlate well with the general uses of marking by the two sexes. The routine urinary marking done by dominant males, or territory holders, serves to notify nearest neighbors and transients of occupancy, as well as to attract and to endocrinologically stimulate females. Females, on the other hand, probably mark in a field population only to stimulate the release of LH in male recipients. It is important

Table 1. Sexual differences in marking and response functions in
          the mouse's pheromonal cueing system.

Phenomenon	Male	Female
Marking Behavior	Gonadally-dependent	Gonadally-independent
LH-Releasing Potency of Urine	Gonadally-dependent	Gonadally-independent
Marking Behavior	Habituates rapidly	Doesn't habituate
LH Response to Urine	Habituates rapidly	Habituates slowly
Elicitation of Marking	Non-specific response to novelty	Specific response to male or male urine

for a dominant male to conserve urine for marking. In the laboratory the male mouse excretes about 1.5 ml of urine per day, enough
for a few hundred to a few thousand marks. Nevertheless, males
often deplete their urine supply during prolonged testing. By
marking only in response to the most novel set of stimuli, e.g.,
those that occur at the edges of the home range or territory,
and those derived from the appearance of transient and nearest
neighbor odors (regardless of sex), the male can effectively serve
his need for communication with the limited amount of urine he
excretes.

Similarly, the habituation characteristics of the LH responses
of the two sexes (Table 1) probably correlate best with the uses of
these responses. The rapid, efficient induction of puberty, for
example, requires the synergism of both tactile and urinary cues
from the male and, physiologically, a prolonged elevation in serum
LH. Thus young females probably would not react to randomly encountered urine marks when dispersing; cohabitation with a dominant
male would seem required before puberty could result. Continuous
cohabitation, however, would be unrealistic given the energy needs
of this small homeotherm, thus the slow or absent habituation of
the female's LH response makes sense. The rapid habituation of the
male's LH response, on the other hand, seems to serve no over-
whelming utility, not surprising since we know absolutely nothing
about resting hormone levels in the field nor about the relationship between episodic surging, habituation, and pheromone synthesis
in the field or in the laboratory.

As a generality, the pheromonal cueing system probably confers
the least direct utility to stable, high density, commensal populations. Since these populations exist for long periods of time at
carrying capacity, one should be concerned here with the suppression of reproduction rather than with the converse. The only
suppressive aspect of the pheromonal cueing system, as conceptualized
here, is the female-female inhibition of the pubertal ovulation.
Gross inhibition of sexual maturation among young females is a
common observation in experimental populations when emigration is
prevented. No attempts have been made in Mus to explore directly
a possible olfactory basis for this phenomenon, but the existing
evidence certainly indicates that this is a non-specific response
to agonistic stimuli (Christian et al., 1965). Thus if dispersal
doesn't occur in a stable commensal population, a stress mechanism
would severely retard maturation, thereby providing a more enduring
and efficient means to this end than could ever be provided by
pheromonal cueing.

Despite the above argument, dispersal is universal in mouse
populations and this process has played an important role in the
global spread of Mus. The period of a few days to a few weeks
during which the pubertal ovulation can be either accelerated or

delayed could be extremely important in the colonization process.
It is obviously an advantage to a dispersing young female to breed
as early as possible, but only after having established a home.
Puberty in young female mice can be induced at a surprisingly early
age, even by the father of the young female. The complete antagonism
of the male's action by odors from female litter mates offers real
protection against pregnancy before dispersal. The need for
synergism between a male's tactile and urinary cues should prevent
rapid induction of puberty via random encounters with male urinary
marks during the actual dispersal itself. Only after having established a home in an adult male's territory, with cohabitation
and consequent tactile and urinary stimulation, will the young female attain her pubertal ovulation efficiently. Given these conditions, however, she can ovulate and become pregnant in 36-60
hours, which is advantageous to a prey species. From this stand-
point then, it is interesting that several researchers have
commented that dispersing young female mice often seem to become
pregnant just before or (more probably) just after establishing a
home (see Myers, 1974; Lidicker, 1976).

Continuing from the perspective of colonization, it is obvious
that both sexes are required for the successful establishment of a
new population. Young males arriving in an exploitable but
unoccupied habitat undoubtedly are in poor reproductive condition.
Thus the mutual feedback component of the pheromonal cueing system
could be critical at this time for rapid pregnancy. Finally, for
a new population to grow rapidly, the young female must lose her
sensitivity to the ovulation-inhibiting influences of other females.
As noted previously, that is precisely what happens at puberty.
Thus from the viewpoint of both the male and the female, and from
the expected sequence of events and conditions before, during and
after dispersal, the mouse's pheromonal cueing system seems ideally
suited to enhance the probability of successful colonization by
young dispersants.

The regulation of ovulation by priming pheromones also could
play a meaningful role in adults in low density, feral populations.
The mutual stimulation component of the cueing system seems ideally
suited for the promotion of reproductive efficiency between widely-
spaced animals, particularly given the high rate of mortality and
the instability in these populations. Indeed, if one looks at the
pregnancy rate of many feral mouse populations, non-pregnancy is
surprisingly common (e.g., DeLong, 1967). Undoubtedly this is
correlated in part with population instability and it is under
such opportunistic and/or chaotic conditions that the pheromonal
cueing system could play a role in adults similar to that which is
of probable benefit to dispersing young. As noted previously,
feral adults often shift home ranges to utilize new habitats as
they become available. It would be advantageous to pheromonally

time ovulation and pregnancy with regard to these movements. It is unfortunate that we have so little hard data from feral populations with which to explore this possibility further. Nevertheless, this concept is supported by the results of corn rick studies, in which it is reported that females entering ricks tend not to be pregnant (Rowe et al., 1963; see also Newsome, 1969).

In summary, if an opportunist is defined as one capable of exploiting favorable habitats in an unstable environment, then the house mouse indeed is an unexcelled opportunist. Furthermore, Mus has demonstrated a truly outstanding history of successful colonization, including a world-wide spread from southeastern Russia. The pheromonal system for cueing reproduction, as conceptualized herein, seems of probably adaptive value in both cases, but particularly in regard to colonization. Aggressive behavior in all Mus populations ensures a high rate of dispersal. The metabolic adaptations that allow this species to live in arid climates also has allowed them to live commensally with man and to be carried around the world in man's routinely dry transports. The extreme plasticity of the ambient cueing of reproduction in Mus allows potential colonizers to breed under highly diverse conditions. The pheromonal cueing system probably has its greatest adaptive value in being still another adaptation supporting colonization, in this case by synchronizing ovulation with dispersal. This cueing system undoubtedly also has value in timing ovulation with the opportunistic movements of adults in feral populations, but little conclusive data are available either to support or to argue against this possibility.

REFERENCES

Adamczyk, K. and Walkowa, W. 1971. Compensation of numbers and production in a Mus musculus population as a result of partial removal. An. Zool. Fennici 8: 145.

Anderson, P.K. 1970. Ecological structure and gene flow in small animals. Symp. Zool. Soc. Lond. 26: 299.

Anderson, P.K. and Hill, J.L. 1965. Mus musculus: experimental induction of territory formation. Science, N.Y. 148: 1753.

Anderson, P.K., Dunn, L.C. and Beasely, A.B. 1964. Introduction of a lethal allele into a feral house mouse population. Amer. Nat. 98: 57.

Bartke, A.R., Steele, E., Musto, N. and Caldwell, B.V. 1973. Fluctuations in plasma testosterone levels in adult male rats and mice. Endocrinology 92: 1223.

Batty, J. 1978. Acute changes in plasma testosterone levels and their relation to measures of sexual behavior in the male house mouse (Mus musculus). Anim. Behav. 26: 349.

Berry, R.J. 1968. The ecology of an island population of the house mouse. J. Anim. Ecol. 37: 445.

Berry, R.J. 1970. The natural history of the house mouse. Field
    Stud. 3: 219.
Berry, R.J. 1977. The population genetics of the house mouse.
    Sci. Prog., Oxford 64: 341.
Berry, R.J. and Jakobson, M.E. 1974. Vagility in an island popula-
    tion of the house mouse. J. Zool., Lond. 173: 341.
Berry, R.J. and Jakobson, M.E. 1975. Adaptation and adaptability
    in wild-living house mice (Mus musculus). J. Zool. Lond. 1976:
    319.
Berry, R.J., Peters, J. and Van Aarde, R.J. 1978. Sub-antarctic
    house mice: colonization, survival and selection. J. Zool.
    Lond. 184: 127.
Breakey, D.R. 1963. The breeding season and age structure of
    feral house mouse populations near San Francisco Bay,
    California. J. Mammal. 44: 153.
Bronson, F.H. 1968. Pheromonal influences on mammalian reproduction
    In Reproduction and Sexual Behavior (M. Diamond, Ed.) pp. 341-
    361, Indiana Press, Bloomington.
Bronson, F.H. 1971. Rodent pheromones. Biol. Reprod. 1: 344.
Bronson, F.H. 1976a. Urine marking in mice: causes and effects
    In Mammalian Olfaction, Reproductive Processes, and Behavior
    (R.L. Doty, Ed.) pp. 119-141, Academic Press, New York.
Bronson, F.H. 1976b. Serum FSH, LH and prolactin in adult
    ovariectomized mice bearing Silastic implants of estradiol:
    responses to social cues. Biol. Reprod. 15: 147.
Bronson, F.H. 1979. The reproductive ecology of the house mouse.
    Quart. Rev. Biol., in press.
Bronson, F.H. and Desjardins, C. 1974. Circulating concentrations
    of FSH, LH, estradiol and progesterone associated with acute,
    male-induced puberty in female mice. Endocrinology 94(6): 1658.
Bronson, F.H. and Maruniak, J. 1975. Male-induced puberty in
    female mice: evidence for a synergistic action of social cues.
    Biol. Reprod. 13: 94.
Bronson, F.H. and Maruniak, J. 1976. Differential effects of male
    stimuli on follicle stimulating hormone, luteinizing hormone,
    and prolactin secretion in prepubertal female mice. Endo-
    crinology 98(5): 1101.
Bruce, H.M. 1959. An exteroceptive block to pregnancy in the
    mouse. Nature, Lond. 184: 105.
Bruce, H.M. 1963. Olfactory block to pregnancy among grouped
    mice. J. Reprod. Fert. 6: 451.
Bruce, H.M. 1966. Smell as an exteroceptive factor. J. Anim.
    Sci. 25: 83.
Caldwell, L.D. 1964. An investigation of competition in natural
    populations of mice. J. Mammal. 45(1): 12-30.
Champlin, A.K. 1971. Suppression of oestrus in grouped mice:
    the effects of various densities and the possible nature of
    the stimulus. J. Reprod. Fert. 27: 233.

Chipman, R.K. and Fox, K.A. 1966. Oestrus synchronization and pregnancy blocking in wild house mice (Mus musculus). J. Reprod. Fert. 12: 233.

Christian, J.J., Lloyd, J.A. and Davis, D.E. 1965. The role of endocrines in the self-regulation of mammalian populations. Rec. Prog. Horm. Res. 21: 501.

Colby, D.R. and Vandenbergh, J.G. 1974. Regulatory effects of urinary pheromones on puberty in the mouse. Biol. Reprod. 11: 268.

Crowcroft, P. 1966. Mice All Over. G.T. Foulis and Company, London.

Crowcroft, P. and Rowe, F.P. 1963. Social organization and territorial behavior in the wild house mouse (Mus musculus). Proc. Zool. Soc. Lond. 140: 517.

Davis, D.E. 1958. The role of density in aggressive behavior of house mice, Anim. Behav. 6: 207.

DeLong, K.T. 1967. Population ecology of feral house mice. Ecology 46: 611.

Desjardins, C., Maruniak, J. and Bronson, F.H. 1973. Social rank in house mice: differentiation by ultra-violet visualization of urinary marking patterns. Science 182: 939.

Drickamer, L.C. 1974a. Contact stimulation, androgenized females and accelerated sexual maturation in female mice. Behav. Biol. 12: 101.

Drickamer, L.C. 1974b. Sexual maturation of female house mice: social inhibition. Devel. Psychobiol. 7(3): 257.

Eibl-Eibesfeldt, I. 1950. Beitrage zur Biologie der Haus-und der Ahrenmaus nebst einigen Beobachtungen an anderen Nagern. Z. Tierpsychol. 7: 558.

Eisen, E.J. 1973. Genetic and phenotypic factors influencing sexual maturation of female mice. J. Anim. Sci. 37(5): 1104.

Fox, K.A. 1968. Effects of prepubertal habitation conditions on the reproductive physiology of the male house mouse. J. Reprod. Fert. 17: 75.

Haines, H. and Schmidt-Nielsen, K. 1967. Water deprivation in wild house mice. Physiol. Zool. 40(4): 424.

Harrington, J.E. 1976. Recognition of the territorial boundaries by olfactory cues in mice. Z. Tierpsychol. 41: 295.

Hoppe, P.C. and Whitten, W.K. 1972. Pregnancy block: imitation by administered gonadotropin. Biol. Reprod. 7: 254.

Justice, K.E. 1961. A new method for measuring home ranges of small mammals. J. Mammal. 42: 462.

Justice, K.E. 1962. Ecological and genetical studies of evolutionary forces acting on desert populations of Mus musculus. Arizona-Sonora Desert Museum Inc: Tucson, Arizona.

Kirchhof-Glazier, D.A. 1979. Reproductive and behavioral effects of cross-fostering female albino house mice (Mus musculus) to prairie deermice (Peromyscus maniculatus bairdii). Ph.D. Thesis, Cornell University.

Lidicker, W.Z. 1966. Ecological observations on a feral house mouse population declining to extinction. Ecol. Monogr. 36: 27.

Lidicker, W.Z. 1975. The role of dispersal in the demography of small mammals In Small Mammals, Their Productivity: Population Dynamics" (F.B. Golley, K. Petrusewicz, and L. Ryszkowski, Eds.) pp. 103-128, Cambridge University Press.

Lidicker, W.Z. 1976. Social behavior and density regulation in house mice living in large enclosures. J. Anim. Ecol. 45(3): 677.

Lloyd, J.A. and Christian, J.C. 1969. Reproductive activity of individual females in three experimental freely growing populations of house mice Mus musculus. J. Mammal. 50: 49.

Lombardi, J.R. and Vandenbergh, J.G. 1977. Pheromonally induced sexual maturation in females: regulation by the social environment of the male. Science 196: 545.

Macrides, F., Bartke, Z. and Dalterio, S. 1975. Strange females increase plasma testosterone levels in male mice. Science 189: 1104.

Maruniak, J. and Bronson, F.H. 1976. Gonadotropic responses of male mice to female urine. Endocrinology 99(4): 963.

Maruniak, J.A., Coquelin, A. and Bronson, F.H. 1978. The release of LH in male mice in response to female urinary odors: Characteristics of the response in young males. Biol. Reprod. 18: 251-255.

Myers, J.H. 1974. Genetic and social structure of feral house mouse populations on Grizzly Island, California. Ecology 55: 747.

Newsome, A.E. 1969. A population study of house-mice permanently inhabiting a reedbed in South Australia. J. Anim. Ecol. 38: 361.

Newsome, A.E. and Corbett, L.K. 1975. Outbreaks of rodents in semi-arid and arid Australia: causes, preventions, and evolutionary considerations In Rodents in Desert Environments (P. Gosh, Ed.) pp. 117-153, The Hague: Junk.

Petrusewicz, K. and Andrezjewski, R. 1962. Natural history of a free-living population of house mice (Mus musculus L.) with particular reference to groupings within the population. Ekol. Polska A10: 85.

Reimer, J.D. and Petras, M.L. 1967. Breeding structure of the house mouse, Mus musculus, in a population cage. J. Mammal. 48: 88.

Rowe, F.P., Taylor, E.J. and Chudley, A.H.J. 1963. The numbers and movements of house mice (Mus musculus L.) in the vicinity of four corn-ricks. J. Anim. Ecol. 32: 87.

Rowe, F.P., Taylor, E.J. and Chudley, A.H.J. 1964. The effect of crowding on the reproduction of the house mouse (Mus musculus L.) living in corn-ricks. J. Anim. Ecol. 33: 477.

Ryan, K.D. and Schwartz, N.B. 1977. Grouped female mice: demonatration of pseudopregnancy. Biol. Reprod. 17: 578.

Schwarz, E. and Schwarz, H.K. 1943. The wild and commensal stocks of the house mouse, Mus musculus L. J. Mammal. 24: 59.

Selander, R.K. 1970. Behavior and genetic variation in natural populations. Am. Zool. 10: 53.

Southern, H.N. and Laurie, E.M.O. 1946. The house mouse (Mus musculus) in corn ricks. J. Anim. Ecol. 15(7): 134.

Southwick, C.H. 1958. Population characteristics of house mice living in English corn ricks: density relationships. Proc. Zool. Soc. Lond. 131: 163.

Stiff, M.E., Bronson, F.H. and Stetson, M.E. 1974. Plasma gonadotropins in prenatal and prepubertal female mice: disorganization of pubertal cycles in the absence of a male. Endocrinology 92(2): 492.

Stueck, K.L. and Barrett, G.W. 1978. Effects of resource partitioning on the population dynamics and energy utilization strategies of feral house mice (Mus musculus) populations under experimental field conditions. Ecology 59(3): 539.

Vandenbergh, J.G. 1967. Effect of the presence of a male on the sexual maturation of female mice. Endocrinology 81: 345.

Vandenbergh, J.G. 1973. Acceleration and inhibition of puberty in female mice by pheromones. J. Reprod. Fert. 19: 411.

Vandenbergh, J.G. 1975. Hormones, pheromones and behavior In Hormonal Correlates of Behavior, Vol. II (B.E. Eleftheriou and R.L. Sprott, Eds.) pp. 551-584, Academic Press, New York.

Vandenbergh, J.G., Drickamer, L.C. and Colby, D.K. 1972. Social and dietary factors in the sexual maturation of female mice. J. Reprod. Fert. 28: 397.

Wheeler, L.L. and Selander, R.K. 1972. Genetic variation in populations of the house mouse, Mus musculus, in the Hawaiian Islands. Stud. Genetics 7: 269.

Whitten, W.K. 1956. Modifications of the oestrus cycle of the mouse by external stimuli associated with the male. J. Endocrinol. 13: 399.

Whitten, W.K. 1959. Occurrence of anoestrus in mice caged in groups. J. Endocrinol. 18: 102.

Whitten, W.K. and Bronson, F.H. 1970. Role of pheromones in mammalian reproduction In Advances in Chemoreception (J.W. Johnson, D.G. Moulton and A. Turk, Eds.) pp. 309-325, Appleton-Century-Crofts, New York.

Whitten, W.K. and Champlin, A.K. 1973. The role of olfaction in mammalian reproduction In Handbook of Physiology (R.O. Greep, Ed.) II(7): 109. Waverly Press, Inc., Baltimore.

Young, H., Strecker, R.L. and J.T. Emlen, Jr. 1950. Localization of activity in two indoor populations of house mice Mus musculus. J. Mammal. 31: 403.

THE MAJOR HISTOCOMPATIBILITY COMPLEX AS A SOURCE OF ODORS IMPARTING

INDIVIDUALITY AMONG MICE

Kunio Yamazaki, Masashi Yamaguchi, Edward A. Boyse
and Lewis Thomas

Memorial Sloan-Kettering Cancer Center
1275 York Avenue, New York, NY 10021

For us, belonging, as we do, to a species in which olfactory
recognition of any kind seems little more than a luxury, it requires
an effort of imagination to envisage how crucial a part the recognition of identity by scent may play in the evolution and survival
of a species like the mouse. The very word 'envisage' betrays our
reliance on sight. How little the mouse depends on vision is
illustrated by the fact that mutations causing gross retinal defects
in laboratory strains may pass unnoticed, and that it is sometimes
difficult even to determine whether the loss of sight caused by such
mutations is partial or complete; so slight, apparently, is the
disability of blind mice.

It is the purpose of this essay to suggest ways in which the
genetic hallmarking of individuals by scent may be a crucial factor
in the fitness of species. We refer here exclusively to chemical
olfactory individuality that is genetically determined. All manner
of non-genetic circumstances, such as health, diet and commensal
flora may be the source of odors that distinguish particular mice
at particular times, but such transitory or fortuitous factors have
no place in this discussion. On the other hand, odors whose quality
or quantity is genetically determined, no matter how tortuous the
route from gene to chemical signal, may form part of a sensory input
that influences reproductive behavior and physiology and that must
thus come to bear in greater or lesser degree on the genetic structure of the species and hence on its fitness and adaptability.

The first requirement of a laboratory model for studying genetic individuality conferred by scent is a gene that not only is
concerned in the production of an odor or odors, but is available
in different allelic forms that change the quantity or chemistry

267

of the odors in a manner that members of the species can detect.
The Major Histocompatibility Complex (MHC) of the mouse (Klein,
1975), commonly called H-2, is the only locus to meet this criterion so far, and it is certainly most provocative that this particular gene complex should prove to be involved in olfactory
individuality. For one reason, this string of many linked genes is
involved in several aspects of immunity, i.e., in the precise
distinction by lymphocytes of a vast diversity of chemical structure. The only other example of such prowess in chemical recognition
is the olfactory system. Although admittedly it is not yet known
to what extent chemical discrimination by olfactory neurons depends
on the same principle of receptor diversity used by lymphocytes, as
opposed to combinational recognition of the kind operating in color
vision, the evident parallel has stimulated discussion of possible
evolutionary implications.

The first intimation that MHC genes have anything to do with
the scents of individual mice came from a fortunate observation
made in the Congenic Mouse Production Unit at Sloan-Kettering. The
value of genetically uniform inbred strains of mice is well known,
and of recent years congenic inbred strains which differ from their
established partner strains at single genetic loci have added
further refinements of genetic analysis. Because the genetic composition of congenic strains and their inbred partner strains is
identical except for allelic differences at the selected locus, any
difference in the characteristics of the two strains must be
ascribed, with certain due reservations, to genes at that locus
(for brief descriptions and illustrations of the derivation of
congenic mouse strains, see Boyse, 1977).

The observation in question was made in breeding tests where
males were being mated with females of the same strain together
with females of a congenic strain differing only at the MHC/H-2
locus. It appeared that pairs of different H-2 types were more
socially inclined to one another than pairs of the same H-2 type.

In order to translate this finding into a more definitive
experimental model, a procedure was devised in which males of a
given H-2 type were offered two females in estrus, the trios then
being observed continuously until successful mating of one of the
females had been confirmed by the presence of a vaginal plug.
In some 5,000 trials of this type, it transpired that males are
typically more prone to mate with females of one of the H-2 types
than with females of the other. Thus it was established that mice
are able to sense each other's H-2 types (Yamazaki et al., 1976;
Andrews and Boyse, 1978; Yamaguchi et al., 1978; Yamazaki et al.,
1978), most probably by olfaction, although proof of this came
later (Yamazaki et al., 1979). Secondly, these studies indicated
that the sensing of H-2 types played a significant part in the choice

of a mate, and this potential could influence the genetic structure of the species.

Ways in which this recognition system might act to the advantage of the species can be illustrated by a simple hypothetical example. In most cases the mating bias was towards matings between males and females of different H-2 types. Of the many genes included in the chromosomal region known as the MHC are several that affect the level of immune responses to particular antigens. These are known as Ir (Immune response) genes, and each is generally dominant for high response to a particular antigen. The mouse that is heterozygous in the H-2 region should therefore have a larger repertoire of high responses to antigens than a mouse which is homozygous for H-2, and should therefore have a wider range of effective immunity. Thus heterozygosity for H-2, fostered by a propensity to mate with partners of a different H-2 type, could be particularly important in a species like the mouse, where inbreeding is a special hazard because of the cyclic derivation of large populations from a few pairs or even a single pair.

Although this experimental model, which we may call the 'mating preference system', was important in implying a natural reproductive bias based on genetic differences, it involves highly laborious procedures, and is ill suited to analysis of the sensory communication system involved. For that reason we began to investigate the use of a Y maze whose two arms are differentially scented by air currents passed through chambers containing congenic mice of different H-2 types. It was decided to use only male mice as the source of odors in the two odor chambers, and these mice were drawn from a large pool according to an appropriate system of randomization to eliminate non-genetic individual variation. The mice to be tested in the maze were deprived of water for 23 hours, the inducement to run the maze being a drop of water.

Four mice of the C57BL/6 (B6) strain were trained in this manner in the Y maze, two females and two males (strain B6 has the H-2 type H-2^b. The males of the congenic strains used as sources of odor were of strain B6 (H-2^b) and of the congenic strain B6-H-2^k, the latter differing from the former only in the small segment of chromosome 17 in the region of H-2. Two of the four trained mice, one male and one female, were trained by reward to enter the arm scented by B6, the same H-2 type as their own, and the other male and female were trained by reward to enter the arm scented by the B6-H-2^k congenic odor-source mice. The mice were trained first to distinguish between the two odors juniper and cinnamon. Next they were trained to distinguish between the odors of B6 and AKR (H-2^k) mice. These are two unrelated inbred strains, and therefore differ at probably hundreds of loci in addition to H-2^b and H-2^k (Boyse, 1977). When they exhibited highly significant proficiency in this distinction, they were then trained for the final distinction, between B6 and B6-H-2^k mice, in which the difference is confined to the region of H-2.

The success of this study (Yamazaki et al., 1979) established a number of points of interest:

(1) The circumstances of the experiment strongly indicate that olfactory recognition is the only necessary sense required for the distinction of H-2 types.

(2) Without preliminary training, the subject mice showed no H-2-associated discrimination in the Y maze. Thus the H-2-associated bias which mice exhibit in mating, assuming this also to be mainly olfactory, is not a behavioral response to the isolated sensory perception of H-2 types, but response to a sensory input that includes not only odors related directly or indirectly to H-2 but also sensory information received in the circumstances of mating. If it were not so, untrained mice would show the same H-2 bias in the Y maze as in mating. In other words, as might be expected in a highly evolved species like a mammal, the perception of prospective partners as individuals, as in fact genetically different individuals, is one item in the total set of information that determines the response. Indeed it is to eliminate as far as possible the many variables incidental to mating, interesting though these may be in their own right, that the Y maze system is being exploited, because it lends itself better to analysis of H-2-associated olfactory recognition in isolation.

(3) The fact that mice can be trained to discriminate H-2 types in the Y maze, and moreover that they can be trained to select the MHC type that is discordant with their natural mating preference, adds to the argument that response to H-2-associated olfactory output is in no sense the sort of automatic reaction implied in the stricter definitions of a pheromone.

The successful use of the Y maze for demonstrating discrimination of H-2 types has led to the use of the maze for investigating the source of H-2 identifying odors. For this purpose materials have been substituted for living mice in the odor chambers through which the two currents of air are conveyed to the mouse. Urine has proved highly effective as an odor source, and the performance of the trained mice in distinguishing H-2 types in the Y maze has been as high as their performance with living mice as odor sources. This finally disposes of the possibility that forms of communication other than olfaction, of which ultrasound is an example (Whitney et al, 1973), constitute a necessary feature in the recognition of individuality according to H-2 type.

To revert to the main topic of this paper, which concerns ways in which MHC-associated sensory communication may affect processes of reproduction, thus affecting the genetic composition and hence the fitness of the species, we shall mention some further ideas which are the subject of present study. Analysis of the data from the original mating preference study, in which the male is presented with H-2 congenic females, emphasizes the active role of the male rather than of the female in the choice between the two alternative possible matings (Yamazaki et al., 1976). This places the male in the role of recipient of the alternative signals, and the active responder to them. But the more prominent role of the male may well be a feature of the experimental design. Comparable test systems in which a single female is offered a choice of alternative males have been devised (Mainardi, 1964; Yanai and McClearn, 1972), but were never tested in the setting of isolated H-2 disparity. So it is an open question to what extent the female would be an active participant in H-2-associated mating preference under natural conditions. It seems unlikely that a device as evidently powerful as MHC-associated sensory communication would be limited to the mating choice of males. Moreover, with communication in the reverse direction, the female recognizing and responding to the H-2 type of the male, effects more far-reaching than mating enter the picture, because at zygosis the male's active participation in reproduction is ended, whereas the female's has just begun. Nor is the embryo entirely shielded from the effects of chemical sensory information, and the embryo remains susceptible to endocrine disturbances which can be initiated in the mother by chemical signals. By far the best example of such neuro-endocrine perturbation resulting in termination of pregnancy goes by the name of its discoverer, Bruce. The phenomenon described by Bruce (1959) is the abortion of pregnancy in the preimplantation period, which lasts from 5 to 6 days of gestation in the mouse, brought about by exposure of the pregnant female to a strange male, particularly one of a different strain from the stud male. Abortion is attributed to an endocrine disturbance which is said to be reversed by administration of prolactin or progesterone (Dominic, 1966b), and the agent that initiates the neuro-endocrine disturbance that is inimical to implantation is considered to be a pheromone present in male urine, because blocking of pregnancy can be effected with bedding or urine from a foreign male (Parkes and Bruce, 1962; Dominic, 1966a).

While the impact of pregnancy-blocking on population structure in wild mice can be a matter only for speculation, there is at least the potential for selection of particular genotypes of the kind implied by mating preference. It is well recognized that genetics has a hand in pregnancy block, because inbred strains of mice differ in the potential of males to block pregnant pre-implantation females, and females of different inbred strains differ in their susceptibility to blocking (Chapman and Whitten, 1968).

The question that we should ask, in the context of the present discussion, is how genetic diversity may account for these strain differences, because behind this lies the deeper question whether blocking of pregnancy is in fact a device for maintaining genetic diversity. Production of the blocking pheromone is a testosterone-

dependent male characteristic (Hoppe, 1975), and resting levels of
serum testosterone have been shown to bear a relation to H-2 types
(Ivanyi et al., 1972). We have a few preliminary data indicating
that the incidence of pregnancy block in susceptible females may
be greater when the blocking male differs from the stud male at
the H-2 locus than in the case where the blocking male has the same
H-2 type as the stud male.

Granted these facts, the differences exhibited by various mouse
strains in relation to blocking of pregnancy might be interpreted
largely in terms of genetic variation, stemming from H-2 in the
present case, in controlling the output and reception of pheromone.
But our present knowledge that H-2 confers sensory identification,
probably comprising characteristic sets of odors, exhibited and received by both males and females, permits a hypothesis with a broader
biological base.

We have seen that urine is the source of discriminatory H-2-
associated chemical signals, and that these signals are in fact concerned in mating choice. We have seen also that the response to
these signals depends on the context in which they are received;
thus the inexperienced male has a mating bias based on the H-2 type
of females, but does not exhibit any bias in the Y maze, the latter
requires training. Thus response of the pregnant pre-implantation
female to the urine of a strange male, while dependent in the first
instance on pheromonal recognition that the donor is male, may be
conditioned by reaction to the accompanying chemical information
relating to the H-2 identity of the donor. This hypothesis, which
we are now studying, casts a more subtle light on how the genotype
might determine varieties of responses to a uniform chemical signal.

This work was supported in part by grant CA-23676 from the National
Institutes of Health and by grant RF-77042 from the Rockefeller
Foundation. EAB is American Cancer Society Research Professor of
Cell Surface Immunogenetics.

REFERENCES

Andrews, P.A. and E.A. Boyse. 1978. Mapping of an H-2-linked gene
    that influences mating preference in mice. Immunogenetics 6
    265.
Boyse, E.A. 1977. The increasing value of congenic mice in bio-
    medical research. Laboratory Animal Science 27 771.
Bruce, H.M. 1959. An exteroceptive block to pregnancy in the
    mouse. Nature 184 105.
Chapman, V.M. and W.K. Whitten. 1968. The occurrence and inheri-
    tance of pregnancy block in inbred mice. Genetics 61 49.

Dominic, C.J. 1966a. Observations on the reproductive pheromones of mice. I. Source, J. Reprod. Fert. 11 407.

Dominic, C.J. 1966b. Observations on the reproductive pheromones of mice. II. Neuro-endocrine mechanisms involved in the olfactory block to pregnancy. J. Reprod. Fert. 11 415.

Hoppe, P.C. 1975. Genetic and endocrine studies of the pregnancy blocking pheromones of mice. J. Reprod. Fert. 45 105.

Ivanyi, P., R. Hampl, L. Starka, and M. Mickova. 1972. Genetic association between H-2 gene and testosterone metabolism in mice. Nature 238 280.

Klein, J. 1975. Biology of the mouse histocompatibility-2 complex. Springer-Verlag, New York.

Mainardi, D. 1964. Relations between early experience and sexual preferences in female mice. Atti. Ass. Genet. It. 9 141.

Parkes, A.S. and H.M. Bruce. 1962. Pregnancy-block in female mice placed in boxes soiled by males. J. Reprod. Fert. 4 303.

Whitney, G., J.R. Coble, M.D. Stockton and E.F. Tilson. 1973. Ultrasonic emissions: do they facilitate courtship of mice? J. Comp. Physiol. Psychol. 84 445.

Yamaguchi, M., K. Yamazaki and E.A. Boyse. 1978. Mating preference tests with the recombinant congenic strain BALB.HTG. Immunogenetics 6 261.

Yamazaki, K., E.A. Boyse, V. Mike, H.T. Thaler, B.J. Mathieson, J. Abbott, J. Boyse, Z.A. Zayas, and L. Thomas. 1976. Control of mating preference in mice by genes in the major histocompatibility complex. J. Exp. Med. 144 1324.

Yamazaki, K., M. Yamaguchi, P.W. Andrews, B. Peake, and E.A. Boyse. 1978. Mating preference of F_2 segregants of crosses between MHC-congenic mouse strains. Immunogenetics 6 253.

Yamazaki, K., M. Yamaguchi, L. Baranoski, J. Bard, E.A. Boyse, and L. Thomas. 1979. MHC-associated recognition among mice: Evidence from the use of a Y maze differentially scented by congenic mice of different MHC types. J. Exp. Med. submitted.

Yanai, J. and G.E. McClearn. 1972. Assortative mating in mice, I. Female mating preference. Behavior Genet. 2 173.

BEHAVIORAL AND STIMULUS CORRELATES OF VOMERONASAL FUNCTIONING IN

REPTILES: FEEDING, GROUPING, SEX, AND TONGUE USE

Gordon M. Burghardt

Department of Psychology
University of Tennessee
Knoxville, TN 37916

Whatever the reason for the general neglect, until recently, of the vomeronasal organ (VNO), its associated structures, and the behavioral significance of the system, we are now beginning to understand its widespread function in reptiles. Studies of the system are hindered, however, by a lack of anatomical, neurological, and neurophysiological information. But a start has been made and the primary purpose of this paper will be to review recent advances in our knowledge of the function of the reptilian vomeronasal system and to present some new data and speculation. Rather complete reviews are available (Burghardt, 1970a; Madison, 1977) that cover early experimental studies on the role of the chemical senses in reptile behavior.

NASAL MORPHOLOGY

A most thorough review of nasal anatomy in reptiles and other vertebrates is provided by Parsons (1970, 1971). The VNO is absent from crocodilians except for early embryonic stages, present in a relatively unspecialized form in turtles, is a distinct organ still associated with the nasal cavity in the sole living representative of the Rynchocephalia, and is most highly developed in squamate reptiles, the lizards and snakes. Here it is typically mushroom shaped and separate from the nasal cavity, opening into the mouth via a duct in the palate. In snakes, it is present in an advanced and prominent state in all species studied. In the lizards, its degree of development varies between and within different families and genera. Unlike the olfactory epithelium, the vomeronasal sensory cells in reptiles and mammals appear to not possess cilia, although apparent precursors may be present (Graziadei and Tucker,

1970; Kolnberger and Altner, 1971). Wang and Halpern (in press) provide a detailed examination of VNO fine structure in garter snakes (<u>Thamnophis</u>) and an excellent literature review.

NEUROANATOMY

Aspects of the neuroanatomy of the system are reviewed and added to by Halpern (1976) and Northcutt (1978). All lizards and snakes possess a main olfactory bulb that receives nerves from the olfactory epithelium. The main bulb projects to a variety of ipsilateral and contralateral areas such as the anterior olfactory nucleus, olfactory tubercle, and lateral cortex. Nerve endings from the VNO remain separate from the olfactory system and terminate in the accessory olfactory bulb (AOB), which is posterior to the main bulb. The afferents of the AOB project exclusively to the nucleus sphericus on the same side (Halpern, 1976; Kowell, 1974; Northcutt, 1978). The reptilian nucleus sphericus is often considered homologous to nuclei in the mammalian amygdala. Thus the tracing of olfactory and AOB projections beyond the bulb has demonstrated that far from being an "accessory organ" the vomeronasal system (VNS) has CNS pathways that remain discrete (see also Moulton and Beidler, 1967; Johns, this volume; Meredith, this volume).

NEUROPHYSIOLOGY

Recording of VNO responses has been successful in turtles (Tucker, 1963; Graziadei and Tucker, 1970; Tucker, 1971) and in lacertid lizards, where the VNO is a blind sac (Altner and Müller, 1968; Müller, 1971). These studies demonstrated differential responses between olfaction and the VNO using pure reagents with unknown biological meaning. In the turtles and lizards tested, the VNO responds readily to most, but not all, stimuli that activate olfactory receptors. The studies also agree that lower molecular weight fatty acids and alcohols are more effective in the VNO than longer homologues. While no primary receptor recordings in snakes appear to have been published, Meredith and Burghardt (1978) did find evidence for AOB firing to earthworm extract in garter snakes, <u>Thamnophis</u> <u>radix</u>, earthworms being a favorite food of this species. This is the first recording in a reptilian VNS above the primary receptor cell.

VNO STIMULATION: THE ROLE OF THE TONGUE

Since the squamate VNO opens into the roof of the mouth, it has long been thought that the frequently protruded tongue picks up chemical stimuli and transfers them to the organ. By 1950 it

had been shown that amputation of the tongue reduced or eliminated
several responses in snakes and lizards dependent upon chemical
cues. The common textbook statement that the bifid tips of the
tongue are actually inserted into the lumen of the organ was dis-
proved by the fact that in many lizards the opening was of smaller
diameter than the tongue tips, while in snakes severing of only
the tips barely decreased responding (Wilde, 1938). Thus while
insertion may be possible in snakes, it is not necessary. The
temporal and spatial constraints on such insertion in rapidly
tongue flicking snakes also argue against this. Since Wilde (1938)
severed the vomeronasal nerve (VNN), we knew that VNO stimulation
had to occur, and was in fact critical, in feeding of Thamnophis
sirtalis. Sheffield et al. (1968) showed in the same species that
newborn snakes not only tongue flicked prior to prey attack at
either earthworms or water surface extracts, but that tongue contact with the prey actually always took place. But in my review
in 1970, I could point to no evidence that conclusively demonstrated
chemical input to the VNO via the tongue or any other mechanism.

In the last few years the situation has changed. Burghardt
and Pruitt (1975) showed that virtually complete removal of the
tongue was necessary to abolish feeding in newborn garter snakes.
Recovery was possible if any stub could protrude from the jaws and
a completely tongueless adult continued to find and eat earthworms,
albeit abnormally, for five years until her apparently natural
death. Such evidence was still indirect, but the distaste in our
laboratory for tongue-severing experiments led to our abandoning
this approach; a tongueless snake, particularly a baby, is not
only not a "real" snake, its lack of feeding soon makes it no snake
at all.

The careful and ingenious work of John Kubie (1977) and others
working in Mimi Halpern's laboratory soon gave us the evidence
needed. They showed that if a garter snake (Thamnophis radix) was
induced to tongue-flick at earthworm extract laced with tritium
labeled proline and later perfused, radioactivity was concentrated
in the VNO epithelium. Although Kahmann (1932) performed a similar
experiment with carbon black and had similar results, there were
possible artefacts and it was not replicated.

Kubie also devised a method for reversibly eliminating the
VNO. Tongue-removal, VNO cauterization and nerve section were
not the way, even though the VNN does show some regeneration.
Kubie (1977) was able to suture together the lateral ridges of the
fenestra vomeronasalis into which the vomeronasal ducts open.
This abolished prey trailing ability, for example, yet was reversible.

Recording from the AOB in T. radix Meredith and Burghardt
(1978) demonstrated that normal tongue flicking at prey or prey

extracts resulted in more firing than controls; EMG recording from
the tongue retractor muscle was necessary for precise timing of
stimulation. Distel (1978) found that in the green iguana (Iguana
iguana) electrical stimulation of the nucleus sphericus, the target
of fibers from the AOB, as well as other structures elicited tongue
protrusion under diverse circumstances. Taken together these
studies show conclusively that the tongue-VNO relationship is
highly important, if not essential, for adequate VNO functioning
in the species investigated.

FUNCTIONAL SIGNIFICANCE

     Because of the lack of a clear separation between air sampled
for nasal, trigeminal, and vomeronasal stimulation in turtles, no
clear behavioral studies on the specific role of the VNO in turtles
have been performed. In snakes, however, the evidence is growing
that the VNO or, indirectly, tongue flicking, is involved in many
activities including trailing, recognizing, striking, and swallowing
prey, predator recognition, courtship, maternal care, aggregation,
exploration, and habitat selection (references in Burghardt, 1979a;
Chiszar and Scudder, this volume; Chiszar et al., 1976; Weldon and
Burghardt, 1979; and below).

     While all snakes seem to rely on the VNS a great deal, in
lizards we find a great deal of diversity in both structure and
function. Thus, lizards provide ideal material to study the behavioral and ecological correlates of degrees of development in
the VNS (Northcutt, 1978). This means gathering information about
the proximate mechanisms underlying the performance of activities
as well as correlating, across taxa, behavior and ecology with gross
and fine-grained features of VNO anatomy and CNS projections. Before broaching some beginnings in this area, I will review some
recent work indicating the pervasive importance of the VNS in
snakes.

Feeding and its Ontogeny

     The critical importance of chemoreception and the tongue-
vomeronasal system for feeding behavior in snakes is secure,
although the story can be complex in arboreal snakes (e.g.,
Burghardt, 1977b), as well as vipers and pit vipers (see Chiszar
and Scudder, this volume). Here I will chiefly mention some of
the newer developments in work on natricine snakes in the family
Colubridae.
     Using adult Thamnophis radix, Kubie and Halpern (1979) repli-
cated and extended Wilde's (1938) study and demonstrated through
meticulous surgery and histological verification of degeneration
that the VNS is critical in snake feeding. Olfaction seems to play

no role but both VNO nerve cuts and duct suture led to loss of
trailing ability, lowered tongue-flick rates, and eventual refusal
to eat. Halpern and Frumin (1979) presented T. radix with
aqueous surface prey extracts on cotton swabs (see Isolation of
an Effective Volatile for similar testing method) and, eliminating
conditioning possibilities, showed through VNO nerve cuts that the
VNS was essential. Olfactory nerve cuts did not reduce prey
attacks or stimulus discrimination; the latter actually improved,
possibly by reducing the amount of "noise-induced" flicking to
distilled water control swabs. Kowell (1974) has established an
important role for the VNS in other Colubridae (Lampropeltis) and
the Boidae (Boa).

    Working on feeding in newborn, ingestively naive snakes con-
tinues. Species and populational differences in response to prey
surface extracts were shown years ago (Burghardt, 1970b), and more
recently, polymorphic chemical prey preferences were demonstrated
within litters in two species (Burghardt, 1975; Gove and Burghardt,
1975). Arnold (1977, 1979) has documented geographic variation in
both actual food choice and chemical cue preferences based on tongue-
flick rates in West Coast garter snakes, Thamnophis elegans.
Importantly, the actual availability of prey at different sites was
known and the food and chemoreceptive preferences were correlated
and largely matched sympatric occurrence even in inexperienced
snakes. Carr and Gregory (1976) tested young of three species of
Thamnophis found on the same island and found differences among
them.

    The modifiability of chemical cue preferences has led to some
exciting findings. Fuchs and Burghardt (1971) found that prey
attacks and tongue flicking of newborn eastern T. sirtalis increased to extracts prepared from fish or earthworms when the snakes
were fed these prey and declined when they were not. Aversions were
easily induced in one trial (Burghardt, 1969; Burghardt et al.,
1973). Arnold (1978) performed a series of experiments with a
West Coast population of this species utilizing fish, frog, and tad-
pole prey and extracts. Here the results of Fuchs and Burghardt
(1971) were replicated with respect to fish, but the responses to
frog extract were high throughout the experiment regardless of diet,
indicating that experience and innate preference might interact
differentially within the same species.

    While Fuchs and Burghardt (1971) and Arnold (1978) found no
evidence for food imprinting in neonates (interpreted as a special
salience of the first feedings on prey preference) Gove and
Burghardt (1975) found that in contrast the food preferences in
adult Nerodia (= Natrix) sipedon were not readily modified by their
current diet.

Mushinsky and Lotz (submitted) looked at responses to prey extracts in newborn water snakes after detailing the food habits of several water snake species found in a Lousiana swamp. They fed restricted diets to newborn broad-banded water snakes, Nerodia fasciata, and yellowbellied water snakes, N. erythrogaster, and periodically tested prey extract responses. The results confirmed that experience acts differentially on different prey preferences and suggested maturational ontogenetic sequencing of modifiability. N. fasciata eats primarily fish (78%) in the wild; naive snakes as well as those reared on all fish or all frog diets for 6 months preferred fish extracts. At 8 months of age more variability was seen with the fish-reared group showing no preference, the mixed diet group still preferring fish, and the frog-diet group tending to prefer frog. In N. erythrogaster, which eats primarily frogs (85%) in the wild, a different picture emerged. Newborns showed no preference between fish and frog, two to six month old snakes preferred fish regardless of diet, and at 8 months both groups shifted to a frog preference. Further field work suggests that, in contrast to adults, small snakes of this species actually do eat fish primarily and shift to frogs at about 40 cm body length.

Drummond (1979) looked at aquatic predation and its stimulus control in young but experienced N. sipedon. He found that fish models in fishy water are readily attacked, as are pebbles coated with stickem (a nondrying glue) and fish mucus. Visual cues of prey movement are also very important, however, and in some circumstances models were attacked in the apparent absence of all chemical cues. Thus the importance of considering modality inter- action and sequencing cannot be ignored in studies of snake feeding (see also Burghardt, 1969; Chiszar and Scudder, this volume).

Grouping or Aggregation

Finding temperate snakes in groups, clumps or aggregations outside the hibernation and mating seasons is common (Noble and Clausen, 1936; Dundee and Miller, 1968). The above two cited papers also report some laboratory experiments. Noble and Clausen (1936) concluded that vision was the most important modality, but their experiments were flawed. Dundee and Miller (1968) tested adult ringnecked snakes (Diadophis punctatus) in a circular arena with 10 identical cover objects and showed that the snakes did not distribute themselves randomly but favored locales developed by one group of snakes that were also preferred spots for newly introduced animals. Adequate controls demonstrated that the response was mediated by some deposited substance. Our experiments have found similar grouping in newborn snakes in some, but not all, species of Thamnophis, Storeria, and Nerodia tested (Burghardt, 1977a; Burghardt and Spottswood, in prep.). The best evidence for

VNO involvement in grouping comes from a comparison of normal and tongueless newborn T. sirtalis.

Our method was modified from Dundee and Miller (1968) and involved use of a 46 x 106 cm glass aquarium with a gray aquarium gravel substrate. Two Petri dish bottoms containing water were placed along the central long axis of the tank and eight slightly creased 10 x 15 cm paper towels were placed regularly along the outside perimeter; one at each end and three along each side. The temperature was 22-25°C and the relative humidity ranged between 39 and 60 percent. An experiment involving the placement of 20 normal brown snakes (Storeria dekayi) and 20 normal garter snakes (T. sirtalis) into this cage on the day of birth will be recounted. Although some manipulations took place such as stirring the gravel, rotating the tank, and replacing the cover objects, they can be disregarded here.

The position of all the animals in the enclosure was noted 12 times over the first 11 days of life, at least once a day. The number of snakes exposed or under each cover object was recorded as well as any aggregation. The latter was defined as two or more snakes in physical contact, although in groups of more than two, each snake did not have to be in contact with every other one. The snakes were found under some covers significantly more than others (see Burghardt, 1977a) and there were more empty covers and more containing large numbers of snakes than would be predicted. Table 1 gives the frequency distribution for all covers, irrespective of position or day, for the T. sirtalis only. Considering the species separately or together led to the same result: substantial clumping occurred deviating significantly from a Poisson estimate.

The distribution of aggregation sizes, summed over all readings, was also skewed from what might have been expected by chance and was substantiated by comparison with a truncated Poisson distribution (Cohen, 1960) since groups of size zero could neither be counted nor estimated. There were many more snakes found in groups larger than predicted. Therefore, a clumping process is evident in aggregations as well as in distributions under covers. The two species showed partial separation on both measures.

Cheryl Pruitt and I then tested twelve tongueless T. sirtalis in the same situation. The tongues of 12 snakes were removed shortly after birth as described in Burghardt and Pruitt (1975). The snakes were placed in the test enclosure at 16 days of age. Their positions were noted once each day, in the afternoon, for 10 days. One snake died between the fifth and sixth observations. Disturbance to the enclosure was minimized to enhance aggregation or habitat conditioning.

Table 1. Frequency distribution of newborn snakes, <u>Thamnophis</u>
          <u>sirtalis</u>, found under each of eight cover objects.

No. under cover	Normal[a]		Tongueless[b]	
	Observed	Poisson Estimate	Observed	Poisson Estimate
0	39	11.5	22	21.3
1	28	24.4	28	28.2
2	7	25.9	19	18.7
3	8	18.3	6	8.2
4	3	9.7	4	2.7
5	0	4.1	0	0.7
6	0	1.5	1	0.2
7	2	0.4	0	0.0
8	1	0.1	0	0.0
9	0	0.0	0	0.0
>9	8	0.0	0	0.0
	96	95.9	80	80.0

$$G = 151.80 \qquad\qquad G = 4.24$$
$$df = 7 \qquad\qquad\quad df = 5$$
$$p < .001 \qquad\qquad .50 < p < .70$$

[a] 12 observations on 20 animals.

[b] 10 observations on 11-12 animals.

Most of the tongueless snakes were always found under covers (92.2%) but overall use of cover objects was far less skewed than in normal snakes and in only 1 day out of 10 did a non-random distribution occur (G test) while normals showed differential use almost every day. The number of snakes under cover objects, totaled over all 10 readings, was remarkably similar to the Poisson distribution (Table 1). Considering aggregation size, over half of the snakes were found isolated and a truncated Poisson estimate gave values extremely close to the observed (Table 2).

Table 2. Distribution of aggregation sizes in tongueless Thamnophis sirtalis.

Size of Group	Observed Frequency	Truncated Poisson Distribution
1	58	56.0
2	17	20.2
3	5	4.8
4	2	0.9
> 4	0	0.1
	82	82.0

$N =$
$\bar{x} = 1.402$
$s^2 = 0.515$

$\lambda = 0.719$
$G = 1.809$
$df = 3$
$.50 < p < .70$

Thus while tongueless snakes showed some tendency to form groups, the degree of aggregation and selective use of cover objects was far less than in the same species with intact tongues. Both the use of cover objects and aggregation size were predicted remarkably well by the Poisson distribution. Although aspects of procedure differed in the comparisons made here, the common reports of grouping in T. sirtalis under diverse conditions argue against such minor differences accounting for the results and strongly indicate that in T. sirtalis an intact VNS facilitates the gathering and grouping of snakes under the same objects. A variety of mechanism could be posited to explain the aggregations that did occur. These would include 1) a minor role for visual or olfactory cues for distal recognition, 2) residual VNO functioning, 3) random movement until physical contact is made, at which time tactile or other stimuli maintain proximity.

In a series of two-choice preference tests, Porter and Czaplicki (1974) established that laboratory-reared water snakes, Nerodia rhombifera, will tend to avoid the side of a chamber where a con-

specific had defecated, while showing no such discrimination between
a clean area and one soiled by an adult plains garter snake,
Thamnophis radix. Adults of the latter species significantly preferred the side soiled by either themselves or a conspecific. Since
our snakes were not fed prior to or during the experiment, feces are
not the basis for the species preferences and the aggregation we
found with neonates.

Sexual Behavior

Although chemoreception has long been considered an important
modality in male courtship of female natricine snakes, experiments
by Noble (1937) led to his concluding that the VNO was, at most, of
equal importance with olfaction in male courtship. Noble correctly
identified the dorsal skin of females as being the source of
chemicals attractive to males and noted the ubiquitous tongue flicking of this area by males which triggers the courtship sequence.
Devine (1977) gathered more evidence on pheromonal function while
Crews (1976) showed that female sexual attractiveness and male
sexual behavior can be stimulated by hormones. Building on these
results Kubie et al. (1978) found that female T. radix injected
with oestradiol benzoate five days prior to shedding are not
attractive to males until after ecdysis. While the mechanism is
not understood, some change in the chemistry of the skin is almost
certainly responsible. More relevant here is the finding that
uninjected penmates kept with a shedding, injected female became
themselves highly attractive to males and were courted. The
authors conclude that there was transfer of the suddenly released
pheromone to the penmates. Since mass-mating balls of garter
snakes are found in some areas of the country after emergence from
hibernation (e.g., Aleksiuk and Gregory, 1974) such a transfer
process may occur naturally and aid in synchronizing courtship.
Ross and Crews (1977, 1978) found that substances in the seminal
plug deposited after copulation greatly reduce the attractiveness
of females to males.

The absolutely critical importance of the VNO in courtship
and copulation by male garter snakes was established by Kubie et al.
(1978) utilizing the same surgical procedures that demonstrated the
role of the VNO feeding. Unfortunately we still have no information
as to female sensory control of male recognition.

WHY A DUAL SYSTEM? IS VOLATILITY THE ANSWER?

Although the VNS is apparently selective in response to various
chemicals, the preceding evidence demonstrates that in reptiles it
is not basically used for one type of communication or class of
compounds. Thus conspecific, interspecific, and even inorganic

sources may all be utilized. Aspects of intraspecific sources of
signals have been reviewed by Madison (1977) who also gives a con-
cise update of glandular sources (as does Quay, 1977) which are
surprisingly well-developed in reptiles. Eisner et al. (1978)
identified the chemicals responsible for the notorious defensive
odor given by the stinkpot turtle (Sternotherus odoratus) and reviewed other identified glandular secretions in reptiles. On the
face of it, a well-developed VNS in lizards and snakes along with
a more than respectable olfactory system seems redundant. Why
can't a good nose do the job that needs being done?

Graziadei (1977) points out that in the mammalian nose there
are five neural elements that have separate pathways to different
CNS centers. These five components are the main olfactory epithelium, the VNO, the septal organ of Masera, and the terminal endings
of both the nervus terminalis and the nervus trigeminus. Graziadei
states that these five nasal systems show "either an apparent
redundancy in distance chemoreception or a finely tuned apparatus
whose components can detect a variety of subtly different chemical
signals (p. 445-446)." He does not resolve the issue, which also
pertains to reptiles, who have at least four of the systems.
Certainly, however, the main olfactory system and the VNS are the
most prominent in reptiles, and at this stage of our knowledge are
the only ones that can be discussed with at least some functional
data.

A question that can be asked in light of all the recent evidence pointing to a predominant, often exclusive role for the
tongue-VNS in snake behavior is "what does nasal olfaction do?"
It must have some function, given its morphological prominance,
yet present data do not exclude the VNS from postulated olfactory
functions. None of the conflicting predictions concerning differing roles of olfaction and the VNS in general sensitivity and
discrimination ability (e.g., Cowles and Phelan, 1958; Hainer,
1964) have been supported. Our view that olfaction is a distance
arousal receptor mostly, but not exclusively, sensitive to volatiles
with the opposite true of the VNO (Sheffield et al. 1968; Burghardt,
1969) is a relativistic model that can accommodate most data,
although the olfactory end still has little support. In this model
olfactory input may lead to, among other outcomes, the onset of
tongue flicking.

By their very nature non-volatile compounds necessitate close
proximity, if not physical contact, by the organism, perhaps in
solution. If the VNO is a derivative of the aquatic smelling organ
of fish (Broman, 1920), then this makes even more sense. Certainly
the squamate VNO as a mushroom shaped closed sac is superficially
similar to the shark nostril (Parsons, 1971). According to
Graziadei (1977), the VNO lumen is filled with dense mucus, which

"streams towards the only aperture in a direction which is opposite
to the direction that the odor molecules must travel to reach the
receptor surface (p. 446)." Graziadei claims that it is not known
why the VNO "morphological arrangements" are in such contrast with
those thought "to favor the reception of airborne chemical stimuli
(p. 446)." The conflict is resolved if we show that the VNO is not
functioning to detect airborne odors. MacLeod (1971) also specu-
lates from the literature that the VNO senses liquids as a comple-
ment to taste. Beets (1971) on the other hand, suggests that in
studying molecular structure in relation to olfaction the use of
compounds with a molecular weight under 140 should be avoided
because little is known of the vomeronasal and trigeminal responses.
Tucker (1971) concludes that the VNO is probably designed to respond
to substances of low volatility not detected readily by olfaction,
but does, however, allow that species differences could exist.
Further, the neurophysiological evidence reviewed earlier indicates
that the VNO in reptiles responds best to the lower molecular weight
compounds in some common chemical series. What may be involved in
some of these contradictions is a confusion between molecular weight,
volatility, water solubility, and the differential sensitivity to
chemicals of different classes. Certainly the marked differences
among serpent species in preferences to prey stimuli and attraction
to conspecifics must entail the over-riding of strict molecular
and volatility aspects by the biological necessity of responding
to certain chemical cues.

Not only has the VNO been implicated in many snake behaviors
but what evidence there is generally points to nonvolatile chemicals
as being the effective ones. If we make the statement that nonvolatiles are the most important, then much of the evidence fits.

1) Tongue contact or close approach is made by lizards and
snakes when responses apparently controlled by the VNO are
involved.

2) Sheffield et al. (1968) found that tongue contact was
necessary prior to chemically elicited prey-attack in naive garter
snakes. The duration of the flick and maximum extension of the
tongue are greatly reduced when objects are touched with the tongue.

3) Kubie and Halpern (1978) found that trailing in garter
snakes of earthworm extracts was dependent upon the tongue being
able to contact the trail. Perforated plastic 1 cm above the trail
that allowed volatiles to escape readily nonetheless lowered choice
point decisions to chance level.

4) Kubie and Halpern (1978) found that snakes trailed dry
trails well above chance levels but with somewhat reduced accuracy
and tongue flick rates. Burghardt (1966) found that dry swabs

previously dipped in prey extracts prepared with water and several other solvents retained their potency, but again with reduced effectiveness. Sheffield et al. (1968) also found that freeze-dried worm extract powder reconstituted with water was still effective. Together these observations suggest that volatiles released in aqueous solutions are at best facilitory.

Although tongue proximity if not actual contact thus seems present in all cases of VNO function analyzed so far, a clear separation of volatile and nonvolatile components has not been made in prey attack. Sheffield et al. (1968) found that increased flicking and prey-attacks were elicited by nonvolatile earthworm extract substances (> 5000 molecular weight) but they were unsuccessful in finding an effective volatile component. Devine (1977) reported that sexual attractants in Thamnophis are "not very volatile" lipids.

Isolation of an Effective Volatile

At Rockefeller University Alan Singer and I tried again to isolate volatile components from the standard earthworm (Lumbricus terrestris) extract. Our reason for attempting this was our conviction that, in spite of the available evidence, there is reason to believe that the serpent (Thamnophis) VNO is sensitive to volatile chemicals. This belief is occasioned by 1) the ubiquitous tongue-flicking in air by snakes, 2) the dispensability of tongue contact for feeding (Kowell, 1974; Burghardt and Pruitt, 1975) and recognition of whether females have been mated by males (Ross and Crews, 1978) and 3) the potency of air passed over prey in olfactometers, resulting in increased activity, tongue-flicking, and even gaping at the inlet tube; tongue excision reduced responding much more than did occlusion of the nostrils (Burghardt, 1977b; Burghardt and Abeshaheen, in prep.).

Chemical surface substance extracts were prepared by the warm water wash method described elsewhere (e.g., Burghardt and Pruitt, 1975). For distillation, 70 ml. of the Lumbricus extract was stirred in a 100 ml. round-bottom flask. A 30 ml/min stream of nitrogen entered the flask about 1 cm above the liquid level and exited through a 15 cm long reflux condensor (maintained at 10°C) and then through a 2.2 mm I.D. by 15 cm long stainless steel tube packed with Tenax porous polymer resin. The flask was heated at 53°C for 5 hr (see Figure 3 in Claesson and Silverstein, 1977).

For behavioral assays, the Tenax-adsorbed distillate was

transferred to a 30 cm by 1.5 mm I.D. glass capillary by heating
the tube containing the resin to 220°C for 4 min in a 5 ml/min
stream of nitrogen that had been passed over brine. The volatiles,
mainly water, were condensed in the attached dry ice cooled
capillary tube. The condensate was then rinsed from the tube with
750 µl of propylene glycol and diluted to 1.0 ml with water or
propylene glycol to make the behavioral assay solution. A control
sample was prepared by the above procedure from 70 ml of distilled
water. The distillate was analyzed by gas chromatography which
confirmed the presence of several volatile compounds by peaks in
the chromatogram.

Young plains garter snakes (T. radix) born to a wild-caught
mother were received at 10 days of age and fed at 5-7 day intervals
solely on Lumbricus. The animals were maintained, in individual
plastic rodent cages with filter tops and a Petri dish cover holding
water, from Sept. 1976 through the tests which took place in May,
1977. During the period of testing the room temperature fluctuated
between 20.5 and 27.5°C but during testing itself was maintained at
24.0 ± 0.5°C. Of 13 snakes available for testing, 10-12 were used
in a given bioassay procedure.

All tests with the extracts and controls were performed by
dipping a 15 cm cotton swab into the stimulus liquid at room
temperature, pressing the excess off against the side of the glass
vial, and introducing it slowly into the animal's cage to approximately 2 cm from its snout and slightly to one side. The snake
was typically in a resting position at this time. Tongue flicks
directed at the swab, as well as total tongue flicks emitted, were
counted for a 30 sec period beginning when the swab was 2 cm from
the snake. An attack towards the swab immediately terminated a
trial. Four series of bioassay tests were run, each on a different
day, involving three to five stimuli (see Table 3). Each animal
was tested, in order, at approximately 17 min intervals. All tests
were performed blind with coded vials. The snakes were unfed for
5-8 days prior to all testing.

Substances used were earthworm volatiles in H_2O and propylene
glycol, Tenax volatiles in H_2O (resin control), distilled H_2O, propylene glycol, standard earthworm extract, and the earthworm extract
after nitrogen flushing (pot residue). Thus a total of seven different stimuli were tested. In Series I, II and IV, each subject received a different ordering of the stimuli, balanced over all positions. In Series III, two snakes received each of the six possible
orders.

Table 3. Responses to various earthworm extract preparations and controls by garter snakes.

Test Series	Response Measures	Standard worm	Pot residue	Resin control	Volatiles in H_2O	Distilled H_2O	Volatiles in propylene glyol	Propylene glycol
I	No. Attacks	10	10	0	1	1		
n = 10	x̄ flicks (to) (prorated)	28.9	22.0	5.5	9.5	12.6		
	x̄ flicks (total) (prorated)	29.7	22.0	13.1	14.7	15.2		
II	No. Attacks	10	10		2	0	0	
n = 10	x̄ flicks (to) (prorated)	21.5	18.0		10.0	7.4	10.0	
	x̄ flicks (total) (prorated)	22.0	22.9		18.6	13.7	20.4	
III	No. Attacks					0	1	0
n = 12	x̄ flicks (to) (prorated)					3.5	8.9	3.7
	x̄ flicks (total) (prorated)					6.6	16.7	9.7
IV	No. Attacks				0	0	0	0
n = 11	x̄ flicks (to)				3.2	1.8	1.9	1.9
	x̄ flicks (total)				9.5	4.4	7.1	4.5

Results for all four series are combined in Table III. When attacks occurred, the number of tongue-flicks was prorated over the remainder of the 30 sec. interval. After comparing all stimuli tested with a nonparametric ANOVA, the hypothesized greater responses to stimuli containing volatiles as compared to controls were tested with the Wilcoxon matched-pairs signed-rank test (Siegel, 1956). In Series I, the standard extract and pot residue elicited attack on every presentation. The other three stimuli did not differ significantly from each other in terms of tongue-flick rates. The two attacks to these other stimuli were by the same individual and followed attacks to the raw extract and pot residue. Thus the volatiles were not effective here.

In Series II, the standard extract and pot residue again elicited attacks at every presentation; the volatiles in H_2O elicited two attacks. The volatiles in H_2O led to significantly greater total tongue flicks than did distilled water alone ($p < .05$). This was also true for propylene glycol volatiles as compared to distilled water ($p < .05$). Unfortunately, a propylene glycol control was not run. This was rectified in Series III. In addition, the standard extract and pot residue were eliminated to remove the possibility that lingering odors or conditioning were affecting responses to the volatiles and controls. Propylene glycol volatiles were more effective than pure propylene glycol in terms of both total flicks ($p < .03$) and flicks directed at the swab ($p < .01$). By Series IV, the reactivity of the snakes had declined markedly. However, total tongue-flicks to volatiles in H_2O were definitely higher than to H_2O alone ($p < .01$). Although the results for the volatiles in propylene glycol as compared to the control were in the same direction, they were not significant.

Throughout Series II, III, and IV, the volatile components resulted in about 50% more tongue flicks than did the controls. The result for Series I is thus anomolous. However, our previous work has shown novelty effects of distilled water on initial tongue flick rates in experimentally naive subjects (Burghardt, 1969; Burghardt and Abeshaheen, 1971). Although we thus showed significant increased responding to volatile prey stimuli, it must be remembered that both olfaction and the VNS were operative here, and the response increase was modest. There could be a substantially higher concentration of volatiles from real prey (which are lost through the use of aqueous extracts) and this should be tested. Much greater response increases to apparently volatile components are found in olfactometer tests where air is passed over living prey. Perhaps in this situation heavier molecules do "take to the air", perhaps as an aerosol, or by dust particles with adsorbed nonvolatiles. Unpublished experiments on species discrimination by skinks in an olfactometer necessitated extremely high rates of air flow (50 m/sec.) for positive results (Duvall, et al. ms.). Together the findings suggest that the issue of volatility is not simple.

TONGUE USE AS A MEASURE OF VNO FUNCTIONING IN COMPARATIVE STUDIES

       Although the morphological development of a sensory system
may give a general insight to the importance of that system in the
ongoing behavior of an animal, it is at best a gross measure, and
in any event, cannot tell us anything about what behavior patterns
the organ is used for, or to what extent, and how the behavior
mechanisms operate. Detailed experiments involving the use of
models, blocking the peripheral organs, and insult to the nervous
system have, we have seen, shown convincingly that the VNO is important in much snake behavior. Experiments on possible chemical
control of some lizard behaviors (e.g., Ferguson, 1966; Curio and
Möbius, 1978) have been less than convincing, whereas in other
species positive results prevail (see Burghardt, 1970a; Madison,
1977). But with well over 5000 species of lizards and snakes, a
complete experimentally based picture of VNO involvement for
comparative and evolutionary purposes is impractical. Also the
rarity, as well as the sensitivity of many species to captive
conditions, make experimental work arguable on conservation and
ethical grounds. Furthermore, experimental studies of sensory
control, even when well done, are hard to compare across species
because of procedural differences. Probably the major contribution
of my early studies on prey chemical preferences in snakes was
the ability to compare the responses of animals tested on similar
series of prey extracts prior to any contaminating effects of
captive rearing conditions. But even those studies depended upon the
availability of newborn snakes for testing and assumed a comparable
reliance upon chemical cues, similar adaptability to laboratory
conditions, and the appropriateness of the testing system. By and
large we were fortunate.

       But when one wants to extend the comparative scope to animals
in which similar use of the senses is problematic or detailed
experiment is impractical, an easily recorded overt measure of
sensory involvement is invaluable. Such a measure is rarely available for any sensory system. Thus, even today the relative size of
the eye is often taken as a measure of the importance of vision to a
vertebrate, a tactic highly misleading when, as is usual, the
behavioral evidence is anecdotal. For example, black bears have
small eyes and are considered to have poor vision whereas our
experiments show that they have good color and form discrimination
abilities (Burghardt, 1975; Bacon and Burghardt, 1976). One comparative measure with vision might be gaze fixation; ear movement
and orientation to sounds might work in mammals. But in both cases,
assessing and quantifying this measure would prove most difficult,
and morphological differences such as frontal or lateral eyes and
degree of pinna development and movement would be further complica-
tions. With squamate reptiles, however, there is a potential

measure which provides the opportunity for a unique unobtrusive
measure in assessing VNO function: tongue movement outside the
oral cavity.

The correlation of tongue use with VNO reception has been
established rather clearly in snakes, and I think that a comparable
link will be established in lizards. Tongue flicking is ubiquitious
among all snakes observed to date and shares common characteristics.
In captivity it is also rather easily observed and counted except
in very small and burrowing forms. But even in captivity there can
be difficulties. Earlier this year in Panama, I counted tongue
flicks in sea snakes, Pelamis platurus. Even when I observed a
good sized (50 cm) individual at the water surface with my eyes no
more than 10 cm from its head and with nothing between us, I had
difficulty. Although significantly increased tongue flicking was
found to fish prey, the almost translucent tongue was extended not
only infrequently but not very far. On the other hand, in the
field conditions are often favorable for tongue voyeurism; I have
been able to count tongue protrusions in the lizards Agama agama
and Iguana iguana while Drummond (pers. comm.) has records for
foraging Thamnophis elegans.

The study of tongue movements in lizards and snakes is a
growing area of comparative study. Not only do all snakes have a
highly developed VNO, they have a long, bifid, extensible,
frequently protruded tongue, on which no taste buds reside (Payne,
1945). Picking up stimuli for the VNO is its main function, along
with tactile, reception and visual signaling (Gove, 1978). In
lizards the range of development of the VNO is rather well
correlated with the specialization of the tongue for chemoreceptive
functions (Gove, 1978; McDowell, 1972); gustation via some lizard
tongues cannot yet be ruled out (e.g., Duvall, in press).

Some lizards with a reduced or absent VNO, accessory olfactory
bulb, and nucleus sphericus (Parsons, 1970; Northcutt, 1978), such
as those collectively termed the Iguania (Agamidae, Chameleonidae,
Iguanidae), often protrude the tongue in various contexts and
almost always make lingual contact with, and even lick, the substrate
or object (DeFazio, et al., 1977; Gove, 1978; Duvall, in press;
Burghardt, pers. obs.). Thus observational data suggest that use
of the VNO for nonvolatiles is greatest in species in which the
system is least developed, and thus probably least sensitive.

In all snakes, as well as lizards of the Autarcholglossid
families, usually considered more advanced and from which the snakes
evolved, the tongue is often protruded into the air without touching
anything; especially in snakes it may often be waved vertically
(oscillations) and to a lesser extent laterally. In all lizards,
however, the tongue is used for functions not found in snakes (Table
4) such as drinking, food manipulation, and jaw cleaning (Gove, 1978).

Table 4. Types of tongue movements in families of lizards and snakes (from Gove, 1978).

	Tongue Movements						
	Drink with Tongue	Wipe Jaws with Tongue	Manipulate Food	Touch the Substrate	Single-Oscillation Tongue-Flick	Multiple-Oscillation Tongue-Flick	Defensive Tongue-Flick
Lizards							
Gekkonidae	X	X	X	X			
Iguanidae	X	X	X	X			
Lacertidae	X	X	X	X	X		
Scincidae	X	X	X	X	X	X[a]	
Teiidae	X	X	X	X	X		
Anguidae	X	X	X	X	X	X[a]	
Helodermatidae	X	X	X	X	X		
Varanidae	X	X		X	X		
Amphisbaenia	X	?	?	X	X		
Snakes							
Acrochordidae	?			X	X		
Aniliidae				X	X	X	
Boidae				X	X	X	
Tropidophidae				X	X	X	
Colubridae				X	X	X	X
Elapidae				X	X	X	X
Viperidae				X	X	X	X

[a]Rare

Detailed comparisons across lizard families in terms of the function, extension, and sampling volume of the tongue (Gove, 1978; also in Burghardt, 1977) and the frequency of exploratory tongue flicking (Bissinger and Simon, 1979) support the view that the use of the tongue for sensory purposes is more pronounced in the more chemically dominated lizard families. But I would have us focus on the fact that some types of lingual protrusion are seen in virtually all lizards.

In the more "primitive" families of lizards we see mostly tongue touching, and although some of this may be visual signaling (e.g., defensive tongue flicks, Table 4) or displacement behavior (Greenberg, 1978, Burghardt, pers. obs.) most is undoubtedly related to chemosensory input. In many other lizards and all snakes tongue movements in air are common. As reptiles lack a diaphram and may have difficulty sampling large volumes of air, using the tongue to "go out" to the odor may have evolved instead (Gove, 1978). Thus the specialized morphology and behavior of the snake tongue are secondarily derived from the tongue-touch and lick system used by early lizards. The multiple oscillation tongue flick found almost exclusively in snakes may serve to concentrate airborne stimuli for use by a system basically developed for reception of nonvolatiles. Interestingly, an oscillation series usually ceases when the tongue contacts any object. The wart snakes (Acrochordidae, Table 4) have not yet been noted to emit multiple oscillation tongue-flicks and they are considered primitive, if not the most primitive, living snakes. They are also highly aquatic and my observations on advanced sea snakes indicate that we may find some regression in the tongue-VNO system in completely (or primarily) aquatic forms.

REPTILES AND MAMMALS COMPARED: LICKS, FLICKS AND SNIFFS

In mammals the use of the VNS is apparently restricted to intraspecific chemicals that the animal approaches close to, or may actually take in through the mouth, as in Flehmen (Estes, 1972; Verberne, 1970, 1976). There are also some suggestions of a functional and chemical separation of the olfactory and VNO systems in mammals. Mykytowycz (1977) points out that the latter is particularly sensitive to compounds of low volatility and seems to be "a special receptor for sex hormones (p. 213)." Powers and Winans (1975) and Winans and Powers (1977) found that in hamsters only the olfactory system was involved in sniffing and attraction to vaginal secretions while both it and the VNS were involved in mating behavior. They also postulated a distal-proximal role separation between olfaction and the VNO very similar to my 1969 model on snake feeding. Nonvolatiles and the VNO have been clearly implicated in Guinea pig sexual behavior (Beauchamp, this volume). Urine-induced ovulation in rats has been shown to be greatly facilitated by both a functioning VNO and contact with soiled bedding rather than volatiles emanating from it (Johns, this volume).

The restriction of VNO function in mammals to sexual behavior, however, seems too narrow. For example, Fleming et al. (1979) found that the normal inhibition of maternal behavior in virgin female rats to foster pups is mediated by both the olfactory and vomeronasal systems, with licking solely under control of the VNS. I expect data implicating the VNS in mammalian behavior to become increasingly common and a nonpheromonal role to be uncovered in at least some macrosmotic mammals. Edwin Gould (pers. comm.) has observed bats (Leptonyzteris Sanborni) emitting increased tongue-flicks in the presence of food odor (banana).

I conclude that while the VNS can be sensitive to airborne chemicals, it is specialized for proximal stimuli of low volatility, where its discriminating ability, at least in snakes, is superb. If the primitive stage in lizards was to use the tongue-VNO as a contact chemical sensing mechanism, then licking, common in mammals, is probably the analogous response. The secondarily added and ubiquitous aerial movements of snake and many lizard tongues arose as the organ became more sensitive and may also have a mammalian counterpart. Sniffing seems absent in snakes and rare, at best, in lizards; perhaps this is due to the limitations mentioned earlier. Sniffing in mammals is olfactory and involves rapid inspiration, concentration, and sampling of volatiles. Where Flehmen occurs, the VNS and the tongue become involved, perhaps much more so than olfaction. Flehmen seems to involve liquids, non-volatiles, and the VNO, although there seems to be no reason why the VNO cannot be important in species without Flehmen. I therefore suggest that tongue flicking in reptiles may represent the use of a single system to accomplish sniffing, Flehmen, and licking functions in a most graceful and efficient manner. The takeover of the sniffing function by the VNS may help explain the often greater relative size of the VNS as compared to the main olfactory system in reptiles. In few mammals does the VNS rival or supercede the olfactory system in development, as it does in many squamates. A detailed comparison of the behavioral patterns used by reptiles and mammals to stimulate the olfactory and vomeronasal organs would be most illuminating.

REFERENCES

Aleksiuk, M., and Gregory, P.T. 1974. Regulation of seasonal mating behavior in Thamnophis sirtalis parietalis. Copeia 1974: 681-689.

Altner, H. and Müller, W. 1968. Elektrophysiologische und elektronenmikroskopische Untersuchungen an der Riechschleimhaut des Jacobsonschen Organs von Eidechsen (Lacerta). Z. Vergl. Physiol. 60: 151-155.

Arnold, S.J. 1977. Polymorphism and geographic variation in the feeding behavior of the garter snake Thamnophis elegans. Science 197: 676-678.

Arnold, S.J. 1978. Some effects of early experience on feeding responses in the common garter snake, Thamnophis sirtalis. Anim. Behav. 26: 455-462.

Arnold, S.J. (in press). The microevolution of feeding behavior. In Foraging Behavior (A. Kamil, Ed.). Garland STPM Press, New York.

Bacon, E.S. and Burghardt, G.M. 1976. Learning and color discrimination in the American black bear. In Bears - Their Biology and Management (M.R. Pelton, J.W. Lentfer, and G.E. Folk, Eds.). pp. 27-36, IUCN, new series, No. 40, Morges, Switzerland.

Beets, M.G.J. 1971. Olfactory response and molecular structure. In Handbook of Sensory Physiology, Vol. 4, Chemical Senses Part 1, Olfaction (L.M. Beidler, Ed.). pp. 257-321, Springer-Verlag, Berlin.

Bissinger, B.E. and Simon, C.A. 1979. Comparison of tongue extrusions in representatives of six families of lizards. J. Herpet. 13:133-139.

Broman, I. 1920. Das organon vomero-nasale Jacobsoni-ein Wassergeruchsorgan. Anat. Hefte, Abt. 1, 58: 137-197.

Burghardt, G.M. 1966. Prey attack in naive snakes: A comparative and analytic study on the role of chemical perception. Unpublished Ph.D. dissertation, University of Chicago.

Burghardt, G.M. 1969. Comparative prey-attack studies in newborn snakes of the genus Thamnophis. Behaviour 33: 77-114.

Burghardt, G.M. 1970a. Chemical perception in reptiles. In Communication by Chemical Signals (J.W. Johnston, Jr., D.G. Moulton and A. Turk, Eds.). pp. 241-308, Appleton-Century-Crofts, New York.

Burghardt, G.M. 1970b. Intraspecific geographical variation in chemical food cue preferences of newborn garter snakes (Thamnophis sirtalis). Behaviour 36: 246-257.

Burghardt, G.M. 1975a. Behavioral research on common animals in small zoos. In Research in Zoos and Aquariums. pp. 103-133, National Academy of Sciences, Wash. D.C.

Burghardt, G.M. 1975b. Chemical prey preference polymorphism in newborn garter snakes Thamnophis sirtalis. Behaviour 52: 202-225.

Burghardt, G.M. 1977a. Of iguanas and dinosaurs: Social behavior and communication in neonate reptiles. Am. Zool. 17: 177-190.

Burghardt, G.M. 1977b. The ontogeny, evolution, and stimulus control of feeding in humans and reptiles. In The Chemical Senses and Nutrition. (M.R. Kare and O. Maller, Eds.). pp. 253-275, Academic Press, New York.

Burghardt, G.M. and Abeshaheen, J.P. 1971. Responses to chemical stimuli of prey in newly hatched snakes of the genus Elaphe. Anim. Behav. 19: 486-489.

Burghardt, G.M. and Pruitt, C.H. 1975. Role of the tongue and
    senses in feeding of naive and experienced garter snakes.
    Physiol. Behav. 14: 185-194.
Burghardt, G.M., Wilcoxon, H.C. and Czaplicki, J.A. 1973. Con-
    ditioning in garter snakes: Aversion to palatable prey in-
    duced by delayed illness. Anim. Learn. Behav. 1: 317-320.
Carr, C.M. and Gregory, P.T. 1976. Can tongue flicks be used to
    measure niche sizes? Can. J. Zool. 54: 1389-1394.
Chiszar, D., Carter, T., Knight, L., Simonson, L. and Taylor, S.
    1976. Investigatory behavior in the plains garter snake
    (Thamnophis radix) and several additional species. Anim.
    Learn. Behav. 4: 273-278.
Claesson, A. and Silverstein, R.M. 1977. Chemical methodology
    in the study of mammalian communication. In Chemical Signals
    in Vertebrates (D. Müller-Schwarze and M.M. Mozell, Eds.).
    pp. 71-93, Plenum, N.Y.
Cohen, A.C. 1960. Estimating the parameter in a conditional Poisson
    distribution. Biometrics 16: 203-211.
Cowles, R.B. and Phelan, R.L. 1958. Olfaction in rattlesnakes.
    Copeia 1958: 77-83.
Crews, D. 1976. Hormonal control of male courtship behavior and
    female attractivity in the garter snake (Thamnophis sirtalis
    sirtalis). Horm. Behav. 7: 451-460.
Curio, E. and Möbius, H. 1978. Versuche zum Nachweis eines
    Riechvermogens von Anolis l. lineotopus (Rept., Iguanidae).
    Z. Tierpsychol. 47: 281-292.
DeFazio, A., Simon, C.A., Middendorf, G.A. and Romano, D. 1977.
    Iguanid substrate licking: A response to novel situations
    in Sceloporus jarrovi. Copeia 706-709.
Devine, M. 1977. Chemistry and source of sex-attractant pheromones
    and their role in mate discrimination by garter snakes.
    Unpublished Ph.D. dissertation, University of Michigan.
Distel, H. 1978. Behavioral responses to the electrical stimula-
    tion of the brain in the green iguana. In Behavior and
    Neurology of Lizards. (N. Greenberg and P.D. MacLean, Eds.).
    pp. 135-147, NIMH, Rockville, MD.
Drummond, H.M. 1979. Stimulus control of amphibious predation in
    the northern water snake (Nerodia s. sipedon). Z. Tierpsychol.
    (in press).
Duvall, D. In press. Western fence lizard (Sceloporus occidentalis)
    chemical signals. I. Conspecific discriminations and release
    of a species-typical visual display. J. Exp. Zool.
Duvall, D., Herskowitz, R. and Trupiano, J. Responses of five-
    lined skinks (Enmeces fasciatus) and ground skinks (Scincella
    lateralis) to conspecific and interspecific chemical cues.
    Unpublished manuscript.
Eisner, T., Cowner, W.E., Hicks, K., Dodge, K.R., Rosenberg, H.I.,
    Jones, T.H., Cohen, M., and Meinwald, J. 1977. Stink of
    stinkpot turtle identified: w-Phenylalkanoic acids. Science
    196: 1347-1349.

Estes, R.D. 1972. The role of the vomeronasal organ in mammalian
    reproduction. Mammalia 36: 315-338.
Ferguson, G.W. 1966. Releasers of courtship and territorial be-
    havior in the side-blotched lizard Uta stansburiana. Anim.
    Behav. 14: 89-92.
Fleming, A., Vaccarino, F., Tambosso, L. and Chee, P. 1979.
    Vomeronasal and olfactory system modulation of maternal
    behavior in the rat. Science 203: 372-374.
Fuchs, J.L. and Burghardt, G.M. 1971. Effects of early experience
    on the responses of garter snakes to food chemicals. Learn.
    Motiv. 2: 271-279.
Gove, D. 1978. The form, variation and evolution of tongue-flicking
    in reptiles. Unpublished Ph.D. dissertation, University of
    Tennessee, Knoxville.
Gove, D. and Burghardt, G.M. 1975. Responses of ecologically
    dissimilar populations of the water snake, Natrix s. sipedon,
    to chemical cues from prey. J. Chem. Ecol. 1: 25-40.
Graziadei, P.P.C. 1977. Functional anatomy of the mammalian
    chemoreceptor system. In Chemical Signals in Vertebrates
    (D. Müller-Schwarze and M.M. Mozell, Eds.). pp. 435-454,
    Plenum, N.Y.
Graziadei, P.P.C. and Tucker, D. 1970. Vomeronasal receptors in
    turtles. Z. Zellforsch. 105: 498-514.
Greenberg, N. 1978. Ethological considerations in the experi-
    mental study of lizard behavior. In Behavior and Neurology
    of Lizards (N. Greenberg and P.D. MacLean, Eds.). pp. 203-
    226, NIMH, Rockville, MD.
Hainer, R.M. 1964. Some suggested critical experiments in
    olfactory theory. Annals of the New York Academy of Sciences
    116: 477-481.
Halpern, M. 1976. The efferent connections of the olfactory bulb
    and accessory olfactory bulb in the snakes, Thamnophis
    sirtalis and Thamnophis radix. J. Morphol. 150: 553-578.
Halpern, M. and Frumin, N. 1979. Roles of the vomeronasal and
    olfactory system in prey attack and feeding in adult garter
    snakes. Physiol. Behav. 22:1183-1189.
Kahmann, H. 1932. Sinnesphysiologische Studien an Reptilien I.
    Experimentelle Untersuchungen über das Jakobson 'sche
    Organ der Eidechsen und Schlangen. Zoologisches Jahrbuch
    51: 173-238.
Kolnberger, I. and Altner, H. 1971. Ciliary-structure precursor
    bodies as stable constituents in the sensory cells of the
    vomero-nasal organ of reptiles and mammals. Z. Zellforsch.
    118: 254-262.
Kowell, A.P. 1974. The olfactory and accessory olfactory bulbs
    in constricting snakes: Neuroanatomic and behavioral con-
    siderations. Unpublished Ph.D. dissertation, University of
    Pennsylvania.

Kubie, J.L. 1977. The role of the vomeronasal organ in garter snake prey trailing and courtship. Unpublished Ph.D. dissertation, School of Graduate Studies, Downstate Medical Center, New York.

Kubie, J.L., Cohen, J. and Halpern, M. 1978. Shedding enhances the sexual attractiveness of Oestradiol treated garter snakes and their untreated penmates. Anim. Behav. 26: 562-570.

Kubie, J.L. and Halpern, M. 1978. Garter snake trailing behavior: effects of varying prey-extract concentration and mode of prey-extract presentation. J. Comp. Physiol. Psychol. 92: 362-373.

Kubie, J.L. and Halpern, M. 1979. Chemical senses involved in garter snake prey trailing. J. Comp. Physiol. Psychol. (in press).

Kubie, J.L., Vagvolgyi, A. and Halpern, M. 1979. Roles of the vomeronasal and olfactory systems in courtship behavior of male garter snakes. J. Comp. Physiol. Psychol. 93: 648-667.

MacLeod, P. 1971. Structure and function of higher olfaction centers. In Handbook of Sensory Physiology, Vol. 4, Chemical Senses. Part 1 Olfaction (L.M. Beidler, Ed.). pp. 182-204, Springer-Verlag, Berlin.

Madison, D.M. 1977. Chemical communication in amphibians and reptiles. In Chemical Signals in Vertebrates (D. Müller-Schwarze and M.M. Mozell, Eds.). pp. 135-168, Plenum, N.Y.

McDowell, S.B. 1972. The evolution of the tongue of snakes and its bearing on snake origins. Evol. Biol. 6: 191-273.

Meredith, M. and Burghardt, G.M. 1978. Electrophysiological studies of the tongue and accessory olfactory bulb in garter snakes. Physiol. Behav. 21: 1001-1008.

Moulton, D.G. and Beidler, L.M. 1967. Structure and function in the peripheral olfactory system. Physiol. Rev. 47: 1-52.

Müller, W. 1971. Vergleichende elektrophysiologische Unter-suchungen an den Sinnesepithelien des Jacobsonschen Organs und der Nase von Amphibien (Rana), Reptilien (Lacerta) und Säugetieren (Mus). Z. Vergl. Physiol. 72: 370-385.

Mushinsky, H.R. and Lotz, K.H. (Submitted). Responses of two sympatric water snakes to the extracts of commonly ingested prey species: Ontogenetic and ecological considerations. J. Chem. Ecol.

Mykytowycz, R. 1977. Olfaction in relation to reproduction in domestic animals. In Chemical Signals in Vertebrates. (D. Müller-Schwarze and M.M. Mozell, Eds.). pp. 207-224, Plenum, N.Y.

Noble, G.K. 1937. The sense organs involved in the courtship of Storeria, Thamnophis, and other snakes. Bull. Am. Mus. Nat. Hist. 73: 673-725.

Noble, G.K. and Clausen, H.J. 1936. The aggregation behavior of Storeria dekayi and other snakes with especial reference to the sense organs involved. Ecol. Monogr. 6: 269-316.

Northcutt, R.G. 1978. Forebrain and midbrain organization in lizards and its phylogenetic significance. In Behavior and Neurology of Lizards (Greenberg, N. and MacLean, P.D., Eds.). pp. 11-64, NIMH, Rockville, MD.

Parsons, T.S. 1970. The nose and Jacobson's organ. In Biology of the Reptilia, Vol. 2 Morphology B. (C. Gans and T.S. Parsons, Eds.). pp. 99-191, Academic Press, London.

Parsons, T.S. 1971. Anatomy of nasal structures from a comparative viewpoint. In Handbook of Sensory Physiology, Vol. 4 Chemical Senses, Part 1 Olfaction (L.M. Beidler, Ed.). pp. 1-26, Springer-Verlag, Berlin.

Payne, A. 1945. The sense of smell in snakes. J. Bombay Nat. Hist. Soc. 45: 507-515.

Porter, R.H. and Czaplicki, J.A. 1974. Responses of water snakes (Natrix r. rhombifera) and garter snakes (Thamnophis sirtalis) to chemical cues. Anim. Learn. Behav. 2: 129-132.

Powers, J.B. and Winans, S.S. 1975. Vomeronasal organ: Critical role in mediating sexual behavior of the male hamster. Science 187: 961-963.

Quay, W.B. 1977. Structure and function of skin glands. In Chemical Signals in Vertebrates (D. Müller-Schwarze and M.M. Mozell, Eds.). pp. 1-16, Plenum Press, N.Y.

Ross, Jr.,P.and Crews, D. 1978. Stimuli influencing mating behavior in the garter snake, Thamnophis radix. Behav. Ecol. Sociobiol. 4: 133-142.

Ross, Jr., P. and Crews, D. 1977. Influence of the seminal plug on mating behaviour in the garter snake. Nature 267: 344-345.

Sheffield, L.P., Law, J.H. and Burghardt, G.M. 1968. On the nature of chemical food sign stimuli for newborn snakes. Comm. Behav. Biol. Part A, 2: 7-12.

Siegel, S. 1956. Nonparametric Statistics for the Behavioral Sciences. McGraw-Hill, New York.

Tucker, D. 1963. Olfactory, vomeronasal and trigeminal receptor responses to odorants. In Olfaction and Taste (Y. Zotterman, Ed.). pp. 45-69, Pergamon, Oxford.

Tucker, D. 1971. Nonolfactory responses from the nasal cavity: Jacobson's Organ and the trigeminal system. In Handbook of Sensory Physiology. Vol. 4 Chemical Senses. Part 1 Olfaction (L.M. Beidler, Ed.). pp. 151-181, Springer-Verlag, Berlin.

Verberne, G. 1970. Beobachtungen und Versuche über das Flehmen katzenartiger Raubtiere. Z. Tierpsychol. 27: 807-827.

Verberne, G. 1976. Chemocommunication among domestic cats, mediated by the olfactory and vomeronasal senses. II The relation between the function of Jacobson's organ (vomeronasal organ) and Flehmen behaviour. Z. Tierpsychol. 42: 113-128.

Wang, R.T. and Halpern, M. In press. Light and electron microscopic observations on the normal structure of the vomeronasal organ of garter snakes. J. Morphol.

Weldon, P.J. and Burghardt, G.M. 1979. The ophiophage defensive
    response in crotaline snakes: Extension to new taxa. J.
    Chem. Ecol. 5: 141-151.
Wilde, W.S. 1938. The role of Jacobson's organ in the feeding
    reaction of the common garter snake, Thamnophis sirtalis
    sirtalis (Linn.). J. Exp. Zool. 77: 445-465.
Winans, S.S. and Powers, J.B. 1977. Olfactory, and vomeronasal
    deafferentation of male hamsters: Histological and be-
    havioral analysis. Brain Res. 126: 325-344.

ACKNOWLEDGEMENTS

    The research reported here was supported by NIMH research
grant MH 15707 and NSF research grants BNS 75-02333 and BNS 78-
14196. Work at the Rockefeller University took place during the
tenure of a John Simon Guggenheim Foundation Fellowship. I am
grateful for the collaborative efforts of Drs. Alan Singer and
Cheryl Pruitt. O. Greg Brock provided some of the snakes tested.
Thanks also to those who provided unpublished material. Hugh
Drummond, Doris Gove, Mimi Halpern, John Kubie, and Alan Singer
made thoughtful comments on the manuscript.

THE VOMERONASAL ORGAN AND ACCESSORY OLFACTORY SYSTEM IN THE HAMSTER

Michael Meredith

The Worcester Foundation for Experimental Biology
222 Maple Avenue
Shrewsbury, MA   01545

The vomeronasal organ in the hamster became a source of great interest when Powers and Winans reported in 1975 that vomeronasal nerve damage produced some deficits in male mating behavior but that almost complete destruction of the olfactory epithelium did not. At about the same time Singer, O'Connell and coworkers showed that a pure chemical, Dimethyl disulfide (DMDS), could be isolated from hamster vaginal discharge (HVD), and accounted for much of the attractiveness of the discharge. We were therefore interested in studying the electrophysiological responses of the vomeronasal system to components of hamster vaginal discharge and other chemicals. Adrian (1955) and Tucker (1963) had failed to elicit electrical activity in the vomeronasal nerve and accessary olfactory bulb (AOB) of the rabbit when odors were simply blown over the entrance to the vomeronasal duct. This was also our experience with the hamster, presumably because no odor molecules could reach the receptor epithelium sequestered inside the vomeronasal organ. When the odor can reach the receptors, odor responses can be evoked. Figure 1 shows the response of a single unit recorded in the accessory olfactory bulb to odor blown over the surgically exposed vomeronasal epithelium in the hamster. The diagram at the right shows the experimental setup. At the top left is a peristimulus time histogram for four repetitions of amyl acetate stimulation. These data are replotted below as firing rate above or below prestimulus spontaneous rate (zero on ordinate). There is a clear response to amyl acetate which starts within the first second and peaks within the first five seconds.

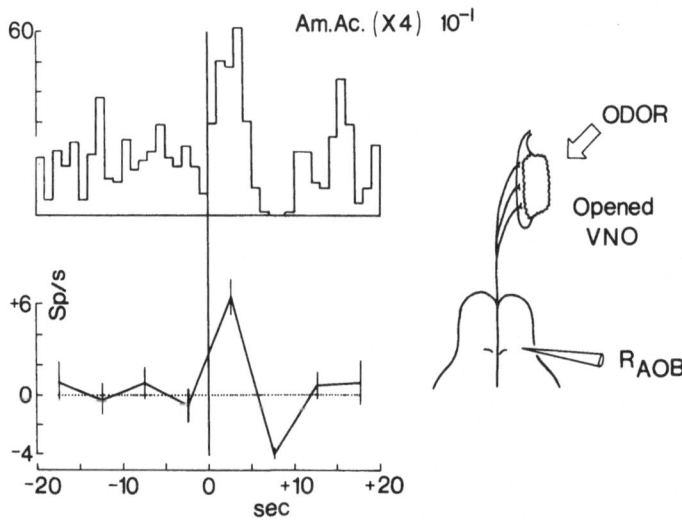

Figure 1. Response of a single unit recorded in the accessory
           olfactory bulb (R_{AOB}) to odor passed over the surgically
           exposed vomeronasal epithelium. Peristimulus time
           histogram (1 sec bins) at top left. The same data are
           replotted below as firing rate above or below the
           average rate for the prestimulus 20 sec (zero on the
           ordinate). Vertical bars are 2 SE of the mean firing
           rate.

As Adrian had previously shown, activity was produced with an
intact vomeronasal organ only by mechanical pressure on the sidewall of the organ. This method probably caused suction of room air
into the lumen when the pressure was released and this may account
for the activity.

Mechanism of Stimulus Access

The question remains, however, how do substances normally get
into the organ, since it consists of a long narrow tube which opens
by a very narrow pore at one end only. Herzfield (1888) suggested
a vasomotor pumping mechanism relying on the cavernous vascular
tissue lying alonside and behind the lumen of the organ. In other
species, especially where the vomeronasal duct opens into the nasopalatine canal, there have been other suggestions for access
mechanisms and these will be discussed later. In the hamster we

have now demonstrated Herzfield's vascular pumping mechanism.   In
this species the organ opens at its anterior end into the nasal
cavity only and has no connection with the nasopalatine canal.   It
consists of a long tube lined on one side with olfactory epithelium
containing the primary sensory neurons (Fig. 2).   The organ is
enclosed inside a bony capsule and enclosed with it are a network
of blood vessels which are capable of constriction and dilation.
The autonomic <u>efferent</u> nerves controlling the vasomotor movements
of these vessels run in the nasopalatine nerve.   The <u>afferent</u> vomeronasal nerves (Fig. 3) contain axons of the receptor cells and end
on the dendrites of mitral cells in the AOB.   The nasopalatine
nerve (Fig. 3) is a branch of the maxillary trigeminal nerve but it
also carries the <u>efferent</u> autonomic fibers.   Stimulating this nerve
causes constriction of the blood vessels in the vomeronasal organ,
which reduces the blood volume within the rigid capsule, causing

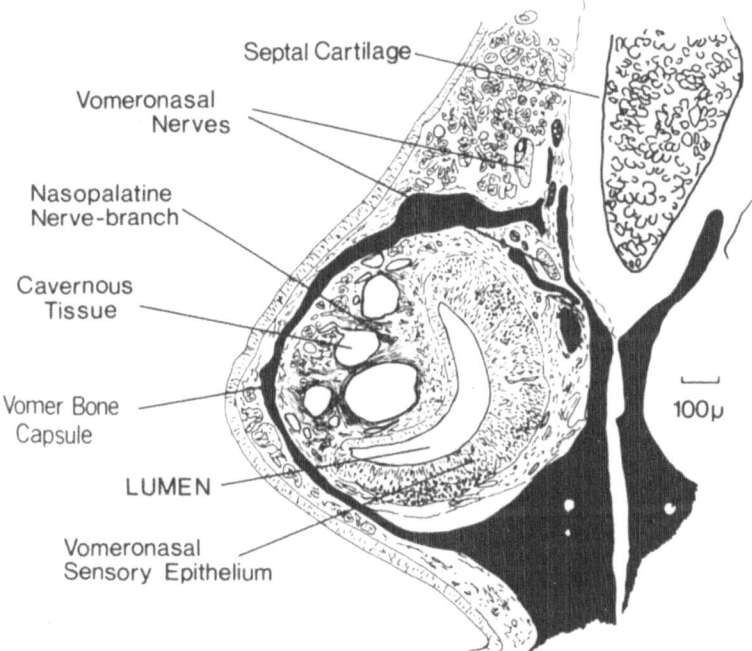

Septal Cartilage

Vomeronasal
Nerves

Nasopalatine
Nerve-branch

Cavernous
Tissue

Vomer Bone
Capsule

LUMEN

Vomeronasal
Sensory Epithelium

100μ

Figure 2.   Camera lucida drawing of a coronal section through the
hamster vomeronasal organ and the base of the nasal
septum.   The fibers running radially between the vessels
of the cavernous tissue had previously been labeled as
elastic fibers on the basis of the descriptions of
previous workers.   This could not be confirmed using
stains specific for elastin, though a longitudinal band
of elastic fibers described by Klein (1881) running
under the non-sensory epithelium was confirmed.

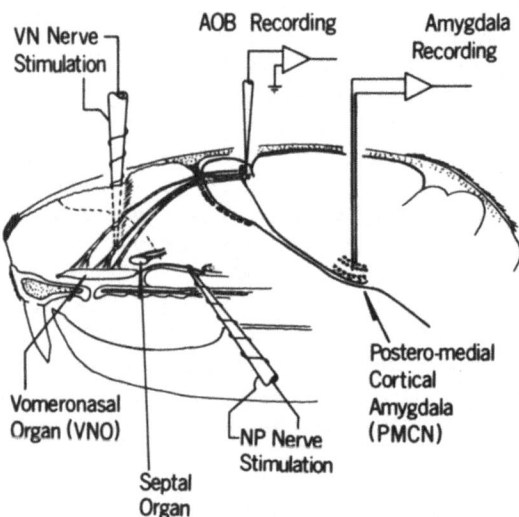

Figure 3. Diagrammatic sagital section through the hamster head
            to show the locations of the vomeronasal organ and its
            anterior pore. The course of the afferent VN nerves
            across the nasal septum and of the efferent NP nerve
            along the dorsal wall of the nasopharynx are indicated
            together with the locations of the suction electrodes
            used to stimulate these nerves. Note that the NP canal
            connecting the nasal and oral cavities is posterior to
            the VN pore. The floor of the nasal groove between
            these points is lined with non-ciliated squamous
            epithelium.

the lumen to enlarge and fluid to be sucked in through the vomeronasal duct. When the blood vessels relax or actively dilate, this
fluid is forced out again through the pore. This fluid movement
can be clearly seen under the dissecting microscope in anesthetized
animals but it cannot be easily measured.

     The suction movements of the organ produced by stimulation of
the nerve were monitored indirectly by recording the movement of
the vomeronasal sidewall using the apparatus shown in Fig. 4. The
anesthetized animal was placed on its back and the vomeronasal organ and the nasopalatine nerve exposed through the palate. A small
window was cut in the vomeronasal capsule so that a probe could
rest on the connective tissue sidewall of the organ. The probe
moved a lever and this movement was detected by a photocell. The
nasopalatine nerve was stimulated with a suction electrode either
with the nerve intact or after cutting.

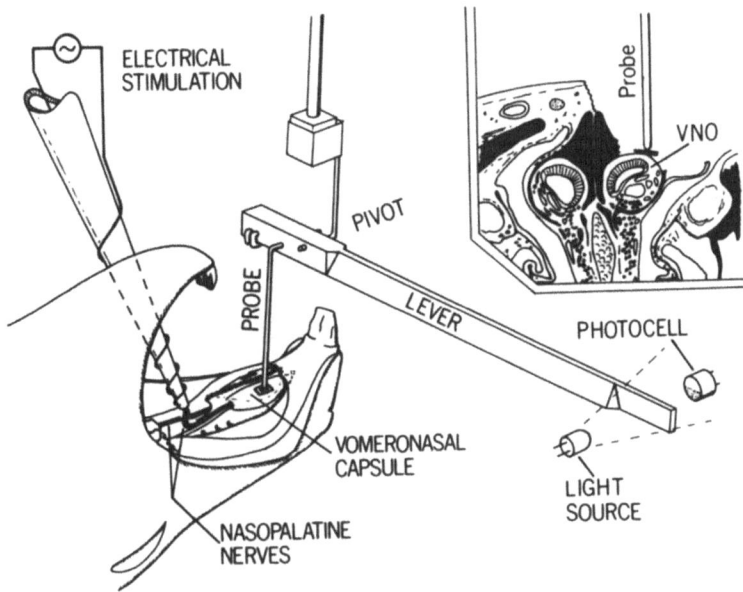

Figure 4. Diagram of the apparatus for measuring vomeronasal organ
response to NP stimulation (see text). The inset shows
in section the probe resting on the vomeronasal sidewall (dorsal is down).

The output of the recording device is shown in Fig. 5. A
typical response to nasopalatine stimulation is shown at the left.
Downward movement corresponds to contraction of the sidewall and
is associated with filling of the lumen. Upward movement corresponds to dilation of the vessels of the sidewall and emptying of
the lumen. The typical time course consists of an initial brief
contraction then a rapid dilation followed by a longer slower
contraction and relaxation back to the baseline.

Actual records are shown at the right with three traces superimposed. The stimulus train consisted of ten pulses, each of one
ms duration, at 30 ms intervals and each of 100 μA. The timing of
the stimulus train is shown below. The time course of the movements appeared to be relatively consistent. Changing the number
of pulses or the pulse current changed the amplitude of the movement but the time course remained approximately the same.

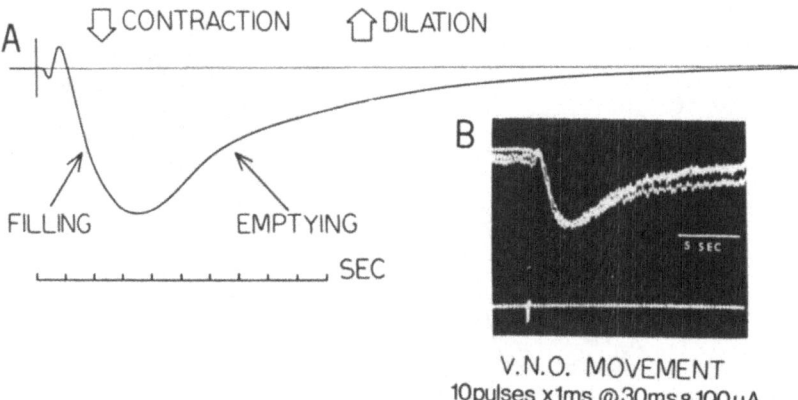

Figure 5.   A.   Typical time course of vomeronasal movement follow-
            ing NP stimulation.   B.   Three superimposed traces of
            actual responses photographed from the oscilloscope.

        To investigate the source of nerve fibers controlling the
vomeronasal movements, two additional preparations were tested.
In one the superior cervical sympathetic ganglion was stimulated
and in the other this ganglion was removed on one side.   Figure 6
shows the effects of sympathetic stimulation.   The experimental
setup is diagrammed at the upper right, the response at the left.
With the normal stimulus parameters a large contraction occurs with
no evidence of an initial dilation.   The contraction precedes the
change in heart rate produced by the sympathetic stimulation (see
EKG trace).   When the pulse train is reduced to one pulse (lower
left) the contraction is produced in the absence of any change in
heart rate.   The movement is therefore unlikely to be due to any
systemic effect on blood flow.   Cutting the sympathetic chain
central to the superior cervical ganglion did not affect the response to stimulation of this ganglion.   After cutting the nasopalatine nerve, stimulation of the peripheral end of the cut nerve
still produced the normal response (lower right), but sympathetic
stimulation now produced only a slow dilation (presumably caused
by the shunting of blood through the vomeronasal vessels due to
constriction of vessels elsewhere in the nose).

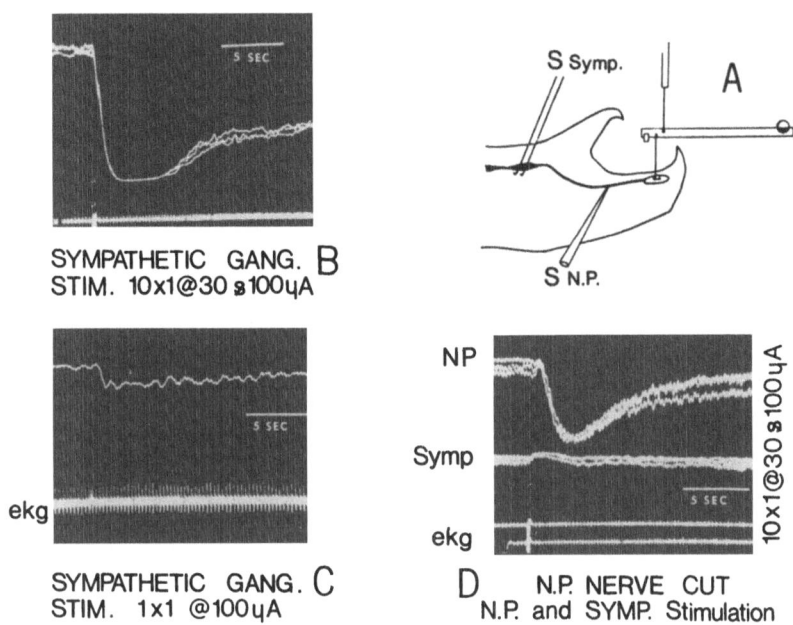

SYMPATHETIC GANG. B
STIM. 10x1@30 s100µA

S Symp.   A

S N.P.

NP

Symp

ekg

10x1@30 s100µA

SYMPATHETIC GANG. C
STIM. 1x1 @100µA

D   N.P. NERVE CUT
N.P. and SYMP. Stimulation

Figure 6.   VN response to sympathetic ganglion stimulation (A) with
a train of 10 pulses (B) or 1 pulse (C). The stimulus
parameters are presented as in Fig. 5, but abbreviated.
(D) The contraction following ganglion stimulation
(S_{SYMP}) but not that following NP stimulation (S_{NP})
disappears after cutting NP nerve central to S_{NP}.

Following unilateral cervical sympathectomy (Fig. 7), six
or more days were allowed for sympathetic fibers to degenerate.
When the appropriate nasopalatine nerves were stimulated a normal
response was produced on the intact side. On the sympathetomized
side, the contraction was eliminated and only the early rapid
dilation movement occurred. These experiments suggest that the
constrictor nerves are adrenergic sympathetic fibers. Injection
of epinephrine into the common carotid artery caused a prolonged
contraction, as would be expected if this were the case.

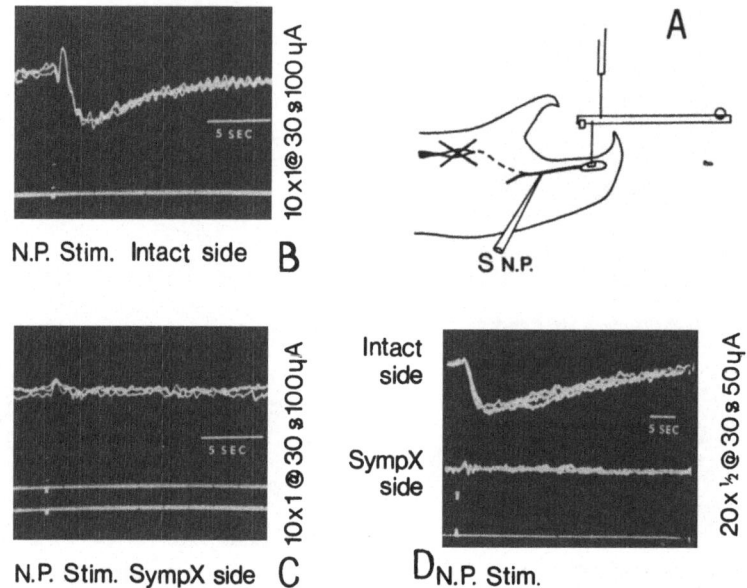

Figure 7. VN response to NP stimulation after degeneration of
          sympathetic fibers on one side.   C. six days after
          sympathectomy the contraction but not the early rapid
          dilation had disappeared on the sympathetomized side.
          D. another animal tested 10 days after sympathectomy
          showed the same result.

     The experiments also suggest that the contraction movement is
independent of the dilation movement (Fig. 8).   The early rapid
dilation tends to reach the same level independently of the degree
of contraction from which it starts.   This is shown for repetitive
nasopalatine stimulation at the left (A).   At the right (B) the
early rapid dilation elicited during an epinephrine contraction
superimposes exactly on the dilation elicited from a different
level of contraction before the epinephrine injection.   It should
also be remembered that sympathetic stimulation produces no early
rapid dilation (Fig. 6, B, C) and that sympathectomy eliminates
the contraction but not the early rapid dilation (Fig. 7 C).

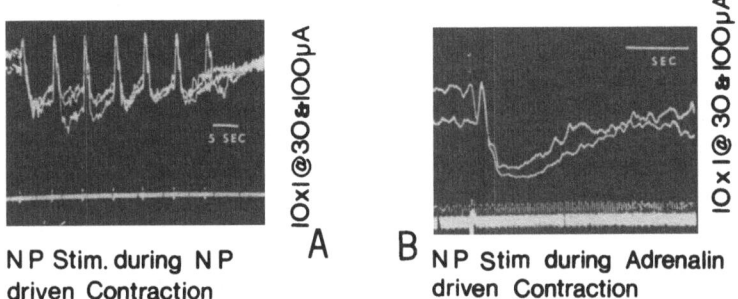

N P Stim. during N P
driven Contraction

N P Stim during Adrenalin
driven Contraction

Figure 8.   Evidence for the independence of the contraction and
dilation movements. See text.

    Vasodilation in the general nasal mucosa is probably controlled
by parasympathetic fibers but the early rapid dilation could not
be blocked by massive injections of atropine, a parasympathetic
blocking agent. Atropine also did not produce noticeable slowing
of the heart, so it is possible that the hamster may be one of
those herbivores which are relatively insensitive to atropine.
The atropine experiments therefore do not rule out the possibility
that the dilator fibers might be cholinergic parasympathetic. It
is also possible however that they could use some other transmitter;
vasoactive intestinal peptide (VIP) for example is present in both
the trachea and the nasal mucosa of several species, it has been
shown to relax smooth muscle and may cause vasodilation (Uddman et
al., 1978).

    The experiments described above show that a pumping mechanism
exists. To show that it is adequate to provide stimulus access to
the organ, the nasopalatine nerve was stimulated to deliver stimulus molecules to the receptor epithelium, while recording the responses of single units in the AOB. The experimental setup is
shown at the bottom right of Fig. 9. The single unit shown at the
top right was driven with almost constant latency with respect to
the field potential evoked by vomeronasal nerve stimulation (S_{VN}).
This suggests that this unit was a second order vomeronasal neurone.
The vomeronasal suction mechanism was operated at approximately 2
min intervals by stimulating the nasopalatine nerve (S_{NP}) while air
or odor was blown continuously over the vomeronasal pore. Peristimulus time histograms for this unit are shown at upper left.
The histograms show activity summed over six repetitions of nasopalatine nerve stimulation during pure air delivery and during
dimethyldisulfide (DMDS) delivery. These data, for both air and

odor, are replotted below on a single axis, as firing rate above or
below prestimulus spontaneous activity. There is a significant
response in both cases but the activity following activation of
the pumping mechanism in the presence of DMDS is clearly greater
than that when only air was present over the pore. The response
with air may be due to odors remaining in the mucus near the pore
since the odor and no-odor conditions were alternated.

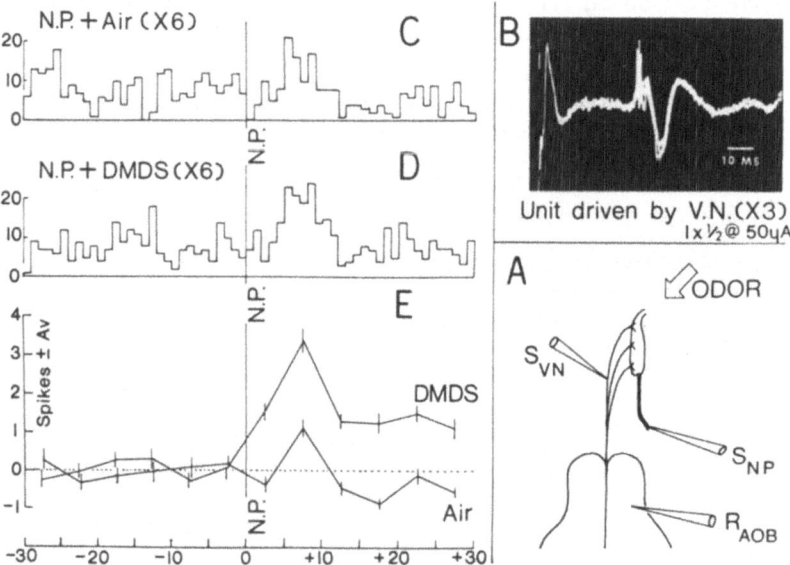

Figure 9.   Response of a single unit recorded in the accessory
             olfactory bulb (R_{AOB}) to: 1, electrical stimulation of
             the vomeronasal nerve (S_{VN}), see B; 2, odor stimulation
             of the vomeronasal organ via suction of odors following
             nasopalatine nerve stimulation (S_{NP}), results shown in
             C-E. The peristimulus time histogram data (C, D) are
             replotted onto one axis in E as firing rate above or
             below the average rate for 30 secs before NP stimula-
             tion.

        Units in the AOB could be excited following suction of odors
into the vomeronasal organ as here, or they could be suppressed;
they could respond to arbitrarily selected odorants such as amyl
acetate as well as to behaviorally relevant substances such as the
sex attractant DMDS. We are not yet in a position to say whether
there is any difference in selectivity or sensitivity between responses recorded in the vomeronasal system and those recorded in
the main olfactory system in the hamster.

There is some evidence that non-volatile substances may be responsible for some behavioral responses mediated by the vomeronasal system (see below). Electrophysiological experiments in other species (Tucker, 1963, 1971; Muller, 1971) suggest a greater response of the VN system to the lower members of homologous series which are usually more volatile. These same substances are also more water soluble and this has dual implications. On one hand such substances might be concentrated in natural secretions and reach the vomeronasal system via routes suitable for truly non-volatile substances. On the other hand the apparent sensitivity to water soluble substances in electrophysiological experiments may be due to their preferential accumulation in the mucus at the proximal end of an airborne stimulus flow path (see Mozell, 1964; Hornung et al., 1978). In Tucker's (1963) experiments on the tortoise this could have resulted in a higher local concentration of these substances over the (anterior) vomeronasal epithelium than over the (posterior) olfactory epithelium. In Muller's (1971) EOG slow potential recordings from Lacerta the VN epithelium gave greater relative amplitudes of response to the more water soluble alcohols but the sensitivity differences between the systems for the other substances were not striking. The mouse VN epithelium did appear to be more sensitive to water soluble substances when compared to the olfactory epithelium of Lacerta but this comparison is difficult to interpret. The relative sensitivity of the vomeronasal and olfactory receptors is therefore not well established in any species, even for the most commonly used odorants.

In any case, the demonstration that a non-volatile or highly water soluble stimulus is necessary for a behavioral response should not be taken as a priori evidence that the vomeronasal system is involved in its detection. Such stimuli may reach and stimulate olfactory receptors or alternatively, they may be detected by the gustatory system or by their systemic effects following absorption into the vascular system.

Alternative Mechanisms of Stimulus Access

In most, if not all, mammalian species with a vomeronasal organ there are large blood vessels enclosed in the vomeronasal capsule. Constriction and dilation in these vessels should produce some pumping action, even though the vessels do not always form cavernous tissue as they do in the hamster. In most species where the vomeronasal duct opens into the nasopalatine canal and particularly in the large ungulates such as antelope, it has been suggested that stimulation of the vomeronasal organ depends on the facial grimace called "Flehmen". Estes (1972) reported that in many species, particularly in the males, this behavior appears after the animal has sniffed female urine. Figure 10, from Estes (1972) shows his interpretation of the stimulation produced during

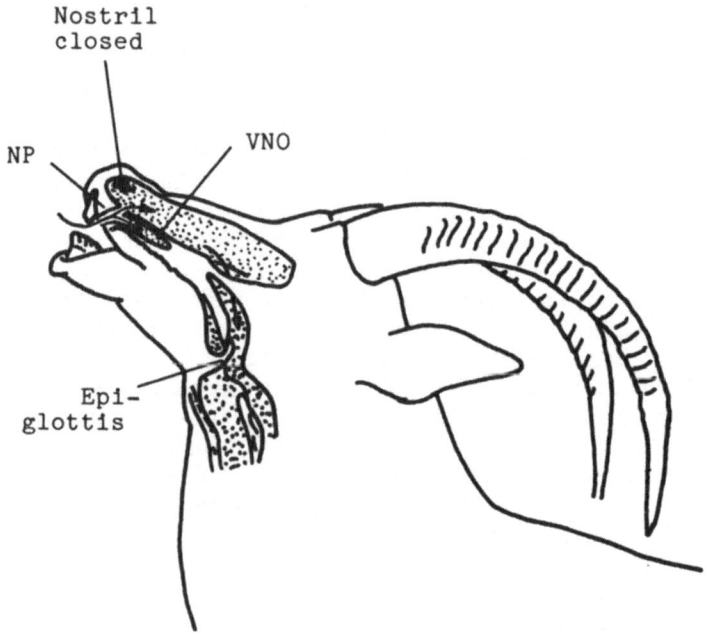

Figure 10. Diagram of Flehmen in a Sable Antelope. (From Estes, 1972).

Flehmen. The animal typically holds its head raised with the upper lip curled back. The nostrils are apparently closed by the lip retraction and the head position should close the epiglottis. Inspiration will therefore cause air to be pulled through the nasopalatine ducts past the vomeronasal entrance. Estes concluded that the venturi effect produced would cause emptying of the VNO, which would then take up some odorized air or mucus when the inspiration stopped and normal pressure in the duct was restored. An alternative hypothesis was proposed for the bat, Phyllostomus by Mann (1961) who observed a retraction of the upper lip during sexual encounters. He suggested that the movement should cause the nasopalatine canal (also known as Stenson's duct) to enlarge because the Stenson's cartilege running along-side the canal was connected to the nasal cartilege and would be pulled up by the lip. Man does not report whether the same maneuver closes the nostrils. Neither Mann's nor Estes' hypothesis has been tested experimentally to see if the movements do in fact deliver stimulus molecules to the vomeronasal organ. The necessary connection between Flehmen and vomeronasal stimulus-access via the nasopalatine canal is not well

established, however. Estes reports some species lacking a patent
nasopalatine connection with the mouth which do show Flehmen and
others where the canal is present and the vomeronasal duct opens
into it but Flehmen is never observed. Verberne (1976) has
shown that interference with the vomeronasal duct or cauterization of the vomeronasal nerves in the cat leads to a reduced incidence of Flehmen, suggesting that vomeronasal stimulation leads to Flehmen rather than vice-versa. It is possible,
of course, that both may occur.

Of all species having a vomeronasal organ, some part of the
mechanism for delivery of substances to the vomeronasal receptor
epithelium has been demonstrated experimentally in only five. These
data are summarized in Table 1. 1) In the turtle, the receptors
are exposed to the nasal airflow (Tucker, 1971) so no special
mechanism is necessary. 2) In the garter snake, stimulus materials
are transported to the roof of the mouth by the tongue. Radiolabelled tracers mixed with stimulus extracts do reach the vomeronasal epithelium and activity can be recorded from the accessory
olfactory bulb following tongue flicks but the mechanism of transport through the vomeronasal ducts had not been established. (For
a detailed comparison of chemoreception in reptiles and mammals,
see the paper by G.M. Burghardt in this volume). 3) In the Guinea
pig, Beauchamp and Wysocki (this volume) showed that a fluorescent
tracer mixed with the urine stimulus appeared in the vomeronasal
organ and in the nasopalatine canal after the animal had sniffed
and licked the stimulus. In preliminary experiments this did not
occur on one side when the nostril on that side was occluded.
This suggests that non-volatile materials may enter the nasal
chamber through the nostril. This route or the NP canal route is
implied in the diagram (4) for the hamster by the arrow connecting
"stim" to "odor-laden mucus". Such an access route is not ruled
out by our demonstration that airborne chemicals can be sucked
into the organ, either directly or after solution in the mucus.
5) In unpublished experiments with Lemur fulvus albifrons Schilling
placed a urine stimulus mixed with an X-ray opaque material on the
tongue of a lightly anesthetized animal. X-ray cinematography
suggested that the mixture was forced into the nasopalatine canal
by the tongue and then sucked up into the VNO lumen.

Neural Pathways

Regardless of the means of stimulus access, it seems likely
that the activity elicited in the receptor cells ultimately excites
a very similar neuronal pathway in all species with an accessory
olfactory bulb (Fig. 11). The vomeronasal pathways up to the third
order vomeronasal areas in the amygdala are completely distinct from
those areas receiving direct main olfactory bulb projections in all

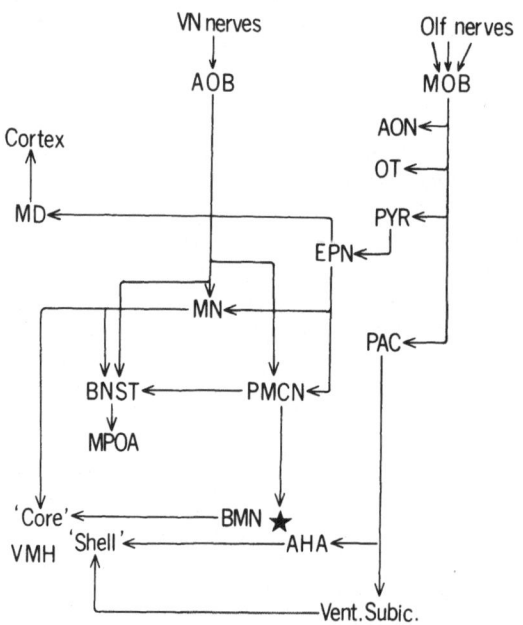

Figure 11. Diagram of the central neuronal projections of the
            olfactory and vomeronasal systems. Taken mainly from
            the descriptions of Kretteck and Price (1978a, b) but
            all projections are not indicated. "Star" explained
            in text. AHA; Amygdalo-Hippocampal Area; AON, Anterior
            Olfactory Nucleus; BMN, Basomedial amygdaloid Nucleus;
            BNST, Bed Nucleus of Stria Terminalis; EPN, Endo-
            pyriform Nucleus; MD Medio-Dorsal thalamic nucleus;
            MN Medial amygdaloid Nucleus; MPOA, Medial Pre Optic
            Area; OT, Olfactory tubercle; PAC Peri-Amygdaloid
            cortex; PMCN, Postero-Medial Cortical amygdaloid
            Nucleus; Pyr, Pyriform Cortex; VMH Ventro-Medial
            Hypothalamus.

species so far studied (Winans and Scalia, 1970; Scalia and Winans, 1975; Broadwell, 1975; Skeen and Hall, 1977; Kretteck and Price, 1978a, b; Davis et al., 1978). More recent work by Kretteck and Price (1978a, b) however shows that in the rat and cat the main olfactory projections may converge with the accessory olfactory pathway at the level of the amygdala. The pyriform cortex, a main olfactory projection area, projects to the endopyriform nucleus and this in turn projects to the medial and postero-medial cortical nuclei, which are accessory bulb projection areas. Kretteck and Price, however, note that the accessory system seems to have more direct inputs to the medial hypothalamus, and thus as suggested by Scalia and Winans (1976) the vomeronasal system has a more direct input to the diencephalic areas shown to be important in sexual behavior.

An additional intra-amygdaloid projection, not resolved by Kretteck and Price has recently been suggested by our electro-physiological recordings in the postero-medial amygdala of the hamster. This is a projection from PMCN to the BMN/AHA region deep to it and is indicated by a star in Fig. 11. Figure 3 shows the experimental setup. The vomeronasal nerve was stimulated electrically and recordings were made in the AOB and in the PMCN. Figure 12 shows a sequence of vomeronasal nerve evoked field potentials on an electrode track (track 3) through PMCN. These are taken at 300μ intervals of depth and each has 3 superimposed traces. In a relatively narrow lamina deep to the superficial PMCN nucleus a negative going potential, the B potential, appears at a latency of approximately 7 msec from the first positive peak. It is followed by a second positive peak, the C potential. All three potential components (A-C) invert as the electrode passes towards the ventral surface (not shown here), suggesting a laminar arrange-ment of elements. Double and repetitive shocks to the vomeronasal nerve at short intervals caused the B and C potentials to disappear leaving a slightly prolonged A potential in isolation. These find-ings and a preliminary, simple form of current density analysis suggested that elements in PMCN are activated by AOB input and that these then activate, synaptically, the deeper elements responsible for the B potential. Both single unit and multiunit recordings suggest a relatively narrow lamina of cells, excited by vomeronasal nerve stimulation, at the level of the peak B potential. The lo-cation of the B potential was marked by passing current through the electrodes. These marks consistently show that the B potential is maximal deep to layer III of the PMCN. Cells in this region; that is, the most posterior part of the baso-medial nucleus (BMN), where it joins the amygdalo-hippocampal area (AHA) project to the ventro-medial hypothalamus (VMH) in the hamster (C. Malsbury, pers. comm.). If they have the same projections as some cells of the AHA in the rat, they may project to the shell region of VMH which does not receive other input from the accessory pathway (Kretteck and Price, 1978a).

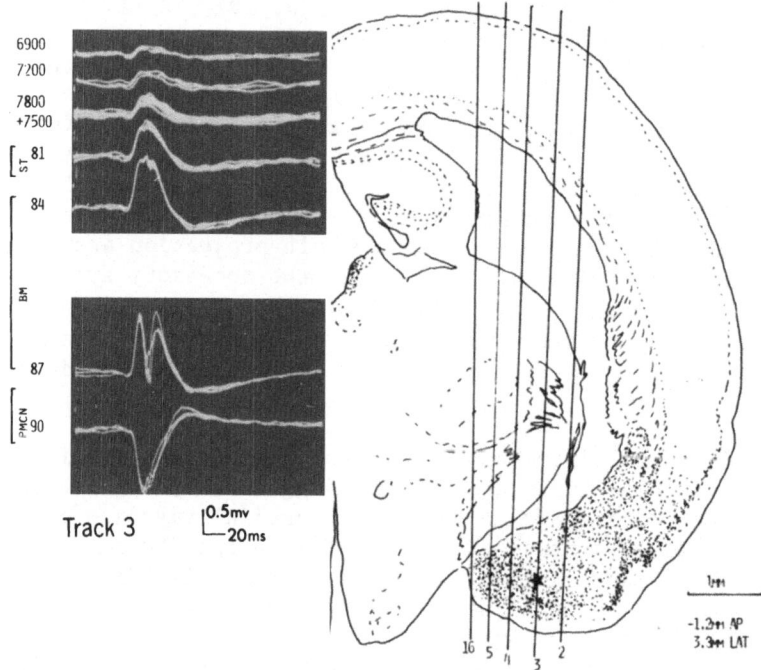

Figure 12. Field potential records from an electrode track (3)
            through PMCN. Depths of successive sets of traces
            and approximate locations of the nuclei on the track
            are shown at the left in µM (zeros omitted from some).
            A camera lucida drawing of the track with the approxi-
            mate location of the peak of the B potential marked by
            a star is shown at the right.

Behavioral Analysis

     The final element of this still very incomplete jigsaw puzzle
returns us to behavior. Powers and Winans (1975; Winans and
Powers, 1977) had shown that lesions of the afferent VN nerves can
cause deficits in sexual behavior in hamsters. We were interested
in extending those experiments to investigate the effect of lesions
of the nasopalatine efferent nerves and to determine the sensory
pathway involved in the attraction of male hamsters to female
hamster vaginal discharge (Meredith, O'Connell, Marques and Stern,
unpublished). Three types of lesions were made:

     1) The NP nerves were cauterized through an opening in the
palate,
     2) the VN nerves were cut with fine scissors where they passed
between the bulbs, and
     3) the olfactory mucosa was destroyed with zinc sulphate

using the same concentration and method of infusion as Powers and
Winans.

Male hamsters were tested in three situations:

1) in their home cages with odor or air diffusing up through
the bedding (the attraction bioassay),
2) with an anesthetized male smeared on the anogenital area
with hamster vaginal discharge (HVD) (scented male test),
3) with a naturally cycling behaviorally estrus female
(cycling female test).

Of the individual treatments given alone, only zinc sulfate produced
significantly reduced attraction to HVD odor suggesting that this
attraction is mediated by the main olfactory system. There was some
reduction in sniffing and digging at the odor port in the VN lesion-
alone group, which may indicate some input from the VN nerves as
suggested recently by Powers et al. (1979). However, neither VN nor
NP lesions alone produced any statistically significant reduction in
attraction. When olfactory lesions were added to the VN and NP
lesions attraction was significantly reduced in both experimental
groups, as shown in Fig. 13, confirming that attraction is mediated
by the main olfactory system.

Preference for the rear of scented males and mounting of scented
males with pelvic thrusts (interpreted as intromission attempts;
Fig. 14) were significantly reduced in groups with combined NP and
zinc sulfate lesions as well as in groups with combined VN and zinc
sulfate lesions. There were significant differences between the
deficits of the two groups but both were significantly different
from the zinc sulfate group which showed no deficits. In the cycling
female test, as shown by Powers and Winans, combined olfactory and
VN lesions produced severe deficits: In our experiments eleven out
of thirteen animals showed no intromissions in a ten minute test
(Fig. 15). The NP plus zinc sulfate group also showed significantly
fewer intromissions although the reduction was not as severe as in
the VN plus zinc sulfate group. VN lesions were also the only lesions
which, when used alone produced significant changes in any of the
sex tests. Attraction for the rear of scented males and mounting of
scented males with intromission attempts were both significantly
reduced by vomeronasal cuts alone. Intromissions with females were
not significantly reduced by this single lesion.

In the scented male test the effects of combined NP plus zinc
sulfate lesions were statistically indistinguishable from those of
the combined VN plus zinc sulfate lesions. In the cycling female
test both produced significantly reduced intromission. Although the
effects of NP lesions did not exactly mimic those of VN lesions,
when combined with zinc sulfate treatment the two lesions produced
similar effects; they differed only in degree of severity. Since

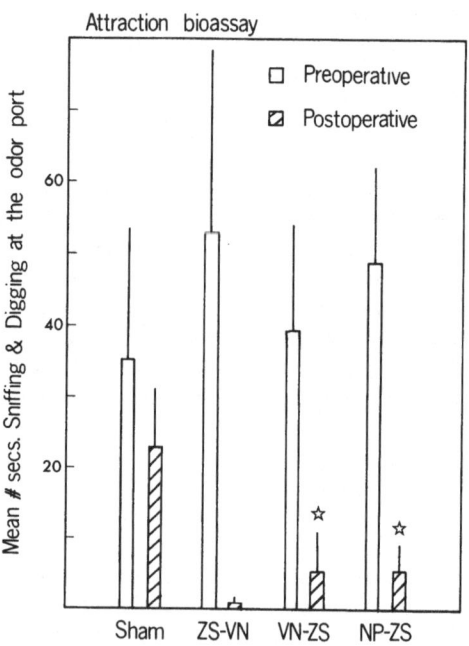

Figure 13. Performance of 4 groups of male hamsters in a task re-
quiring location of the odor of female hamster vaginal
discharge (HVD). NP-ZS; sequential nasopalatine nerve
lesions and zinc sulphate (ZS) treatment. ZS-VN; zinc
sulphate followed by vomeronasal nerve lesions. VN-
ZS; the same lesions applied in reverse order. In
Figs. 13-15 solid stars indicate that the difference
between pre- and postoperative performance is significantly different at the 5% level from the difference
seen in shams. Open stars indicate the same at the 1%
level.

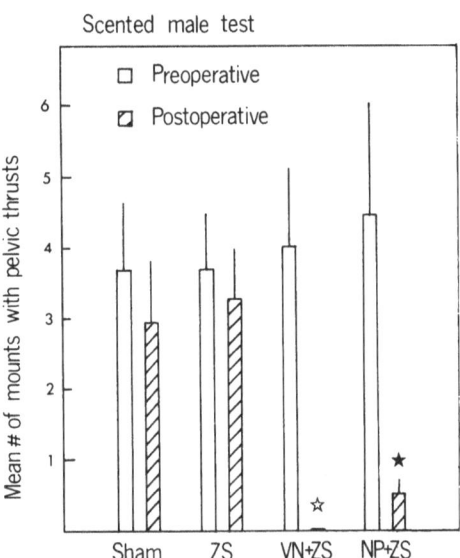

Figure 14. Sexual behavior of male hamsters with surrogate females: Mounting of anesthetized animals scented with HVD during a 5 min test. For abbreviations, see Fig. 13.

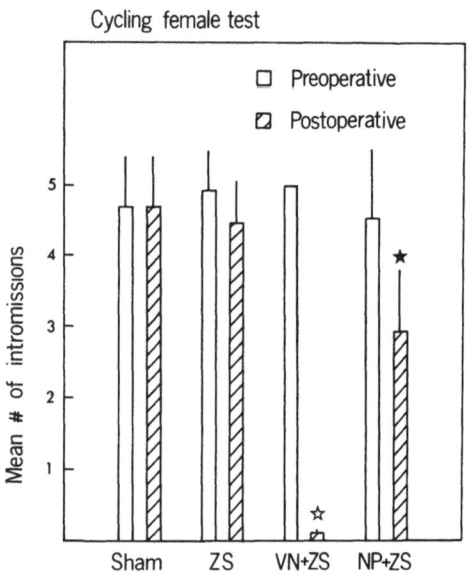

Figure 15. Sexual behavior of male hamsters with behaviorally
            estrus, naturally cycling females: numbers of
            intromissions during a 10 min test (animals limited
            to 5 intromissions). For abbreviations, see Fig. 13.

the NP lesions should disable the VNO pump, these results suggest
that the pumping mechanism is in fact necessary for normal vomeronasal function in the behaving animal.

The residual differences between the effects of VN and NP
lesions are not trivial but might be accounted for by one of the
following: 1) There could be some diffusion into the VN lumen
during a ten minute test; the mount and intromission latencies for
NP group animals were somewhat though not significantly longer
than those in other groups showing continued behavior. 2) Denervation
hypersensitivity of the adrenergic targets in the VNO could make
the system sensitive to circulating levels of epinephrine. Any
fluctuation in these levels during the ten minute test would tend
to operate the pump.

This brings us back to the question of the normal operation
of the VN pump and to the question of stimulus access in vapor or

Table 1.   Apparent access routes for substances reaching the vomero-
           nasal organ. Heavy arrows indicate the most probable route,
           light arrows possible routes which were not ruled out by
           experiment. Asterisks indicate sites where there was
           evidence for the presence of the stimulus during or after
           the experiment.

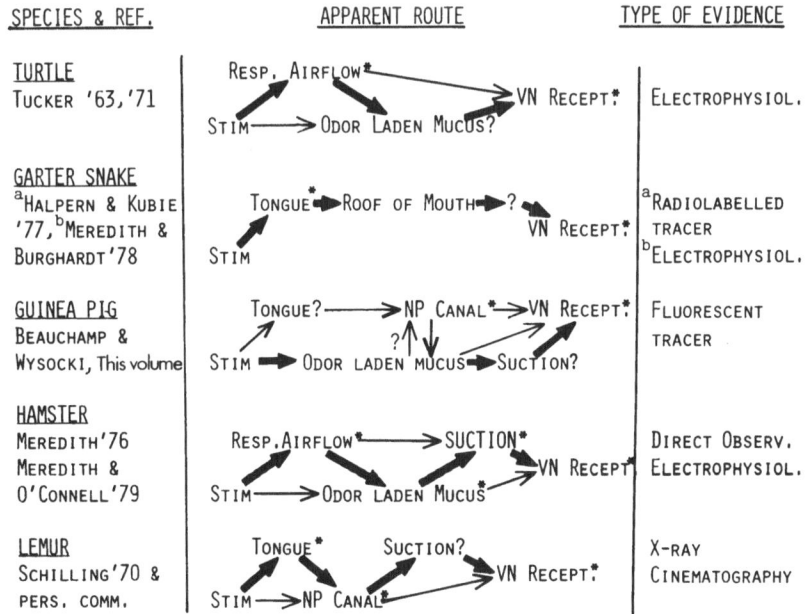

in liquid form.   On the first question, three possibilities present themselves: 1) that the pump is operated voluntarily: this
cannot be tested in my anesthetized preparation but it seems unlikely
since autonomic efferents are involved; 2) that the pump is operated continuously, perhaps by the fluctuations in autonomic activity
which normally accompany respiration.   However, no fluid flow
without overt stimulation was observed in anesthetized animals but
this might be attributed to the effect of anesthetic.   The
possibility which seems most likely is, 3) that the pump is operated
reflexly following sensory input via the main olfactory system or
other sensory system.   This also could not be demonstrated in
anesthetized animals.

On the question of vapor or liquid access, it seems likely
that both can enter the lumen.   In the hamster we have now observed
bubbles at the pore after vomeronasal contraction when the mucus
surrounding the pore had been allowed to dry.   This suggests that
air had been sucked in during the contraction.   There is some
evidence in garter snakes (Kubie and Halpern, 1978) and rats
(Johns et al., 1978) that physical contact with the stimulus is
necessary for activation of the vomeronasal organ.   In the garter
snake, in the Guinea pig and in the lemur there is some evidence
that nonvolatile liquids may pass into the lumen of the organ
(Table 1), and this may be the normal mode of stimulation.   It
should not be forgotten, however, that water soluble substances
would rapidly dissolve in nasal mucus and thus appear in relatively
high concentration at the vomeronasal pore in those animals with
a nasal opening of the vomeronasal ducts.

ACKNOWLEDGEMENT

This work was supported in part by Grant No. NINCDS NS 14453
and Fellowship No. NSO 5849.

REFERENCES

Adrian, E.D. 1955. Synchronized activity in the vomeronasal
    nerves with a note on the function of the organ of Jacobson.
    Pflügers Archiv. ges. Physiol. 260: 188-192.
Broadwell, R.D. 1975. Olfactory relationships of the telenephalon
    and diencephalon in the rabbit I. An autoradiographic study
    of the efferent connections of the main and accessory
    olfactory bulbs. J. Comp. Neurol. 163: 329-346.
Davis, B., Macrides, F., Youngs, W., Schneider, S. and Rosene, D.R.
    1978. Efferents and centrifugal afferents of the main and
    accessory olfactory bulbs in the hamster. Brain. Res. Bull.
    3: 59.
Estes, R.D. 1972. The role of the vomeronasal organ in mammalian
    reproduction. Mammalia 36: 315-341.

Halpern, M. and Kubie, J. 1977. Transport of ^{3}H proline to
vomeronasal epithelium by tongue flick in garter snakes.
See J. Kubie Thesis, Downstate Medical Center SUNY. Dept.
Anat., 1977.

Herzfield, J. 1888. Über das Jacobsonsche Organ des Menschen
und der Säugetiere. Zool. J. (Anat) III, 551.

Hornung, D., Lansing, R.D. and Mozell, M.M. 1975. Distribution
of butanol molecules along the bullfrog olfactory mucosa.
Nature 254: 617-618.

Johns, M.A., Feder, H.H., Komisaruk, B.R., and Meyer, A.D. 1978.
Urine-induced reflex ovulation in anovulatory rats may be a
vomeronasal reflex. Nature 272: 446-448.

Klein, E. 1881. The Organ of Jacobson in the Rabbit. Quart. J.
Microscop. Sci. 21: 549-570.

Kretteck, J.E. and Price, J.L. 1978a. Amygdaloid projections to
subcortical structures within the basal forebrain and brainstem in the rat and cat. J. Comp. Neurol. 178: 225-254.

Kretteck, J.E. and Price, J.L. 1978b. A description of the
amygdaloid complex in the rat and cat with observations on
intra-amygdaloid axonal connections. J. Comp. Neurol.178:
255-280.

Kubie, J.L. and Halpern, M. 1978. Garter snake trailing behavior.
Effects of varying prey extract concentration and mode of
prey extract presentation. J. Comp. Physiol. Psych. 92:

Mann, G. 1961. Bulbus olfactorius accessorius in chiroptera. J.
Comp. Neurol. 116: 135-144.

Meredith, M. 1978. Amygdala response to electrical stimulation
of vomeronasal nerve. Neurosci. Abs. 4: 89.

Meredith, M. and O'Connell, R.J. 1979. Efferent control of
stimulus access to the hamster vomeronasal organ. J. Physiol.
286: 301-316.

Mozell, M.M. 1964. Evidence for sorption as a mechanism of the
olfactory analysis of vapors. Nature 203: 1181-1182.

Müller, W. 1971. Vergleichende elektrophysiologische Untersuchungen
an den Sinnesepithelien des Jacobsonschen Organs und der Nase
von Amphibien (Rana), Reptilien (Lacerta) und Säugetieren(Mus).
Z. Vergl. Physiol. 72: 370-385.

Powers, J.B. and Winans, S.S. 1975. Vomeronasal Organ: Critical
role in mediating sexual behavior of the male hamster. Science
187: 961-963.

Powers, J.B., Fields, R.B. and Winans, S.S. 1979. Olfactory and
vomeronasal system participation in male hamsters' attraction
to female vaginal secretions. Physiol. Behav. 22: 77-84.

Raisman, G. 1972. An experimental study of the projection of the
amygdala to the accessory olfactory bulb and its relationship
to the concept of a dual olfactory system. Exp. Br. Res. 14:
395-408.

Scalia, F. and Winans, S.S. 1975. The differential projections
of the olfactory bulb and accessory olfactory bulb in mammals.
J. Comp. Neurol. 161: 31-56.

Scalia, F. and Winans, S.S. 1976. New perspectives on the mor-
    phology of the olfactory system: Olfactory and vomeronasal
    pathways in mammals. In Mammalian Olfaction, Reproductive
    Processes and Behavior. R.L. Doty (Ed.), A.P. 1976, New York.

Schilling, A. 1970. L'Organe de Jacobson du Lemurien Malagache
    Microcebus murinus (Miller 1777). Mem. Mus. Nat. D'Hist.
    Natur. A51: 203-280.

Singer, A.G., Agosta, W.C., O'Connell, R.J.,Pfaffman, C., Bowen,
    D.V. and Field, F.H. 1976. Dimethyl disulphide: An
    attractant pheromone in hamster vaginal secretion. Science
    191: 948-950.

Skeen, L.C. and Hall, W.C. 1977. Efferent projections of the main
    and the accessory olfactory bulb in the tree shrew (Tupaia
    glis). J. Comp. Neurol. 172: 1-36.

Tucker, D. 1963. Olfactory, vomeronasal and trigemminal responses
    to odorants. In Olfaction and Taste I. Proc. 1st Int. Symp.
    Y. Zotterman (Ed.). Oxford: Pergamon.

Tucker, D. 1971. Non-olfactory responses from the nasal cavity:
    Jacobson's organ and the trigemminal system. In Handbk. of
    Sens. Physiol. Vol. IV: Chemical Senses 1. Olfaction, L.M.
    Beidler (Ed.). Berlin Springer-Verlag.

Uddman, R., Alumets, J., Densert, O., Hakanson, R. and Sundler, F.
    1978. Occurrence and distribution of VIP nerves in the nasal
    mucosa and tracheobronchial wall. Acta Otolaryng. 86: 443-448.

Verberne, G. 1975. Chemocommunication among domestic cats,
    mediated by the olfactory and vomeronasal senses. II Relation
    between the function of Jacobson's organ (vomeronasal organ)
    and Flehmen behavior. Z. Tierpsychol. 42: 113-128.

Winans, S.S. and Scalia, F. 1970. Amygdaloid nucleus: New
    afferent input from the vomeronasal organ. Science 170:
    330-332.

Winans, S.S. and Powers, J.B. 1977. Olfactory and vomeronasal
    deafferentation of male hamsters. Histological and Behavioral
    Analysis. Brain Res. 126: 325-344.

CHEMICAL COMMUNICATION IN THE GUINEA PIG: URINARY COMPONENTS

OF LOW VOLATILITY AND THEIR ACCESS TO THE VOMERONASAL ORGAN

Gary K. Beauchamp[1,2], Judith L. Wellington[1], Charles J. Wysocki[1], Joseph G. Brand[1,3], John L. Kubie[4], and Amos B. Smith, III[1,5]

[1]Monell Chemical Senses Center, 3500 Market Street Philadelphia, PA 19104
[2]Department of Otorhinolaryngology and Human Communication, School of Medicine, University of Pennsylvania
[3]Department of Biochemistry, School of Dental Medicine, University of Pennsylvania
[4]Department of Physiology, University of Pennsylvania
[5]Department of Chemistry, University of Pennsylvania

Guinea pigs (<u>Cavia</u> sp.) utilize chemical information from conspecifics in the regulation of social behavior (Beauchamp et al., 1977; King, 1956; Rood, 1972). We have been investigating a number of aspects of this communication system. Here, we will present observations and evidence from behavioral and chemical studies that suggest that information transfer among individuals is mediated at least in part by compounds of low volatility. If compounds of low volatility are used, the receptor organ(s) and the means by which the compounds reach the organ(s) are called into question. Volatile compounds can easily reach the olfactory receptor cells during inspiration; non-volatile substances presumably must reach appropriate receptors by some other means. We have hypothesized that the vomeronasal organ is involved in responses to chemical signals, particularly those of low volatility. We also will describe studies investigating access of non-volatile material to the vomeronasal organ.

BEHAVIORAL RESPONSES TO URINE

When presented with conspecific female urine, sexually mature male guinea pigs exhibit considerable interest, investigating the

urine by sniffing, licking and headbobbing (Beauchamp, 1973;
Beauchamp et al., 1977). During headbobbing the animal places his
nose very near, usually in contact with, a stimulus and then vigorously moves his head up and down at 2-4 cycles/sec. Headbobbing
accounts for the majority of the investigation time in tests of responses to urine. Males rarely exhibit extensive headbobbing in
response to substances from other species or to pure odorants.

Two-choice tests have been used to determine the information
content of guinea pig urine. If urine samples from one class of
individuals (e.g., females) are preferred to urine samples from
another class (e.g., males) then we conclude that information is
available in urine which differentiates the classes (i.e., females
from males). In these two-choice tests individually housed male
guinea pigs are presented with two 0.2 ml samples of urine on
glass plates. The plates are placed approximately 10 cm apart for
a period of 4 minutes. The amount of time the animal spends with
his nose within approximately 1 cm of each sample is recorded by
an observer. During the investigation period the animals invariably contact the urine samples. Through an extensive series
of such tests we have shown that guinea pig urine transmits information relevant to the genus, species, sex, individual identity
and current diet of the donor animal (Beauchamp, 1973; Beauchamp
and Berüter, 1973; Beauchamp, 1976; Beauchamp et al., in press;
Beauchamp, unpublished).

In addition to establishing what information is available in
urine, two-choice tests provide data on attractiveness of a sample.
For male guinea pigs, urine collected from female conspecifics
elicits more investigation time than any other substance tested
(Beauchamp, 1973). We suggest that this interest, as measured by
the duration of investigation and contact, is indicative of a
hedonic quality of the urine sample. Dilution of the urine reduces
contact duration (Fig. 1) which indicates that contact time is
not a measure of difficulty in obtaining information or in de-
ciphering a confusing signal.

Domestic male guinea pigs exhibit diminished discrimination
between male and female urine when the urine is some distance from
the male. When male and female urine samples are placed into
another male's home cage, there is no trend for the male to approach
or contact the female sample first. This suggests that he does not
discriminate between the samples from a distance. In a series of
tests, urine samples (either wet or after being dried for 3-5 hours)
were placed either 1.0 cm below a wire mesh screen or on top of
the screen prior to their presentation to male subjects (Fig. 2).
When urine was beneath the wire screen the overall response was
greatly attenuated. These data suggest that some important aspects of the stimulus urine are lacking when guinea pigs are denied
contact with the urine. A male's response to urine at a distance

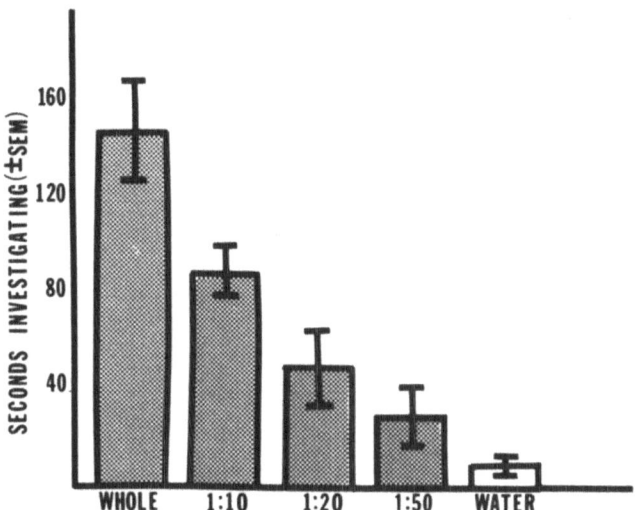

Figure 1. Effects of dilution of female urine on investigation
time of male guinea pigs. Each bar represents the
results (mean seconds investigating ± the standard
error of the mean) of 2 separate 2-min tests with 8
sexually experienced adult male guinea pigs. In each
test, each male was given a single glass plate containing 0.2 ml of female guinea pig urine or water and
investigation time was recorded (see Beauchamp and
Berüter, 1973, for a full description of this method).
Fresh urine was collected from four females, pooled,
then divided into 4 aliquots, three of which were
diluted with deionized water at 1 ml urine diluted to
10 ml, etc. The urine was then divided into separate
samples and frozen until needed for testing.

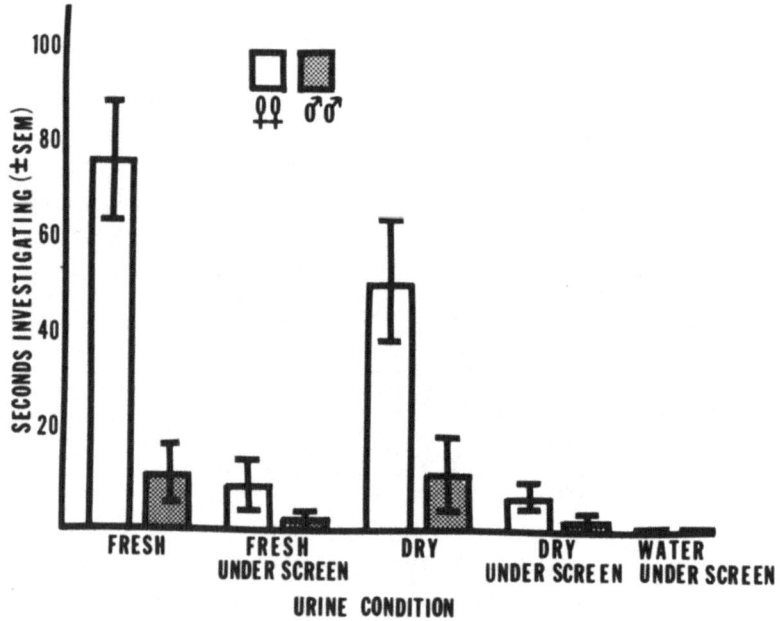

Figure 2.   Responses of male guinea pigs to fresh or dry 0.2 ml urine
             samples placed on top of or 1 cm beneath a wire mesh
             screen.   Nine males were each given 2-choice tests
             comparing urine sample from males and females or water
             as described in the text (see also Beauchamp, 1973).
             Fresh urine was presented in a liquid form; dry urine
             had been placed on the glass plates 3 hours prior to
             use and the water had evaporated.   Female urine was
             preferred to male urine in all cases (p < .05).   Total
             investigation time was greatly attenuated when the
             urine was beneath the screen, unavailable for contact
             (p < .01).

is in some ways similar to his response to diluted urine: investigation time is diminished. However, one distinct difference is
observed. When urine is placed beneath the screen males often
intensively scratch at the screen. This behavior appears as if it
is directed toward making contact with the stimulus urine. If this
interpretation is correct, then perhaps stimulation by some particularly attractive urinary component can only be attained by
direct contact with the urine. This hypothesis would implicate
substances of low volatility.

Further evidence supporting this hypothesis is found in our
study of the stability of the attractive components in urine
(Beauchamp, Wellington, Criss and Smith, in prep.). Urine samples
were placed on frosted glass plates and allowed to dry for lengths
of time ranging from 5 hr to 90 days. Male guinea pigs were then
presented with samples of male and female urine which had been
allowed to dry for equal lengths of time. The animals sniffed,
licked and headbobbed to the urine samples much more extensively
than they did to water and the preference for the female sample
relative to the male sample was evident for samples dried for 60
days or less.

CHEMICAL STUDIES

Chemical studies also suggest that compounds of low volatility
can mediate the attraction to urine in guinea pigs. Separation
of urinary constituents on the basis of volatility was achieved
by sublimation of frozen urine. Domestic males were able to distinguish between the residues of male and female urine, spending
significantly more time with the latter. Interestingly they also
distinguished between the volatiles of male and female urine but
in this case more time was spent with the male volatiles in contrast to their usual preference for whole female urine relative
to whole male urine (Berüter et al., 1973).

Volatility is dependent upon a number of parameters including
molecular size. Most compounds over a molecular weight of 400 g/
mole are considered to be non-volatile. The molecular weights of
the components of guinea pig urine were determined using gel filtration columns. Initially the columns were standardized with
compounds of known molecular weight and then unknowns were determined by comparison of their elution volumes to those of the
standards.

Specifically, domestic male and female guinea pig urine samples were each chromatographed on a Sephadex G15 gel filtration
column which fractionates polypeptides with molecular weights under
1500 g/mole. The chromatograms from male and female urine were
similar (Fig. 3). Glutathione and reduced glutathione were

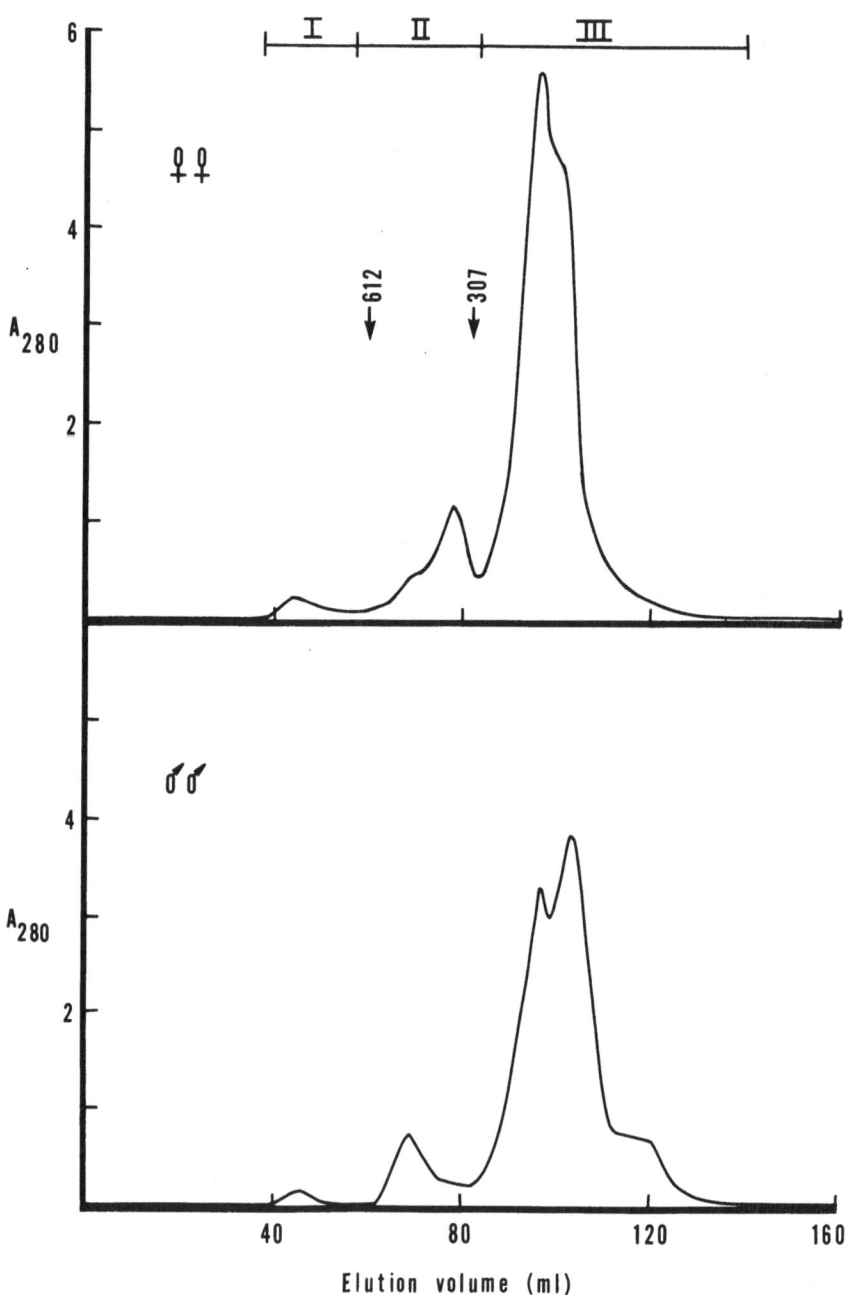

Figure 3.   Elution profiles for 3 ml of centrifuged female and male
            guinea pig urine eluted from a 55 cm x 1.5 cm (i.d.)
            Sephadex G15 column with water. The profiles record the
            absorbance of each fraction at 280 nm. Elution volumes
            of standard peptides are indicated with arrows. Pooled
            fractions I, II, and III were tested with the bioassay.
            Qualitatively similar profiles were observed by monitor-
            ing 210 nm absorbence.

chromatographed in the same column as molecular weight standards (molecular weight = 612 and 307 g/mole respectively) and their elution volumes are indicated with arrows in the figure.

Urinary G15 column fractions were combined into pools as indicated in Figure 3, lyophilized, and diluted to original concentration with deionized water. The corresponding fractions from male and female urine were tested against each other in the previously described two-choice assay. The results (Fig. 4) indicate that test males discriminate between the male and female fractions I and III. The combined peaks I, II, and III resulted in somewhat greater investigation times than did either peak I or III alone.

It is emphasized that the attractiveness of urine and the communication of sexual identity is transmitted, at least in part, by components in peak I which have a molecular weight of greater than 600 g/mole. From the foregoing behavioral and chemical data we cannot rule out the possibility that this communicant is a small volatile molecule(s) bound to a larger molecule(s) eluting in peak I. However the following evidence suggests that this hypothesis is not necessary: non-volatile molecules do reach the chemosensory vomeronasal organ.

SENSORY PERCEPTION OF URINE: THE VOMERONASAL ORGAN

Male guinea pigs whose olfactory bulbs have been removed exhibit no sniffing, licking or headbobbing behavior in response to conspecific urine (Beauchamp et al., 1977). From these data we conclude that some sensory component, which is disrupted by bulbectomy, is necessary for perception of the behaviorally relevant substances. Olfactory bulbectomy eliminates sensory input from the main olfactory system and the accessory olfactory system (as well as several other possible sensory components). The accessory olfactory system has often been suggested to play a role in the perception of sexual signals (see particularly Planel, 1953; Estes, 1972; Powers et al., 1979). Furthermore, it has often been suggested that the vomeronasal organ may be specialized to perceive non-volatile substances (see Wysocki, in press, for a comprehensive review of the literature).

In the guinea pig, the bilateral vomeronasal organ opens into the rostral portion of the nose near the external nares. Caudal to the organ, the nasopalatine ducts connect the nasal and the oral cavities. Thus, there are at least two possible pathways for a non-volatile substance to reach the organ. In the first, the substance could go directly through the external naris, while in the second, the substance would enter through the mouth and pass through the nasopalatine ducts into the nasal cavity and then rostral to the opening of the organ.

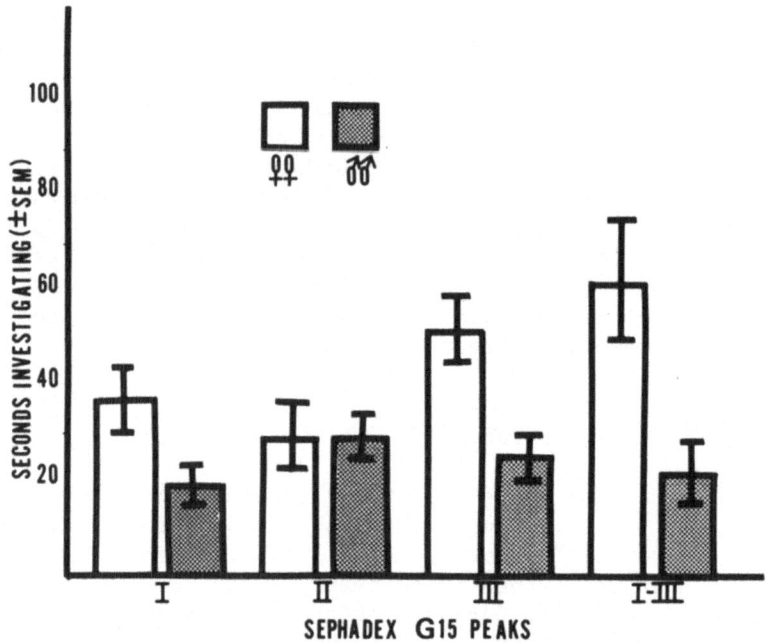

Figure 4.   Responses of male guinea pigs to Sephadex G15 column
            fractions. Ten males were given 2-choice tests
            comparing fractions from the chromatographic separation
            of female urine with the corresponding fractions from
            the separation of male urine. Fractions I and III,
            as well as the pooled fractions (I, II and III) from
            female urine were preferred to the corresponding
            fractions from male urine (p < .05).

To determine whether non-volatile substances do have access
to the vomeronasal organ and further to investigate what pathway(s)
might be employed, we have used fluorescence microscopy (Wysocki,
Wellington and Beauchamp, submitted). Male guinea pigs were presented with urine or other substances which had been adulterated
with rhodamine (B or 6G) hydrochloride, a completely ionized, non-
volatile fluorescent dye. The animals were sacrificed immediately
following the presentation of urine. After gross observations
for the location of the dye, the vomeronasal organs and occasionally
the olfactory epithelium, nasopalatine ducts, and the external nares
were removed, frozen, sectioned and viewed under fluorescence
microscopy.

The results (Table 1) clearly demonstrate that the non-volatile
fluorescent dye does reach the vomeronasal epithelium. All intact
animals which were exposed to dye-adulterated urine had dye bi-
laterally. This was not due to artifacts associated with preparation of the tissue as the control conditions demonstrate. While
the dye was found in the vomeronasal organ, it was not found on
the olfactory epithelium. Since a non-volatile dye in urine reaches
the vomeronasal organ, we conclude that other non-volatile components in urine also reach this organ. Thus during contact with
the urine, non-volatile substances can be transported to the
vomeronasal organs, and based upon our results, this appears to be
a normal occurrence.

While we have emphasized access of non-volatile urinary compounds, non-volatile substances from other sources may also reach
the vomeronasal epithelium. Dye fluorescence was seen in at least
one animal which had consumed dye-adulterated drinking water.
Perhaps the sensory functions of the vomeronasal organ encompass
more then mediation of sexual and social information (see discussion in Cooper and Bhatnagar, 1976). For example, as is true
for snakes, chemical stimuli in foods may be sensed by the vomeronasal organ as well as by taste, olfactory and trigeminal systems.

Following the animal's contact with dye-adulterated urine,
its mouth, nasopalatine ducts, external nares, and esophagus
were extensively stained. Several experiments (Table 1) have been
conducted to investigate the route(s) of access of non-volatile
stimuli to the vomeronasal organ. When one nostril was sealed
prior to exposure to dye-adulterated urine, dye was not observed
ipsilateral to the closed nostril but was seen contralaterally.
Thus, sealing the nostril prevents access of the fluorescent dye
to the vomeronasal organ. While it is likely that the nostril
provides the access route this cannot be concluded with certainty.
Nostril closure could serve to prohibit access in other ways. For
example, substances may accumulate behind the sealed nostril which
prevent access of stimuli to the ipsilateral vomeronasal organ
via the nasopalatine duct. More generally, an open route (mouth,

Table 1. Presence (+) or absence (o) of rhodamine fluorescence in vomeronasal organ sections following several experimental treatments.[1]

Treatment	Rhodamine Fluorescence
1. Removal from home cage (1 male) or contact with unadulterated female urine (3 males)	o
2. Contact with female urine adulterated with rhodamine (7 males)	+ bilateral
3. Contact with male urine adulterated with rhodamine (7 females)[2]	+ bilateral
4. Contact with female urine adulterated with rhodamine following unilateral nasal closure (4 males)	+ contralateral to closure o ipsilateral to closure
5. Contact with drinking water mixed with rhodamine (2 males)	+ bilateral - one male o one male
6. Water (1 female, 1 male) or male urine (1 female) adulterated rhodamine flushed through mouth of anesthetized animal[3]	o

[1]Sample size and sex indicated in parenthesis.

[2]Prior to the presentation of male urine, 6 of the 7 females had undergone one of a variety of surgical attempts to block the nasopalatine duct unilaterally. None of the attempts were success- ful. However, in 4 of the females, the amount of fluorescence in the vomeronasal organ ipsi- lateral to the damaged nasopalatine duct subjectively reduced relative to the contralateral organ.

[3]During dissection, the vomeronasal organ of the male was damaged unilaterally; a surgical implement penetrated the capsule wall. Subsequent fluoresence microscopy revealed rhodamine fluorescence in this caudal region of damage. No dye fluorescence was seen in the rostral portion near the ostium of the vomeronasal organ.

nasopalatine duct, naris) may be required for stimulus transport and an interruption at any point along this route may be disruptive.

If the route of access of the non-volatile fluorescent dye remains somewhat problematical, many other questions remain and are under investigation. We know nothing about speed of transport; furthermore, we are ignorant of what limitations in transport may exist due to size or other characteristics of a molecule. Finally the mechanisms by which the dye reaches the vicinity of the organ are not known although once near, the pumping mechanism described in detail by Meredith (this volume) presumably contributes to delivery of the compounds to the receptors.

Non-volatile and presumably volatile substances in urine can reach the vomeronasal organ. While an increasing number of studies are showing that interruption of vomeronasal function disrupts behavior and/or physiology (see studies described in this volume and Wysocki, in press) clear evidence that chemical signals normally activate the vomeronasal organ is not available. We have thus initiated studies using radiolabelled 2-deoxy-D-glucose (2DG) to map peripheral and central nervous system sites of activity in response to chemical signals. We have found that when males were exposed to conspecific female urine following a 2DG injection, consistent labelling was seen in the accessory olfactory bulbs. However, labelling was also evident in approximately one-half of the control animals' accessory olfactory bulbs. These results are not inconsistent with the hypothesis that the accessory olfactory system plays an important role in mediating responses to chemical communicants but the variability among control animals is so far an enigma. We are continuing to investigate the stimulus parameters involved in the activation of this system.

CONCLUSIONS

Three issues have been emphasized. First, behavioral evidence suggests that some of the urinary substances which transmit information among guinea pigs are of low volatility. Second, chemical studies support this view. These studies indicate that fractions with molecular weights of greater than 600 g/mole contain attractive components. Third, a mechanism exists by which non-volatile materials may be transported into the vomeronasal organ. Additionally, some preliminary evidence implicates this organ as a functioning receptor for urine-based chemical signals.

Rood (1972) described the social behavior of Cavia in considerable detail. He concluded that olfactory stimuli were of particular importance in cavies. For example, cavies frequently pause to investigate a spot where another individual has urinated. Females commonly "spray urine to repulse an approaching male which

then licks up the urine from the ground. Urine sprayed on a female
by a male <u>Cavia</u> during rumping may mark the female, possibly
causing subordinate males to avoid her" (Rood, 1972, p. 72). These
descriptions implicate chemical signals which could be of low volatility. Perhaps for surface-living, highly social species like
cavies which live in stable social groups, volatile chemical
signals which could transmit information over long distances are
not generally necessary.

    Guinea pigs are not unique in establishing contact with con-
specific chemical signals. Licking, mouthing, ingesting and
sniffing from very near the source is extremely common in many
species of mammals. Such behaviors suggest (but surely do not
prove) that compounds of low volatility may be of widespread importance in chemical communication.

    The involvement of low volatility substances in chemical
communication does not prove that the vomeronasal organ is the
receptor. However, the suggestion that this receptor organ may be
particularly responsive to such substances is not new (see Powers
et al., 1979). Receptors within the organ need not be unique for
the organ to have a distinct function; if molecules which are of
very low volatility have access only to these receptors, then a
specific role is assured.

REFERENCES

Beauchamp, G.K. 1973. Attraction of male guinea pigs to con-
    specific urine. Physiol. Behav. 10: 589-594.
Beauchamp, G.K. 1976. Diet influences attractiveness of urine.
    Nature 263: 587-588.
Beauchamp, G.K. and Berüter, J. 1973. Source and stability of
    attractive components in guinea pig (<u>Cavia</u> <u>porcellus</u>) urine.
    Behav. Biol. 9: 43-47.
Beauchamp, G.K., Criss, B. R. and Wellington, J.L. In press.
    Chemical communication in <u>Cavia</u>: Responses of wild (<u>C</u>. <u>aperea</u>),
    domestic (<u>C</u>. <u>porcellus</u>) and F$_1$ males to urine. Anim. Behav.
Beauchamp, G.K., Magnus, J.G., Shmunes, N.T. and Durham, T. 1977.
    Effects of olfactory bulbectomy on social behavior of male
    guinea pigs (<u>Cavia</u> <u>porcellus</u>). J. Comp. Physiol. Psychol.
    91: 336-346.
Berüter, J., Beauchamp, G.K. and Muetterities, E.L. 1973. Com-
    plexity of chemical communication in mammals: Urinary com-
    ponents mediating sex discrimination by male ginuea pigs.
    Biochem. Biophys. Res. Comm. 53: 264-271.
Cooper, J.G. and Bhatnagar, K.P. 1976. Comparative anatomy of the
    vomeronasal organ complex in bats. J. Anat. 122: 571-601.
Estes, R.D. 1972. The role of the vomeronasal organ in mammalian
    reproduction. Mammalia 36: 315-341.

King, J.A. 1956. Social relations of the domestic guinea pig
    living under seminatural conditions. Ecology 37: 221-228.
Planel, H. 1953. Études sur la physiologie de l'organe de
    Jacobson. Arch. Anat. Histol. Embryol. 36: 197-205.
Powers, J.B., Fields, R.B. and Winans, S.S. 1979. Olfactory
    and vomeronasal system participation in male hamster's
    attraction to female vaginal secretions. Physiol. Behav.
    22: 77-84.
Rood, J.P. 1972. Ecological and behavioural comparisons of three
    genera of Argentine cavies. Anim. Behav. Monog. 5: 1-83.
Wysocki, C.J. In press. Neurobehavioral evidence for the involve-
    ment of the vomeronasal system in mammalian reproduction.
    Neurosci. Biobehav. Rev.
Wysocki, C.J., Wellington, J.L. and Beauchamp, G.K. Submitted.
    Access of urinary nonvolatiles to the mammalian vomeronasal
    organ.

ACKNOWLEDGEMENTS

    The work reported here was supported by NSF grant BNS 76-
01642. Other support included NIH 1F32NS05690-01 (J.L.W.) and NIH
5T32NS07068-03 (C.J.W.) and NIH 1F32NS06152-01 (J.L.K.). We
thank Barbara R. Criss for excellent technical assistance.

THE ROLE OF THE VOMERONASAL SYSTEM IN MAMMALIAN REPRODUCTIVE

PHYSIOLOGY

Margaret A. Johns

Division of Endocrinology
Department of Medicine
Mount Sinai Medical Center
New York, NY   10029

GENERAL INTRODUCTION

External chemical stimuli such as those contained in urine and vaginal secretions have been shown to play a significant role in reproductive functions of many mammalian species. Most studies of effects of external chemical stimuli on reproduction have dealt with male behavior or with female physiology. A few studies have demonstrated effects of external chemicals on male physiology and on female behavior. For example, exposure to bedding soiled by adult conspecifics has been shown to affect testicular weight in prepubertal male mice (see Rogers and Beauchamp, 1976), and exposure to vaginal odor has been found to cause an acute increase in plasma testosterone in sexually experienced male hamsters (Macrides et al., 1974). That female reproductive behavior may be affected by external chemical stimuli is suggested by the fact that removal of the olfactory bulbs of female hamsters results in decrements of female courtship behavior (see Murphy, 1976).

Mammalian Chemoreception

The behavioral and physiologic sexual responses to external chemical stimuli might be mediated by any one of five apparently morphologically distinct nasal structures (see Graziadei, 1977): 1) the olfactory neuroepithelium proper; 2) the vomeronasal (VN), or Jacobson's, organ; 3) the septal organ of Rodolfo Masera; 4) portions of the trigeminal nerve (V cranial nerve); and 5) the terminal endings of the nervus terminalis. However, most current

literature emphasizes the role of the olfactory neuroepithelium
proper, part of the "main" or "primary" olfactory system (referred
to simply as the OLF system here).

Role of the "Main" Olfactory System vs. the Vomeron sal System in Mammalian Reproductive Physiology

Although the septal organ of Masera, the terminal endings of
the trigeminal nerve and the terminal endings of the nervus terminalis may be involved in physiologic effects of exteroceptive
chemicals in mammals, we know too little at this point about these
structures and their functions to speculate about their contributions. Therefore, this paper will focus exclusively on the OLF
and the VN systems.

It has traditionally been assumed that the OLF system is
responsible for most, if not all, of the effects of external chemical stimuli on reproductive physiology. There are several possible
reasons for the assumption: 1) previous dramatic demonstrations
of the effects of minute amounts of airborne chemical stimuli,
"pheromones" (Karlson and Lüscher, 1959), on the behavior of insects
suggested that all chemical effects on female and male mammals--
physiologic as well as behavioral--may be mediated by airborne
chemical molecules; 2) there seemed little necessity for looking
beyond the OLF system because airborne chemicals can obviously
affect at least some aspects of mammalian, as well as insect,
behavior (e.g., approach to prey or avoidance of predators);
3) until advanced or refined techniques were available for studies
of olfactory pathways, the possibility that the OLF and VN systems
might be anatomically separate throughout the brain was not fully
appreciated; 4) misinterpretation of neurological effects of
bulbectomy and brain lesions may have resulted in unwarranted
assumptions about functional correlates of central projections of
the OLF system (see Murphy, 1976 for discussion of these assumptions); 5) evidence that the VN dendrites lack cilia, combined with
the assumption that it is the cilia (rather than the microvilli
found on VN dendrites) that contain receptor sites for the odorant
molecules (see Graziadei, 1977), may have contributed to relative
lack of interest in the VN system; 6) although the well known
olfactory grimace, "flehmen" (observed in males of many mammalian
species in response to female urine), has indicated a possible role
for the VN organ in behavior (see Estes, 1972) the relative rarity
of the "flehmen" response in females may have cast doubt on the
ability of the female VN organ even to perceive urinary "pheromones"
(Estes, 1972), much less respond physiologically to chemical molecules contained in the urine; 7) results of an early wind tunnel
experiment that tested the effects of airborne chemical factors
from male mice on estrous cycles of grouped female mice (Whitten
et al, 1968), combined with the suggestion that effective stimuli

might have to be conveyed to the VN epithelium in a liquid vehicle
(Broman, 1920), may have led to the general conclusion that "primer
pheromones" are volatile and probably act through receptors of the
OLF system.

ε

Purpose

The purpose of this paper is 1) to describe experiments that
have established that chemical substances provided by male or female
mammals are capable of affecting the reproductive physiology of
conspecifics, 2) to show why the design of these (and related)
experiments might not clearly distinguish between the functions of
the OLF and VN systems, 3) to propose that under natural conditions, the OLF system serves primarily to orient animals towards
contact with chemical factors that act on the VN system, and that
4) the VN system then mediates (in whole or in part) chemically
stimulated pituitary hormone release responsible for reproductive
maturation, pregnancy block, successful implantation, estrous
cyclicity and reflex ovulation.

EVIDENCE THAT EXPOSURE TO EXTERNAL CHEMICAL STIMULI ALONE CAN AFFECT ENDOCRINE CHANGES

This section provides examples of some studies that have led
to an appreciation of the importance of external chemical stimuli
on endocrine processes in mammals.

Puberty

It has long been known that reproductive maturation in
prepubertal female and male rodents is affected in various ways
by exposure to adult animals (Vandenbergh, 1967; see Rogers and
Beauchamp, 1976). These observations led to a number of studies
that clearly established that the physical presence of adult
animals is not necessary for the effects (reviewed by Bronson,
1974), although it may play a contributory role (Drickamer, 1974).
For example, chemicals released by adults provide stimulation
sufficient to advance or retard sexual maturation in both males
and females. The most common way of testing the effects of
external chemical stimuli has been to place the prepubertal animal
in direct contact with bedding soiled by adult males or females
or with urine from adult males or females. These studies have
demonstrated that 1) puberty is delayed by exposure of immature
female mice to adult female urine (Cowley and Wise, 1972), 2)
maturation is not accelerated by bedding soiled by perinatally
androgenized female mice (Drickamer, 1974), but is accelerated by
exposure of immature female mice to bedding soiled by intact,

adult male mice (Vandenbergh, 1969) or urine from intact, adult
males (Cowley and Wise, 1972; Vandenbergh et al., 1975) in comparison with control females not exposed to chemical stimuli
from males and 3) in male mice, puberty is differentially affected
by direct contact with bedding soiled by adult mice (see Rogers
and Beauchamp, 1976).

Direct contact with male urine has been found to stimulate
estradiol and luteninizing hormone (LH) release as well as prolactin (PRL) release in prepubertal female mice (Bronson and
Maruniak, 1976).

Ovarian Function in the Adult Female

Females of some mammalian species have been thought to require contact with conspecific males in order to ovulate ("reflex
ovulators") while others ovulate "spontaneously". Among animals
in the first group are prairie voles. Ovulation in female voles
is rarely observed in the absence of mating stimuli from male
voles. The inducing stimulus however need not be mating. Although
artificial stimulation (e.g., glass rod) appears generally
ineffective (Kenney et al., 1977), placing females near male
voles, but separated from them by a wire screen, or in direct
contact with male-soiled bedding is sufficient stimulation for
estrus induction (Richmond and Conaway, 1969).

Some animals that are normally "spontaneous" ovulators can
be maintained under conditions that inhibit ovulation. These
animals can also respond to males, or chemical stimuli provided
by males. For example, the female rat normally cycles every 4-5
days. When placed in continuous light (LL) she ceases to ovulate
spontaneously and remains in a state of persistent vaginal estrus
(PE) (Everett, 1942). These females ovulate reflexly as a consequence of receiving male mating stimuli (Dempsey and Searles,
1943). They also ovulate reflexly when simply placed in direct
contact with male-soiled bedding or with male urine (Johns et al.,
1978).

In some instances "spontaneous" ovulators show altered estrous
cycles rather than total inhibition of ovulation. For example,
mice exposed to urine from pregnant or lactating females (see
Methods for Testing Effects of Chemical Stimuli) experience more
total estrous smears than do mice exposed to urine from non-
pregnant, non-lactating females (Hoover and Drickamer, 1979).
Group caging, or exposure to chemical stimuli from other females
via air currents (see Methods for Testing Effects of Chemical
Stimuli) leads to estrus synchrony in rats (McClintock, 1978) and
group caging of females, without males, disrupts normal estrous

cycling in mice (van der Lee and Boot, 1955). The remainder of this section will review some of the classic external chemical effects on reproductive physiology.

The Lee-Boot Effect. Isolated female mice usually have cycles 4-6 days long. If the females are caged in groups, however, the cycle lengths are increased (Champlin, 1971; van der Lee and Boot, 1955) and/or cycles are suppressed altogether (Whitten, 1959). The suppressive effect of female mice on other intact females is abolished by ovariectomy and restored by estradiol 17 β implants (Clee et al., 1975).

As is true for effects of male or female adults on sexual maturation, the physical presence of other females is not required for the Lee-Boot effect. It is sufficient that a female be exposed to soiled bedding or urine of other females for the cycle-lengthening effect to be demonstrated. Direct contact with the chemical substance has again been the usual method for presenting the stimulus. For example, Champlin (1971) showed that estrous cycles of isolated female mice are lengthened by housing the mice in contact with bedding previously soiled by other females.

Although the neuroendocrine mechanism mediating the effects of chemical stimuli on estrus suppression has not been established, inhibition of gonadotropin secretion (Whitten and Champlin, 1972) and a concomitant increase in PRL secretion (see Reynolds and Keverne, in press) may be responsible for suppression of estrous cycles among grouped female mice.

Just as "reflex ovulators" can be induced to ovulate, females showing long cycles, rather than no cycles, should also be able to respond to conspecific males, or chemical stimuli from males, with a shortening of cycle lengths. There are examples of this. The prolonged, or suppressed, cycles observed among grouped female mice can be shortened by exposing the females to male mice. This is known as the Whitten Effect.

The Whitten Effect. After all-female grouping of mice has resulted in lengthened or suppressed estrous cycles, the presence of the male restores normal cycling. This effect of the male is not observed if olfactory bulbs of the female have been removed (Whitten, 1956). Direct contact with urine produced by intact males (Marsden and Bronson, 1964) or females given exogenous androgen in adulthood (Whitten and Bronson, 1970) is followed by cycles of normal length in groups of female mice. Thus, androgen apparently mediates the effects of the urinary substance that induces estrus in mice. Furthermore, group-housed female mice with prolonged estrous cycles show significantly more estrous cycles of normal length than do controls after being placed downwind from males (Whitten et al., 1968).

Stimuli provided by males (which may include chemical stimuli) can also advance ovulation in non-cycling ewes (Morgan et al., 1972) and pigs (Signoret and Mauleon, 1962) with prolonged estrous cycles. Among female rats with 5-day cycles, exposure to urine of castrated male and female, as well as intact male rats, leads to a decrease in rate of progesterone (P) secretion and can shorten the cycle to 4 days (Chateau et al., 1976). This, incidentally, is an interesting exception to the general rule that androgen is responsible for urinary effects on shortening estrous cycles.

Pregnancy

The Bruce Effect. Exposure of a recently inseminated mouse to a strange male conspecific frequently results in pregnancy blockage (Bruce, 1959). This finding was followed by a number of studies that attempted to identify the relevant sensory stimuli. The pregnancy-block effects have been shown to be mediated by direct contact exposure to urine from intact adult males and also from androgen-treated, castrated adult males (Bruce, 1966; Dominic, 1965). Pregnancy blockage in recently inseminated female mice appears to result from a reduction in PRL secretion secondary to an odor-induced stimulation of gonadotropin release (Dominic, 1966).

On the other hand, after insemination, normal development of functional corpora lutea in rats and mice depends on a continued male influence. Male influence is usually understood to mean extended cohabitation accompanied by copulation. Indeed a positive correlation between a large number of preejaculatory intromissions experienced by the female rat during copulation and the probability of her becoming pregnant has been demonstrated (Adler, 1969; Terkel and Sawyer, 1978) and housing inseminated light-induced persistent-estrus (LLPE) rats overnight with a male does lead to a higher incidence of pregnancy than observed after a shorter mating exposure of only 2 hours (Brown-Grant, 1977). Although a series of nightly PRL surges is activated by a certain threshold level of coital stimulation (Terkel and Sawyer, 1978) the contribution of male chemical stimuli to this effect was not examined by these authors.

In mice however, for whom it has also been demonstrated that successful implantation of ova requires exposure of the inseminated female to the male (or to male stimuli) for more than 2 hours, the role of chemical stimuli has been examined. Kranz and Berger (1975) found that removal of the male from the females' cage after two hours and placing him upwind of the female results in successful implantation in a significant number of inseminated females compared with females deprived of continued male stimuli.

Testosterone Levels in the Adult Male

Exposure of sexually experienced male hamsters to the vaginal discharge of an adult female hamster, which has been smeared onto the inner wall of a small glass bottle, is followed by an acute increase in plasma testosterone (Macrides et al., 1974).

Summary and Conclusions

The above experiments are among those that demonstrate that exposure to chemical stimuli can influence timing of puberty, ovarian function and pregnancy. The variety of the known effects of urine (or chemical substances present in other materials), from either intact or castrated males or females, on mammalian reproductive physiology suggests that they are associated with changes in secretion of LH, FSH, estradiol, progesterone and/or testosterone.

Many of these experiments clearly distinguish between endocrine effects of stimulating olfactory receptors from the effects of stimulating non-olfactory receptors (e.g., tactile, auditory, and visual receptors). However, in general, they were not designed to discriminate between one type of odor receptor and another.

EVIDENCE FOR, AND CRITIQUE OF, THE IDEA THAT THE "MAIN" OLFACTORY SYSTEM IS INVOLVED IN CHEMICALLY MEDIATED ENDOCRINE CHANGES

This section discusses studies that seemed to indicate critical involvement of the OLF system in the effects of external chemical stimuli on reproductive physiology.

Methods for Testing Effects of Chemical Stimuli

As suggested in the previous section the usual methods for testing effects of external chemicals in mammals have involved placing an immature or mature animal in direct physical contact with male-or female-soiled bedding or with urine. In some studies, the test animal was separated by a wire mesh screen from the source of the chemical molecules (e.g., a conspecific male) but not from the chemicals themselves. These testing methods were not designed to distinguish between different types of chemosensitive receptors. Thus, the relative importance of the OLF system compared to the VN system in mammalian reproduction cannot be determined from these experiments.

Not all olfaction studies, however, have placed the subject in direct physical contact with effective chemical stimuli. Alternative methods of conveying the stimuli have been employed, with some type of air current used most frequently.

In the first of four air current studies to be discussed the effects of chemical stimuli derived from males on estrous cycles of females was assessed. Whitten et al. (1968) placed a group of female mice 2 m downwind from a group of males and compared them with groups of females caged underneath the males or caged 2 m upwind of the males. For 96 hours an exhaust fan drew air from the male's cages through one of the female's cages. The downwind group (and the group caged below the males) showed simultaneous resumption of estrus. In contrast, the females caged upwind did not. It was concluded that the chemical stimuli from the males were volatile and that they almost certainly acted through the OLF receptors.

The next two air current studies tested effects of chemical stimuli of females on estrous cycles of other females. In the first study, after establishing that groups of female rats show synchronous estrous cycles when they live together, McClintock (1978) housed females in individual isolation cages interconnected in groups of five so that each group of five shared a common air supply for 30 days. The system was designed so that air was drawn from, and fed equally to, each of the five units. This study showed that airborne chemical communication among singly caged female rats is sufficient to produce the same level of estrus synchrony as that found among female rats that are living together in the same cage. McClintock (1978) suggests that the communication system that mediates estrus synchrony is based on a volatile chemical acting on the OLF system although involvement of the VN system has not been ruled out.

In another study, McClintock and Adler (1978) tested effects on estrous cyclicity of airborne chemical communication between grouped females in constant light and grouped females in a fluctuating light/dark environment. Exposure to constant light (LL) results in persistent vaginal estrus (PE) (Browman, 1937). Airtight boxes containing the two groups of females were connected in series by 6 feet of hose. A large fan pulled air from an intake valve in the first box (containing females in constant light) through the second box (containing the females in a fluctuating environment) and directly out of the external exhaust vent of the room. The speed of the fan and intake valve were adjusted to maximize the air flow through the system and within the boxes. It was found that downwind females were in persistent estrus only when they received airborne cues from females in light-induced persistent estrus (LLPE).

The last study of this type tested effects of airborne chemical communication between males and recently inseminated females.

Kranz and Berger (1975) removed male mice from the cages of female mice two hours after insemination of the females. Males were placed in a wind tunnel upwind of the females. Chemical stimuli carried on air currents provided sufficient stimuli to significantly increase the pregnancy rate among these females compared with other inseminated females deprived of the additional male stimuli (see Pregnancy).

These experiments provide clear evidence that airborne chemical communication can affect mammalian reproduction. However, is evidence that the OLF receptors are involved conclusive or only suggestive? Is it possible that air currents produced by the various fans used in these studies could have carried effective compounds of relatively low volatility of the type to which the VN receptors are perhaps most sensitive (see Berüter et al., 1973)? Alternatively, might a reproductively important fraction of the more volatile compounds have acted on the VN receptors as well as, or instead of, acting on the OLF receptors?

In addition to the four studies that assessed the effectiveness of stimuli that can be transported by air currents, other experiments (see Testosterone Levels in the Adult Male) have assessed the effectiveness of chemical stimuli placed in small holders. For example, Monder et al. (1978) placed cotton swabs soaked with male mouse urine (or extract of the urine) inside straight Girton drinking tubes and put the open tubes in cages of female mice. The narrow openings of the tubes separated the females from the cotton by 2-4 mm. Both the tubes containing the whole urine and those containing the lipid portion of urine also blocked a significant number of pregnancies. In another study, Hoover and Drickamer (1979) placed cotton swabs soaked with urine from pregnant or lactating female mice inside perforated plastic capsules and set the capsules in cages of female mice. The urine significantly increased the total number of vaginal estrous smears of the females. As was true for the wind current experiments, methods used here would seem to implicate OLF receptors rather than VN receptors. The females could not actually touch the effective chemical substances on the cotton. However, in the first study the noses of the females might have been as little as 2 mm away from the cotton (H. Monder, personal communication). In the second study, the distance from the cotton might have been as little as 13 mm (L. Drickamer, personal communication). In either case, these rather short distances might not have put the effective stimuli out of reach of the receptors of the VN organ (see Meredith and O'Connell, 1979).

Thus, in none of the above experiments was involvement of the VN receptors in endocrine processes conclusively ruled out.

Methods of Deafferentating the Olfactory System

Just as methods commonly used to test effects of chemical stimuli on reproductive states are unlikely to reveal a specific role for the VN system, the usual methods for producing olfactory deficits have not been designed to differentiate between functions of OLF and VN systems (see Murphy, 1976; Scalia and Winans, 1976 for discussions). For example, surgical removal of the "main" olfactory bulb (MOB), the most frequently used method of producing olfactory impairment (and one which interferes with female reproduction in several species)(Signoret and Mauleon, 1965; Morgan et al., 1972; Sato et al., 1974), almost always includes destruction of the VN nerve fascicles (as well as the terminal nerve and pars rostralis of the anterior olfactory nuclei), even if the accessory olfactory bulb (AOB) remains intact. Moreover, partial damage may be produced in the rest of the anterior ol- factory nuclei, olfactory tubercle, pyriform cortex, and frontal pole. Severing the olfactory peduncle involves cutting the lateral olfactory tracts, including projections from the AOB as well as the MOB. Results of the peripheral impairment of OLF mucosa by surgical (i.e., scraping the cribiform plate) or chemical means (i.e., zinc sulfate infusion into the nasal cavities), by anesthetization of the OLF receptors or by nasal blockage would be no less difficult to interpret than results of bulbectomy (e.g., scraping the cribiform plate cuts the VN nerves as well as the olfactory nerve fascicles). Thus, traditional methods for producing olfactory impairment might not easily distinguish be- tween reproductive effects that are mediated by the OLF system and effects that are mediated by the VN system.

EVIDENCE THAT THE VOMERONASAL SYSTEM IS INVOLVED IN CHEMICALLY MEDIATED ENDORCINE CHANGES

What is the likelihood that the VN system, rather than the OLF system, will be found to have an important role in mammalian reproductive physiology? To begin with, the VN organ is well developed in most mammals, including the human fetus (see Gasser, 1975), although its presence or absence in the adult human is still controversial. Furthermore, there does not appear to be any evidence that excludes participation of the VN system in reproductive maturation, estrous cyclicity, implantation or changes in hormone levels (see previous section).

More importantly, there is evidence that strongly suggests that receptors of the VN organ have access, through the accessory olfactory bulb (AOB), to a system of central nervous connections, throughout the telencephalon and diencephalon, not available to the OLF mucosa (Raisman, 1972; Broadwell, 1975; Heimer, 1975; Scalia and Winans, 1976).

Nuclei of the amygdaloid body that receive input from the AOB seem to be the medial amygdaloid nucleus and the postero medial sector of the cortical amygdaloid nucleus (Scalia and Winans, 1976). This part of the amygdala is not reached by axons arising from the MOB (Heimer, 1975). Instead, nuclei of the amygdaloid body that receive input from the MOB seem to be in the lateral part of the cortical amygdaloid nuclei (Scalia and Winans, 1976). For the most part, however, the MOB projects to the olfactory tubercle and to certain parts of the pyriform lobe (Heimer, 1975). In turn, these areas are further related to the medial forebrain bundle (Raisman, 1972).

The significant point about projections from the AOB to the amygdala is that this area is related to the medial hypothalamus and medial preoptic area, a region involved in the generation of the cyclic preovulatory surge of gonadotropins in rodents (Raisman, 1972). In fact, electrical stimulation of this area leads to ovulation in rats (Bunn and Everett, 1957).

Although it has been suggested that the VN system might have a function in mammalian reproduction (e.g., Whitten, 1963; Winans and Scalia, 1970; Estes, 1972; Raisman, 1972) there are no known behavioral responses or stimulus discriminations that seem to be crucially dependent on the VN system alone (Murphy, 1976), and until very recently no published evidence that it has any role at all to play in reproductive physiology. However, while both the OLF and VN systems serve a function in the sexual behavior of the male hamster, mating can be seriously affected by deafferentation of the VN system. For example, in one of very few attempts to experimentally demonstrate a function for the VN system in reproductive behavior, Powers and Winans (1975) found that after both VN nerves had been severed, 10/26 (38%) of male hamsters stopped mating entirely, or mated only once in a total of 8 subsequent tests over a four-week period. Prior to this study, the only indication of a possible role for the VN organ in males was the apparent involvement of the VN organ in the flehmen response to female urine (see Estes, 1972).

It has been less apparent that the VN system of mammalian females is involved in the perception of urinary "pheromones". However, a role for the VN system in maternal behavior has been experimentally demonstrated by Fleming et al. (1979) and Mayer and Rosenblatt (in preparation), as have roles for the VN system in reflex ovulation (Johns et al., 1978) and in estrus suppression (the Lee-Boot Effect) (Reynolds and Keverne, in press).

Role of the Vomeronasal System in Reflex Ovulation in Light-Induced
Persistent-Estrus Rats

In a recent series of experiments we (Johns et al., 1978) investigated the role of the VN organ in reflex ovulation among
singly-caged, adult, virgin rats whose estrous cycles had been
suppressed by exposure to continuous light. Prior to the experiments involving disruption of the VN system we had discovered that
brief (1/2 hour) contact with bedding soiled by adult male rats
(of the same strain as the females but from a different breeder)
provided stimuli sufficient to trigger reflex ovulation (Johns and
Komisaruk, 1975; Johns et al., 1978). We investigated further the
nature of the stimulus that had induced ovulation and found that
contact with the urine of intact males was responsible for the
effect. Direct contact with urine of male rats of the same strain
as the females (Sprague-Dawley), but from a different breeder, was
markedly less effective than urine of male rats from the same
breeder while contact with clean bedding or with urine of castrated
male or intact female rats had no effect on ovulation at all (Johns
et al., 1977). The experiments with male-soiled bedding or male
urine were the first to demonstrate that an external chemical stimulus is capable of inducing reflex ovulation.

We originally interpreted our results as demonstrating yet one
more effect of external chemical stimuli on female reproductive
physiology that is mediated by the OLF system. The following test
on the effects of direct contact with male-soiled bedding vs those
of no direct contact with male-soiled bedding forced us to question
this interpretation.

Contact vs No Contact with Male-Soiled Bedding. Adult,
light-induced persistent-estrus rats were tested with male-soiled
bedding from male rats of the same strain but from a different
breeder, and examined 19-22 hours later for signs of ovulation.
They were exposed to direct contact (i.e., male-soiled bedding
was placed on metal trays in their home cages) or no direct contact
with the bedding (i.e., females were exposed to the same amount of
male-soiled bedding on trays placed 3 cm under the wire mesh floor
of the home cage).

The results of this test were: direct contact with soiled
bedding: 49% of 29 rats ovulated; direct contact prevented by a
wire mesh floor: 0 of 15 rats ovulated; (Fisher exact probability
test P = 0.001). These results suggested that the VN organ might
be involved in chemical induction of reflex ovulation, because
contact exposure to the stimulus rather than distance exposure was
apparently necessary.

Lesion Experiment. In order to test the possible role of
the VN system, we used electrocautery to occlude the canals leading
to the VN organ. Lesions were made in the mucosal lining of the

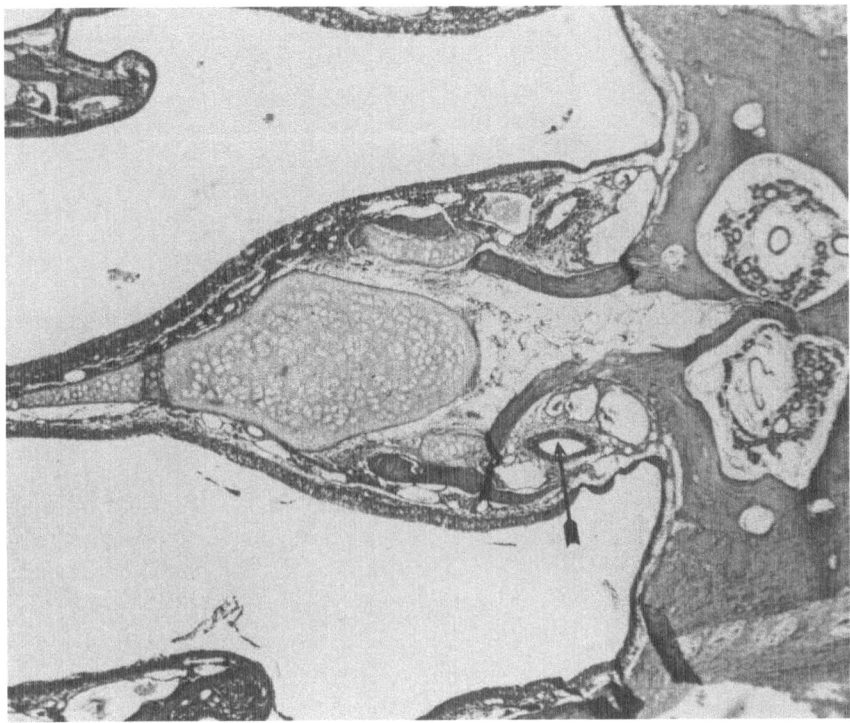

Figure 1b

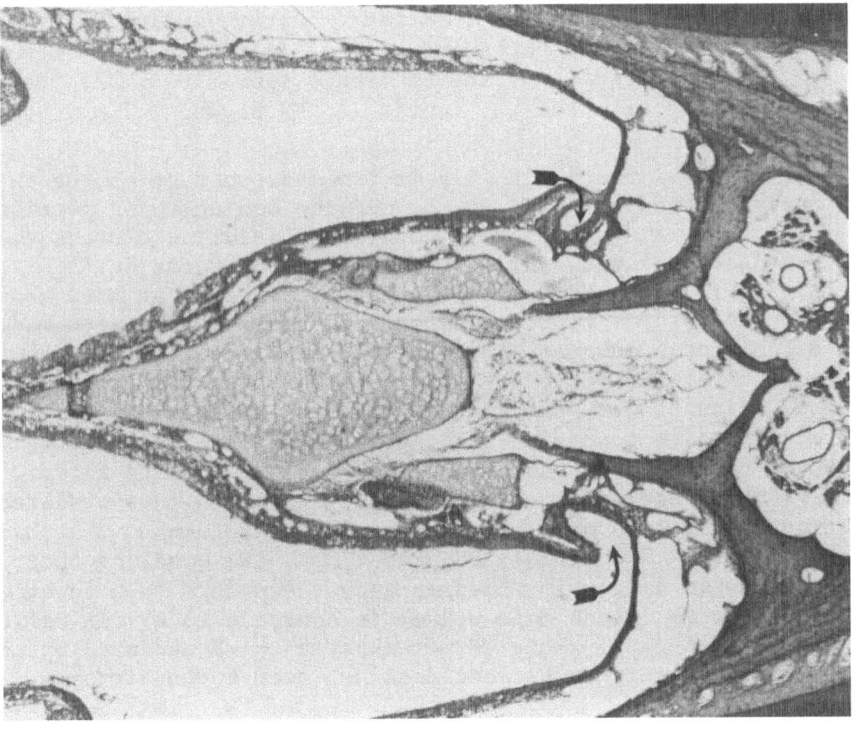

Figure 1a

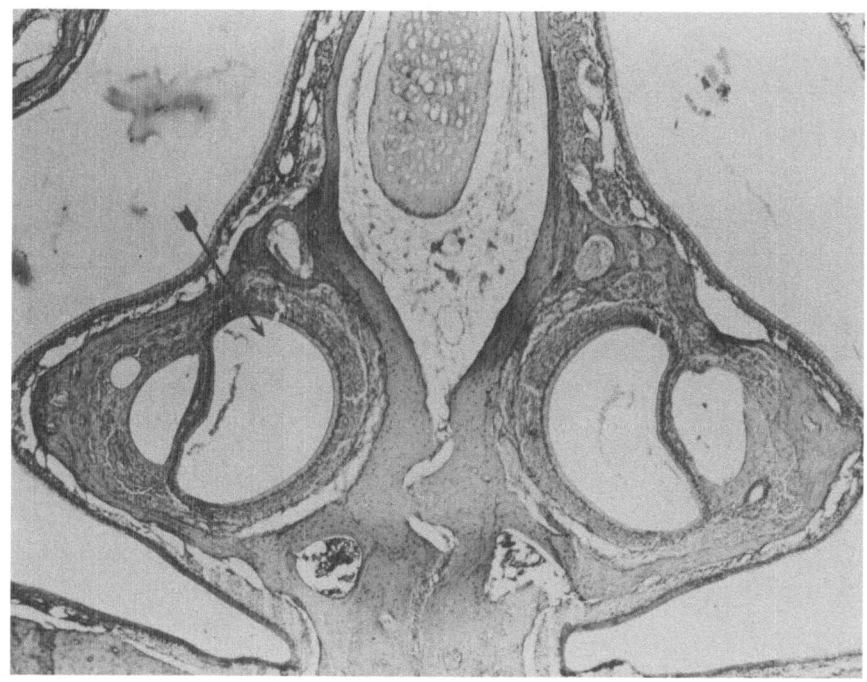

Figures 1a-c. Photomicrographs of frontal sections through the base
              of the nasal septum, 4-6 mm from the surface of the ex-
              ternal nares, of LLPE rats. Arrows indicate the begin-
              ning of the pores and canals leading to the vomeronasal
              (VN) organs and the VN organs themselves. A. Begin-
              ning of pore. B. Canal. C. VN organ.

floor of each nasal vestibule (4-6 mm from the surface of the external nares, approximately over the opening and anterior portion
of the VN canal) using a specially designed lesioning electrode.
A complete description of the technique is in preparation (A.
Mayer and S. Winans). Tissue damage was confined to an area about
2.5 mm long and 0.5 mm wide. Bilateral control lesions were made
in the dorsolateral mucosal lining of the nasal vestibule, with
the same electrode directed away from the VN pore. No lesion
approached areas of OLF epithelium. Eight days later all females
were placed in direct physical contact with male-soiled bedding
(see above) for 1/2 hour and checked for ova 19-22 hours later.
Of females that received control lesions, 54% of 13 rats ovulated.
In contrast, ovulation occurred in only 9% of 22 females that had
VN occlusions (P = 0.006) (see Figures 1, 2). We conclude that
brief exposure to male-soiled bedding (i.e., exposure to a relatively
specific factor in intact male urine) is adequate to elicit reflex
ovulation in LLPE rats. This effect appears to be mediated by the
VN organ and the active substance does not seem to be airborne.

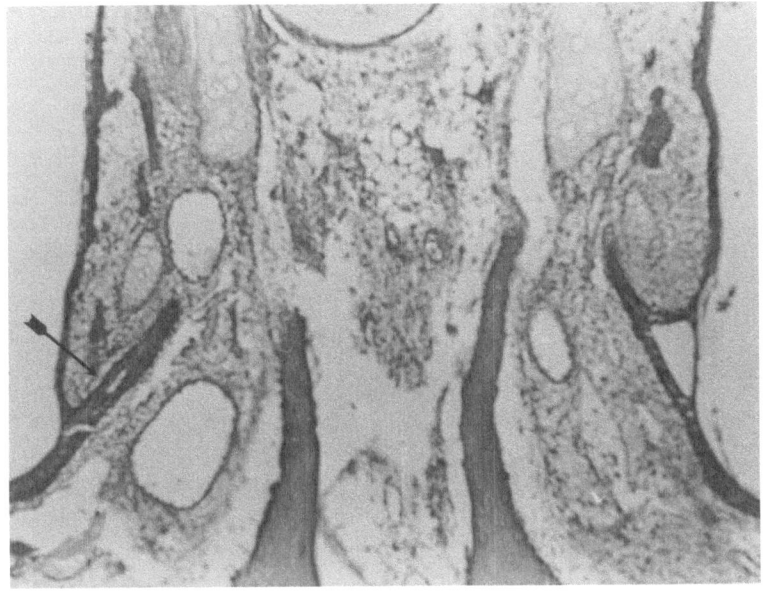

Figure 2a

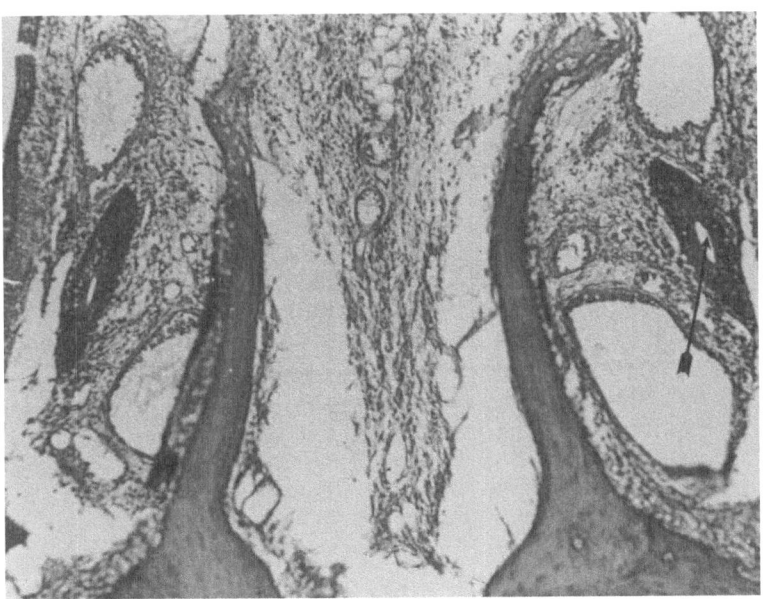

Figure 2b

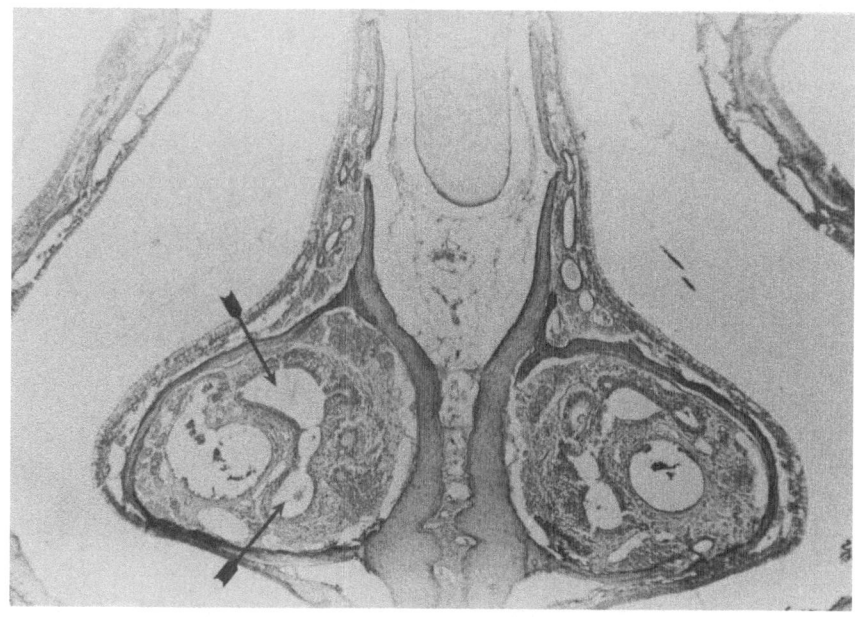

Figures 2a–c. Photomicrographs of frontal sections through the base
              of the nasal septum, 4–6 mm from the surface of the
              external nares, of LLPE rats, nine days after electro-
              cautery to occlude the canals leading to the VN organ.
              Arrows indicate the beginning of the pores and canals
              leading to the VN organs and the VN organs themselves.
              A. Beginning of pore.   B. Canal.   C. VN organ.

In LLPE rats, exposure to males for mating or for mounts without intromission induces a surge in LH while exposure to a novel mating cage (containing male-soiled bedding) apparently induces a smaller rise in LH (Brown-Grant et al., 1973). The possibility that contact with male-soiled bedding, or urine alone, is at least partly responsible for the increase in LH is suggested by the recent finding that (1) such contact is as effective in triggering reflex ovulation in LLPE rats as exposure of the LLPE rat to either male mounts without intromissions or the novel mating cage (Johns et al., in preparation) and (2) in immature female mice, contact with male urine is followed by an immediate (30 min.) elevation in serum LH (Bronson and Maruniak, 1976).

We do not believe that occlusion of the pores leading to VN organs results in hormonal disruption. However, we have not ruled out the possibility that VN pore occlusion may have more widespread effects than simply blocking pathways of the VN system.

Role of the Vomeronasal System in Suppression of Estrous Cycling in Grouped Mice (The Lee-Boot Effect)

Recently, Reynolds and Keverne (in press) investigated the role of the VN organ in suppression of estrus following all-female grouping of mice. Singly caged, adult, virgin mice, showing normal estrous cycling, were grouped in two boxes with seven other females. Three weeks after grouping (in a room isolated from male mouse odors) the 16 mice were again vaginally smeared to establish the absence of estrous cycles. Significantly more females showed suppressed estrus after grouping than before grouping (15/16 vs. 3/16).

Lesion Experiment. In the next experiments, lesions of the VN organ were performed on the grouped females. A small midline incision in the palate of the upper jaw (3-4 mm posterior to the upper incisor teeth) exposed the cartilages enclosing the VN neurosensory epithelium. With the use of a diathermic cutting tool, the VN organ on each side was completely destroyed. Two weeks after surgery, the grouped females (10/13) were cycling normally. In contrast, sham lesions in the lower jaw in eight mice had no effect on the number of females with non-suppressed cycles. Both pre- and post-operatively, 88% were suppressed.

Other females were lesioned to assess the time course of the change. Only four days after lesioning, 11/16 females were beginning to cycle. These findings led Reynolds and Keverne to conclude that "in female mice the VN organs receive pheromonal cues from other females in a group which serves as a block to estrus". Further experiments with dopamine agonists and antagonists suggest that all-female grouping may affect PRL levels via the VN system.

In conclusion, two studies explored the possible role of the
VN system in different aspects of mammalian physiology. By
selectively deafferentating the VN organ, while leaving the OLF
mucosa intact, two quite different functions for the VN system
have now been found: The first study (Johns et al., 1978) reported
that reflex ovulation can be triggered in rats by brief exposure
to male urine and suggested that the VN system is involved in the
effects of urinary chemicals on gonadal function. The second
study (Reynolds and Keverne, in press) provides evidence that the
suppressive effects of all-female grouping on estrous cycling are
mediated by the VN system.

FUTURE DIRECTIONS FOR STUDIES OF THE ROLE OF THE VOMERONASAL
SYSTEM IN REPRODUCTIVE PHYSIOLOGY

The two studies of the role of the VN system in reproduction
in the preceding section leave several issues unresolved. For
example, although it appears that the VN organ mediates effects of
male urine on LLPE rats and effects of all-female grouping of mice,
involvement of the OLF system in these effects has not been ruled
out. In order to demonstrate that the VN system alone mediates
chemical effects on reproductive physiology the OLF system would
have to be rendered non-functional by some method that does not
interfere with normal functioning of the VN system. Only if responses to exposure to male urine or all-female grouping were
shown to be unaffected by such a procedure could participation of
the OLF system in these effects be ruled out completely.

It is also premature to assume that lesions of the VN canal
or VN organ have no effect other than cutting the VN system off
from access to chemical substances. Perhaps other olfactory
systems are affected by the deafferentation procedure. The
septal organ of Masera, portions of the trigeminal nerve and terminal endings of the nervus terminalis are in rather close proximity
to the VN organ and could be damaged. A subtle effect on hormonal
response thresholds could result from the VN lesion procedure
itself or neurotransmitter concentrations might be modified in some
way (see Meredith and O'Connell, 1979). On the other hand, after
deafferentation of the VN organs, LLPE females ovulated in response
to exposure to male mounts without intromission (M. Johns, unpublished data) and estrous cycling in grouped female mice with VN
lesions was resuppressed by a drug that elevates PRL levels. These
findings suggest that VN deafferentation does not prevent normal
hormonal responses to appropriate stimuli.

Studies that implicate the VN system in female physiology have
not identified the factor(s) in urine that induces reflex ovulation
in LLPE rats nor the factor(s) that suppresses estrous cycles in
mice. Are only non-volatile substances involved in these effects?

In the first study (Johns et al., 1978) the answer would appear
to be "yes" because only direct contact with the urine resulted
in reflex ovulation. In the second study (Reynolds and Keverne,
in press) direct contact with other females was not prevented
and no clue is presently available as to whether only non-volatile
chemical substances are involved in estrus suppression.

Not only do we need to know more about the role of the VN
system in mediating changes in female reproductive state (e.g.,
the role of the VN system in puberty, pregnancy block or pregnancy
maintenance, male effects on estrus cycles) but we need to know
more about its role in male reproductive physiology. Furthermore,
controversy over the possible persistence of the VN organ in adult
humans needs to be resolved and variations in VN organ function
among different species of mammal determined.

Finally, we do not know what role the adrenal plays in perception of, and physiologic response to, the urinary chemical
factor that induces reflex ovulation in LLPE rats. Results of
experiments with adrenalectomized rats (M. Johns, H. Feder, B.
Komisaruk, in preparation) showed that adrenalectomized females
tend to ovulate less often than adrenal-intact females after exposure to male-soiled bedding (results did not reach statistical
significance). Is the adrenal involved in other chemically induced
effects of reproductive physiology? Would providing corticosterone
replacement therapy to adrenalectomized rats or ACTH treatment to
adrenal-intact rats affect those responses to external chemicals
that are mediated by the VN system?

SUMMARY AND CONCLUSIONS

The fact that two different functions have now been found for
the VN system, with respect to reproductive neuroendocrinology,
suggests that other functions may soon be found. It therefore may
be more important than it might have seemed previously to attempt to
assess, both at the functional and at the anatomical level, the
relative contributions of the OLF and VN systems to reproductive
physiology and to resolve the present controversy over the presence
or absence of a functioning VN system in adult humans.

At least four techniques have been used to determine whether
the OLF or the VN system mediates chemically-induced changes in reproductive physiology and behavior: 1) occlusion of the pore leading to the VN organ (Johns et al., 1978; A. Mayer and S. Winans,
in preparation); 2) lesion of the VN nerve anterior to the cribiform plate (Planel, 1953) or inside the cranium as it passes between the main OLF bulbs (Winans and Powers, 1974); 3) selective
deafferentation of the OLF system by careful intranasal infusion
of zinc sulfate solution (Winans and Powers, 1974); 4) lesion of

the VN organ itself (Reynolds and Keverne, in press). Such a range
of techniques, along with routine comparisons of direct contact vs.
no direct contact effects of chemical substances on reproduction,
and improved histological and electrophysiological methods, should
increase our understanding of the roles of the VN and OLF systems.

In a paper by Thiessen (1977), it was suggested that in order
for chemicals to act as pheromones they must meet the physical
criteria necessary for transmission at a distance. Still other
writers (Whitten and Champlin, 1972) have expressed the idea that
mammalian pheromones should be volatile. Whether or not the writer
chooses to use the term "pheromone", these views reflect the
emphasis of most current literature. They leave one with the strong
impression that effects of chemical stimuli (e.g. urine) on reproductive physiology as well as behavior are mediated by small quantities
of volatile, low molecular weight substances that act via the distance receptors of the OLF system. However, in women, significant
variations in olfactory sensitivity to involatile esters (but not
to volatile esters) have been observed between ovulation and
menstruation (Mair et al., 1978) and the substance(s) that affects
puberty in female mice (Vandenbergh et al., 1975) and reflex ovulation in LLPE rats (Johns et al., 1978) is not an airborne substance. Moreover, there is evidence that the VN system is involved
in reflex ovulation and estrus suppression, and no evidence that
excludes participation of the VN system in any of the other reproductive functions discussed in this paper. A testable hypothesis
is that, under natural conditions, the OLF system serves to orient
animals toward contact with chemical factors that act on the VN
system. The VN system then mediates chemically stimulated
pituitary hormone release responsible for reproductive maturation,
pregnancy block, successful implantation, estrous cycle modification and reflex ovulation.

REFERENCES

Adler, N.T. 1969. Effects of the male's copulatory behavior on
    successful pregnancy of the female rat. J. Comp. Physiol.
    Psychol. 69: 613.
Berüter, J., Beauchamp, G.K. and Muetterties, E.L. 1973. Com-
    plexity of chemical communication in mammals: urinary compo-
    nents mediating sex discrimination by male guinea pigs.
    Biochem. Biophy. Res. Comm. 53: 264.
Broadwell, R.D. 1975. Olfactory relationships of the telencephalon
    and diencephalon in the rabbit. 1. An autoradiographic study
    of the efferent connections of the main and accessory bulbs.
    J. Comp. Neurol. 163: 239.

Broman, I. 1920. Das organon vomero-nasale Jacobsoni -ein Wassergeruchsorgen. Anat. Hefte 58: 140.

Bronson, F.H. 1974. Pheromonal influences on reproductive activities in rodents, In Pheromones. (M.C. Birch, Ed.), pp. 344-365, American Elsevier, New York.

Bronson, F.H. and Maruniak, J.A. 1976. Differential effects of male stimuli on follicle-stimulating hormone LH and PRL secretion in prepubertal female. Endocrinology 98: 1101.

Browman, L.G. 1977. Light in its relation to activity and estrous rhythms in the albino rat. J. Exp. Zool. 75: 375.

Brown-Grant, K. 1977. The induction of pseudopregnancy and pregnancy by mating in albino rats exposed to constant light. Horm. Behav. 8: 62.

Brown-Grant, K., Davidson, J.M. and Greig, F. 1973. Induced ovulation in albino rats exposed to constant light. J. Endocrinol. 57: 7.

Bruce, H.M. 1959. An extroceptive block to pregnancy in the mouse. Nature (London) 184: 105.

Bruce, H.M. 1965. The effect of castration on the reproductive pheromones of male mice. J. Reprod. Fertil. 2: 138.

Bunn, J.P. and Everett, J.W. 1957. Ovulation in persistent-estrous rats after electrical stimulation of the brain. Proc. Soc. Exp. Biol. Med. 96: 369.

Champlin, A.K. 1971. Suppression of oestrus in grouped mice: The effects of various densities and the possible nature of the stimulus. J. Reprod. Fertil. 27: 233.

Chateau, D., Ross, D., Plas-Roser, S., Roos, M. and Aron, C. 1976. Hormonal mechanisms involved in the control of oestrous cycle duration by the odor of urine in the rat. Acta Endocrinol. 82: 426.

Clee, M.D., Humphreys, E.M. and Russell, J.A. 1975. The suppression of ovarian cyclical activity in groups of mice and its dependence on ovarian hormones. J. Reprod. Fertil. 45: 395.

Cowley, J.J. and Wise, D.R. 1972. Some effects of mouse urine on neonatal growth and reproduction. Anim. Behav. 20: 499.

Dempsey, E.W. and Searles, A.F. 1943. Environmental modification of certain endocrine phenomena. Endocrinology 32: 119.

Dominic, C.J. 1965. The origin of the pheromones causing pregnancy block in mice. J. Reprod. Fertil. 10: 469.

Dominic, C.J. 1966. Observations on the reproductive pheromones in mice. II. Neuroendocrine mechanisms involved in the olfactory block to pregnancy. J. Reprod. Fert. 11: 415.

Drickamer, L.C. 1974. Contact stimulation, androgenized females and the accelerated maturation in female mice. Behav. Biol. 12: 101.

Estes, R.D. 1972. The role of the vomeronasal organ in mammalian reproduction. Mammalia 36: 315.

Everett, J.W. 1942. Certain functional interrelationships between spontaneous persistent estrus, "light estrus" and short-day anestrus in the albino rat. Anat. Rec. 82: 409.

Fleming, A., Vaccarino, F., Tambosso, L. and Chee, P. 1979.
    Vomeronasal and olfactory system modulation of maternal
    behavior in rats. Science 203: 372.
Gasser, R.F. 1975. Atlas of Human Embryos. Harper and Row,
    New York.
Graziadei, P.P.C. 1977. Functional anatomy of the mammalian
    chemoreceptor system, In Chemical Signals in Vertebrates.
    (D. Müller-Schwarze and M.M. Mozell, Eds.), pp. 435–454,
    Plenum Press, New York.
Heimer, L. 1975. Olfactory projections to the diencephalon,
    In Anatomical Neuroendocrinology. (W.E. Stumpf and L.D.
    Grant, Eds.), pp. 30–39, Karger, Basal.
Hoover, J.E. and Drickamer, L.D. 1979. Effects of urine from
    pregnant and lactating female house mice on oestrous cycles
    of adult females. J. Reprod. Fertil. 55: 297.
Johns, M.A. and Komisaruk, B.R. 1975. The role of sensory
    stimuli and the adrenal in mediating ovulation and receptivity
    in persistent-estrus rats, Eastern Conference on Reproductive
    Behavior, Nags Head, N.C. May, Abstract.
Johns, M.A., Feder, H.H. and Komisaruk, B.R. 1977. Male urine
    provokes ovulation in light-induced constant estrus rats.
    Eastern Conference on Reproductive Behavior, U. of Conn.,
    Stoors, Conn., June, Abstract.
Johns, M.A., Feder, H.H., Komisaruk, B.R. and Mayer, A.D. 1978.
    Urine-induced reflex ovulation in anovulatory rats may be a
    vomeronasal effect. Nature 272: 446.
Karlson, P. and Lüscher, M. 1959. "Pheromones": A new term for
    a class of biologically active substances. Nature (London)
    183: 55.
Kenney, A. McM., Lanier, D.L. and Dewsbury, D.A. 1977. Effects
    of vaginal-cervical stimulation in seven species of muroid
    rodents. J. Reprod. Fert. 49: 305.
Kranz, L.K. and Berger, P.G. 1975. Pheromone maintenance of
    pregnancy in Microtus montanus, In Abstracts of the 55th
    Annual Meeting of American Society of Mammologists, p. 66.
McClintock, M.D. 1978. Estrous synchrony and its mediation by
    airborne chemical communication (Rattus norvegicus). Horm.
    Behav. 10: 264.
McClintock, M.D. and Adler, N.T. 1978. Induction of persistent
    estrus by airborne chemical communication among female rats.
    Horm. Behav. 2: 414.
Macrides, F., Bartke, A., Fernandez, F. and D'Angelo, W. 1974.
    Effects of exposure to vaginal odor and receptive females on
    plasma testosterone in the male hamster. Neuroendocrinology
    15: 355.
Mair, R.G., Bouffard, J.A. and Trygg, E. 1978. Olfactory
    sensitivity during the menstrual cycle. Sens. Processes 2: 90.
Marsden, H.M. and Bronson, F.H. 1964. Estrous synchrony in mice:
    Alteration by exposure to male urine. Science 144: 3625.

Meredith, M. and O'Connell, R.J. 1979. Efferent control of stimulus access to the hamster vomeronasal organ. J. Physiol. 286: 301.

Monder, H., Lee, C., Dovonick, P.J. and Burright, R.G. 1978. Male mouse urine extract effects on pheromonally mediated reproductive functions of female mice. Physiol. Behav. 20: 447.

Morgan, P.D., Arnold, G.W. and Lindsay, D.R. 1972. A note on the mating behavior of ewes with various senses impaired. J. Reprod. Fertil. 30: 151.

Murphy, M.R. 1976. Olfactory impairment, olfactory bulb removal, and mammalian reproduction, In Mammalian Olfaction Reproduction Processes and Behavior. (R.L. Doty, Ed.), pp. 95-117, Academic Press, New York.

Planel, H. 1953. Etudes sur la physiologie de l'organ de Jacobson. Arch. Anat. Cytol. Pathol. 36: 197.

Powers, J.B. and Winans, S.S. 1975. Vomeronasal organ: Critical role in mediating sexual behavior of the male hamster. Science 187: 961.

Raisman, G. 1972. An experimental study of the projection of the amygdala to the accessory olfactory bulb and its relationship to the concept of a dual olfactory system. Exp. Brain Res. 14: 395.

Reynolds, J. and Keverne, E.B. Accessory olfactory system and its role in the pheromonally mediated suppression of oestrus in mice. J. Reprod. Fertil. In press.

Richmond, M.E. and Conaway, C.H. 1969. Induced ovulation and oestrus in Microtus ochrogaster. J. Reprod. Fertil., Supplement 6: 357.

Rogers, J.G. and Beauchamp, G.K. 1976. Some ecological implications of primer chemical stimuli in rodents, In Mammalian Olfaction, Reproductive Processes and Behavior (R.L. Doty, Ed.), pp. 181-195, Academic Press, New York.

Sato, N., Haller, E.W., Powell, R.P. and Henkin, R.I. 1974. Sexual maturation in bulbectomized female rats. J. Reprod. Fertil. 36: 301.

Scalia, F. and Winans, S.S. 1976. New perspectives on the morphology of the olfactory system: Olfactory and vomeronasal pathways in mammals, In Mammalian Olfaction, Reproductive Processes and Behavior (R.L. Doty, Ed.), pp. 7-28 Academic Press, New York.

Signoret, J.P. and Mauléon, P. 1962. Action de l'ablation des olfactifs surles mechanismes de reproduction chez la truie. Annal de Biologie Animale, Biochime, Biophysique 2: 167.

Terkel, J. and Sawyer, C.H. 1978. Male copulatory behavior triggers nightly prolactin surges resulting in successful pregnancy in rats. Horm. Behav. 2: 304.

Thiessen, D.D. 1977. Thermoenergetics and the evolution of pheromone communication, In Progress in Psychobiology and Physiological Psychology, Vol. 7, pp. 91-191, Academic Press, New York.

Vandenbergh, J.G. 1967. Effect of the presence of a male on
     sexual maturation of female mice. Endocrinology 81: 345.

Vandenbergh, J.R. 1969. Male odor accelerates female sexual
     maturation in mice. Endocrinology 84: 658.

Vandenbergh, J.G., Whitsett, J.M. and Lombardi, J.R. 1975. Partial
     isolation of a pheromone accelerating puberty in female mice.
     J. Reprod. Fertil. 43: 515.

van der Lee, S. and Boot, L.M. 1955. Spontaneous pseudopregnancy
     in mice. Acta Physio. Pharmacol. Neerl. 4: 442.

Whitten, W.K. 1956. The effect of removal of the olfactory bulbs
     on the gonads of mice. J. Endocrinol. 14: 160.

Whitten, W.K. 1959. Occurrence of anoestrus in mice caged in
     groups. J. Endocrinol. 18: 102.

Whitten, W.K. 1963. Is the vomeronasal organ a sex chemoreceptor
     in mice? Second Asia and Oceania Cong. Endocrinol. Sydney.

Whitten, W.K. and Bronson, F.H. 1970. Role of pheromones in
     mammalian reproduction, In Advances of Chemoreception. (D.
     Johnson, C. Moulton and A. Turk, Eds.), pp. 309-325, Appleton-
     Century-Crofts, New York.

Whitten, W.K. and Champlin, A.K. 1972. Pheromones and olfaction
     in mammalian reproduction. Biol. Reprod. 19: 149.

Whitten, W.K., Bronson, F.H. and Greenstein, J.A. 1968. Estrus-
     inducing pheromone of male mice: Transport by movement of
     air. Science 196: 584.

Winans, S.S. and Powers, J.B. 1974. Neonatal and two stage
     olfactory bulbectomy: Effects on male hamster sexual
     behavior. Behav. Biol. 10: 461.

Winans, S.S. and Scalia, F. 1970. Amygdaloid nucleus: New
     afferent input from the vomeronasal organ. Science 170: 330.

CHEMICAL STUDIES OF HAMSTER REPRODUCTIVE PHEROMONES

Alan G. Singer[1], Foteos Macrides[2] and William C. Agosta[1]

[1]The Rockefeller University, 1230 York Avenue
New York, NY 10021

[2]The Worcester Foundation for Experimental Biology
Shrewsbury, MA 01545

INTRODUCTION

Male golden hamsters respond to the vaginal discharge of females in a number of ways. The vaginal discharge attracts males,
stimulates their sexual behavior, decreases their aggressive behavior, reduces their tendency to scent mark with the flank gland,
and increases their plasma testosterone levels (cf. Johnston, 1977).
It is not clear to what extent these are responses to different
compounds in the discharge and to what extent they are different
aspects of the response to a single compound or mixture. This is
one of the questions which we are attempting to answer as we study
the chemistry of the discharge. This paper is primarily concerned
with our recent work on the substances in vaginal discharge which
stimulate sexual behavior. However, since attraction is an important component of the sexual response, we first will review the results of earlier work on the attractant response.

IDENTIFICATION OF AN ATTRACTANT PHEROMONE

Our work on the chemistry of hamster vaginal discharge began
with the isolation of compounds that would attract hamsters to a
remote odor source. The behavioral assay used to monitor isolation
of the attractant has been described before (Singer et al., 1976).
The substance to be tested is placed in one of two jars under the
ends of a male hamster's home cage. If the test substance is vaginal discharge or an active fraction, the hamster will dig down

through the bedding, apparently attempting to reach the odor source
in the jar, which is isolated from the cage by a wire screen. A
typical response to fresh vaginal discharge is two minutes of vigorous digging and gnawing at the wire covering the mouth of the jar
containing the discharge. The response is not displayed by females,
and food or water in the jars will not produce the response.

Attractant activity was isolated from the vaginal discharge by
distillation. Two active gas chromatographic fractions were detected
with the behavioral assay, and compounds in these fractions were
identified by mass spectrometry. Dimethyl disulfide was identified
in the first active fraction, and dimethyl trisulfide was identified
in the second. Authentic dimethyl disulfide proved to be about half
as active as the fresh vaginal discharge and accounted for the
activity of the first fraction (Singer et al., 1976). It appears
that the activity of the second fraction also is accounted for by
dimethyl disulfide. This fraction always contains some disulfide
produced by thermal disproportionation of the trisulfide during
collection from the gas chromatograph. Since the purest dimethyl
trisulfide that we have been able to prepare by gas chromatography
contains at least 0.07% disulfide, the 2 ng per female of the trisulfide collected in the second fraction must have had at least
1.4 pg of the disulfide. This amount of dimethyl disulfide is
sufficient to account for the attractiveness of the second fraction
(O'Connell et al., 1979).

The difference noted between the activity of fresh vaginal
discharge and dimethyl disulfide seems not to be accounted for by
other single volatile compounds. A number of other gas chromatographic fractions containing one or a few peaks have been tested
in the attractant assay without obtaining consistent activity. On
the other hand, fractions containing a large number of compounds
frequently were active as attractants. Therefore, we next tested
mixtures of twelve of the most abundant volatile compounds in
hamster vaginal secretion, the low molecular weight acids and alcohols listed in Table 1. It is interesting that the mixture of
acids obtained is qualitatively and quantitatively quite similar
to that found in primate vaginal secretions (Michael et al., 1976).
This similarity suggests that these acids are bacterial metabolites
in the hamster, as has been established in the rhesus monkey.

Neither the mixture of acids nor the mixture of alcohols was
active in the attractant assay. About 60% activity was obtained
with a grand mixture of the acids, alcohols, and dimethyl disulfide (i.e., all thirteen compounds in Table 1), and this grand
mixture was not significantly more active than dimethyl disulfide
alone (O'Connell et al., 1978). The presence of a small quantity
of dimethyl disulfide thus could account for most of the attraction
to a remote odor source, even when this compound was assayed in

Table 1. Amounts of some volatile compounds identified in hamster vaginal discharge.[a]

	Micrograms per female
Dimethyl Disulfide	0.005
Acids	
Acetic	130.
Propionic	24.
Isobutyric	1.8
Butyric	24.
α-Methylbutyric	0.51
Isovaleric	3.6
Isocaproic	1.7
Phenylacetic	0.25
Alcohols	
Methyl	0.35
Ethyl	0.95
n-Propyl	0.37
n-Butyl	0.20

[a]Adapted from O'Connell et al., 1978.

the presence of relatively large quantities of inactive biological
odorants. These findings indicate that dimethyl disulfide is an
attractant pheromone which may serve to draw males into the immediate vicinity of females.

CHEMICAL STUDIES OF THE MOUNTING PHEROMONE

     The next phase of our work on the chemistry of hamster re-
productive pheromones is the main subject of this paper, the mounting pheromone. When an anesthetized male hamster is placed in the
cage of a normal adult male, the latter will spend some time investigating this novel object. Usually about two-thirds of this
time is devoted to investigating the head and flanks, and one-third
of the investigation time is devoted to the hindquarters. If the
hindquarters of the anesthetized hamster are scented with the vaginal secretion from an estrous female, the response of the alert
male to this surrogate female is quite different (cf. Macrides
et al., 1977). There is less investigation time at the head and
flanks of the surrogate and much more attention is devoted to the
hindquarters. In addition the tendency of the male to scent mark
with the flank gland is reduced. However, the most striking response of the male hamster to vaginal discharge in this bioassay is
the large increase in sexual behavior. The active male repeatedly
will mount the anesthetized male and exhibit bouts of pelvic
thrusting as if attempting to intromit. The substance or substances in the vaginal discharge stimulating these intromission
attempts (i.e., mounts accompanied by pelvic thrusting) we are
calling the mounting pheromone.

     Typical data from the mounting assay are presented in Table 2.
The males are tested with a surrogate placed in their home cages
for five minutes. The males receive one test every 2 to 3 days,
and their behaviors during the scented and unscented conditions
are compared. The number of intromission attempts is, of course,
our most direct measure of the ability of vaginal discharge or a
chemical fraction of the discharge to elicit overt sexual behavior.
However, the ability of substances to attract males preferentially
to the hindquarters of the surrogate also is of interest. We
consistently have found that the percent of total investigation
time spent at the hindquarters is high when the number of intromission attempts is high, although the percent time at the hindquarters can be increased without an increase in mounting activity
(see below). The typically concomitant increase in these two
responses may indicate that the mounting pheromone also is
attractive to male hamsters, or that certain precopulatory behaviors such as sniffing and licking the hindquarters must be elicited
by an attractant before the mounting pheromone can be effective.

Table 2. Mean responses ($\pm$ S.E.M.) of male hamsters (N = 10) to an unscented surrogate or to a surrogate whose hindquarters were scented with fresh vaginal discharge.

	Unscented	Fresh Discharge
Investigation time at head and flanks (sec)	40 $\pm$ 8	23 $\pm$ 4
Time at hindquarters (sec)	24 $\pm$ 9	177 $\pm$ 24*
% Time at hindquarters	36 $\pm$ 8	85 $\pm$ 4*
Scent marks	2.0 $\pm$ 0.6	0.5 $\pm$ 0.3
Intromission attempts	0.1 $\pm$ 0.1	3.2 $\pm$ 1.0*

*Different from unscented condition ($p < 0.05$; Wilcoxon matched-pairs signed-ranks test).

A portion of the attraction to the hindquarters in the mounting assay can be attributed to the volatile compounds of the vaginal discharge that are extracted by distillation. This is demonstrated by the small but significant increase in the percent time at the hindquarters when dimethyl disulfide is tested in this assay (Table 3). However, much of this investigatory response to the discharge must be stimulated by the relatively less volatile compounds since the residue from distillation of vaginal discharge is inactive in the attractant assay (Singer et al., 1976), while it substantially increases attraction to the hindquarters in the mounting assay (Table 4). The mixtures of volatile compounds that were tested in the attractant assay also were tested in the mounting assay (Table 3). Neither the mixture of acids nor the mixture of alcohols was attractive in the mounting assay. Only dimethyl disulfide and the grand mixture that included dimethyl disulfide had significant effects on the percent time at the hindquarters. The most important finding presented in Table 3, however, is that neither dimethyl disulfide nor the mixtures had any effect on the number of intromission attempts. The attractant pheromone, dimethyl disulfide, is not the mounting pheromone.

Table 3. Mean responses ($\pm$ S.E.M.) of male hamsters (N = 18) to
          mixtures of volatile compounds in the mounting assay.[a]

	% Time at Hindquarters	Intromission Attempts
Unscented	21 $\pm$ 3	0.1 $\pm$ 0.1
Alcohols	20 $\pm$ 3	0.4 $\pm$ 0.3
Acids	25 $\pm$ 4	0.3 $\pm$ 0.2
Dimethyl disulfide	35 $\pm$ 5*	0.1 $\pm$ 0.1
Grand mixture	41 $\pm$ 5*	0.1 $\pm$ 0.1
Fresh discharge	82 $\pm$ 2*	5.7 $\pm$ 1.3*

[a]Adapted from O'Connell et al., 1978

*Different from unscented condition (p < 0.05; Wilcoxon matched-
 pairs signed-ranks test).

Table 4. Distribution of behavioral activity in the mounting assay
          on fractionation of vaginal discharge.[a]

	% Time at Hindquarters	Intromission Attempts
Unscented	21 + 3	0.1 + 0.1
Distillate	53 + 6*	0.1 + 0.1
Aqueous residue	78 + 5*	2.3 + 0.8*
Unscented	20 + 5	0
Ether extract	45 + 6*	0.1 + 0.1
Aqueous residue	62 + 8*	2.5 + 1.0*
Unscented	20 + 3	0
Chloroform/methanol extract	55 + 6*	1.8 + 1.4
Aqueous residue	58 + 5*	3.8 + 1.9*
Unscented	25 + 5	0.6 + 0.3
XAD-2 resin extract	67 + 5*	3.3 + 1.0*
Aqueous residue	76 + 4*	4.2 + 1.1*

[a]Means (+ S.E.M.) are based on the responses of 10 males per assay.

*Different from unscented condition ($p < 0.05$; Wilcoxon matched-
pairs signed-ranks test).

The results of a number of attempts to separate the mounting activity from an aqueous dispersion of vaginal secretion are shown in Table 4. The main conclusion from these results is that it was not possible to get a good recovery of mounting pheromone by distillation or by extraction with organic solvents. In all cases the residue was significantly active and only with the XAD-2 resin was an extract obtained which significantly increased the number of intromission attempts. (XAD-2 is a copolymer of styrene and divinylbenzene that has been used very effectively for extracting prostaglandins, steroids, and steroid conjugates from physiological fluids. To recover absorbed compounds from the polymer it is extracted with methyl or ethyl alcohol (Bradlow, 1968).) A second point is that the mounting activity is not particularly volatile. The distillate did not produce any increase in the number of intromission attempts, and it was possible to concentrate the methanol extract of the XAD-2 resin under vacuum and still retain significant activity. Clearly the extraction with resin is the most promising of these techniques, and we are exploring ways of improving the recovery by this method. We can say that greater quantities of resin in longer columns and repeated extractions are not able to reduce the activity of the residual aqueous material.

Another approach that we have taken to the isolation of this pheromone is based on our observation that the pheromone activity is associated with macromolecules (see below). When we say that mounting activity is associated with a macromolecule, of course it always means that the fraction containing the macromolecule is active. It is not intended, however, to specify the chemical nature of the pheromone. It may mean that the macromolecule itself is active or that a piece of the macromolecule is the active substance, or that another molecule bound to the macromolecule is the active compound, or that the combination of macromolecule and bound molecule is the pheromone. In other words, the macromolecule may be the pheromone itself, a carrier of the pheromone, or a part of the pheromone.

We have found that hamster mounting pheromone activity definitely is associated with high molecular weight compounds. About half of the vaginal secretion and full mounting activity can be dissolved in 1.5 $\underline{M}$ ammonium acetate. When this extract was subjected to gel permeation chromatography on a column of Bio-Gel P-100 the results shown in Figure 1 and Table 5 were obtained. Mounting activity was located in fractions which have retention volumes corresponding to those of globular proteins of molecular weight between 15,000 and 60,000 daltons. Because significant activity was extracted with the XAD-2 resin, it was expected that good activity also would be obtained in the low molecular weight fraction (fraction 4 in Figure 1). However, this was not realized even when the gel chromatography was carried out at 45°C (conditions that have been generally useful for separating small molecules such as

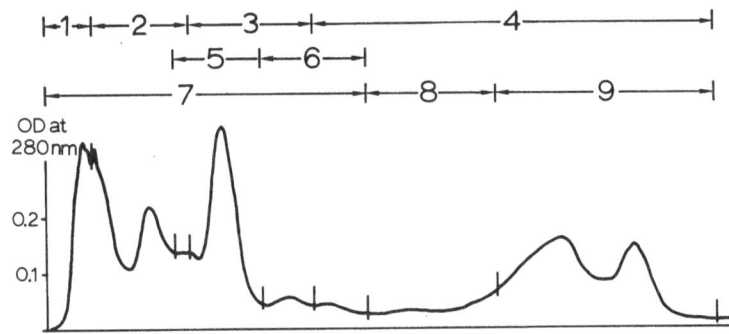

Figure 1. Gel permeation chromatography of hamster vaginal dis-
          charge extract. Material was eluted from the 2.0 cm
          by 40 cm column of Bio-Gel P-100 at 16 ml/hr with 1.5 M
          ammonium acetate at pH 6.5. Fractions 1 to 6 were ob-
          tained from two runs at 22°C; fractions 7 to 9 were
          obtained at 45°C.

Table 5. Behavioral activity of gel permeation chromatography
          fractions described in Figure 1.[a]

	% Time at Hindquarters	Intromission Attempts
Unscented	36 + 8	0.1 + 0.1
Fraction 1	42 + 7	0
Fraction 2	58 + 6	1.3 + 0.7
Fraction 3	71 + 6*	1.6 + 0.8*
Unscented	32 + 5	0.1 + 0.1
Fraction 4	51 + 8	1.0 + 0.6
Unscented	30 + 8	0.5 + 0.3
Fraction 5	64 + 8*	1.6 + 0.9
Fraction 6	34 + 8	0.4 + 0.3
Unscented	32 + 8	0.2 + 0.2
Fraction 7	74 + 5*	1.9 + 0.7*
Fraction 8	45 + 5	0.2 + 0.2
Fraction 9	54 + 10	0.4 + 0.3

[a]Means ($\pm$ S.E.M.) are based on the responses of 10 males per assay.

*Different from unscented condition ($p < 0.05$; Wilcoxon matched-
 pairs signed-ranks test).

steroids from carrier proteins (Westphal, 1969)). Only the macro-molecular fraction (fraction 7 in Figure 1) produced a significant increase in the number of intromission attempts.

At present, our work is concentrating on attempts to isolate and characterize the high molecular weight activity. The results of a crude ion exchange fractionation of the high molecular weight fraction are shown in Table 6. Activity is obtained in the fraction which is bound to DEAE cellulose and eluted with 2.0 M ammonium acetate at pH 6. Although this does not represent much of a purification, since 80% of the sample is in the active fraction, it does indicate that it is possible to recover good activity by ion exchange chromatography. We are now attempting to fractionate the high molecular weight activity by this method, monitoring the fractions with a combination of behavioral assay and gel electrophoresis.

Table 6. Behavioral activity of ion exchange chromatography fractions derived from the high molecular weight fraction of vaginal discharge.[a]

	% Time at Hindquarters	Intromission Attempts
Unscented	32 ± 5	0.1 ± 0.1
Fraction 1	47 ± 8	0
Fraction 2	74 ± 6*	2.1 ± 0.5*
Fresh discharge	76 ± 6*	2.1 ± 0.8*

[a]Fraction 1 was eluted with 0.2 M ammonium acetate and fraction 2 was eluted with 2.0 M ammonium acetate at pH 6.0 from a 1.5 cm by 30 cm column of diethylaminoethyl cellulose. Means ($\pm$ S.E.M.) are based on the responses of 10 males.

*Different from unscented condition ($p < 0.05$; Wilcoxon matched-pairs signed-ranks test).

To sum up our findings thus far, we can say that there is mounting pheromone activity associated with fractions containing proteins of 15,000 to 60,000 daltons and that good activity can be recovered by ion exchange chromatography. From the extractions it appears that there is activity associated with lower molecular weight materials, but we were unable to recover good activity in low molecular weight fractions by our conditions of gel permeation chromatography. It remains to be determined whether non-volatile macromolecules are active themselves, or whether they function as carriers or precursors for a volatile active component.

ACKNOWLEDGEMENTS

This work is supported by NIH grants HD11164 and NS12344 and Rockefeller Foundation grant RF70095.

REFERENCES

Bradlow, H.L. 1968. Extraction of steroid conjugates with a neutral resin, Steroids 11: 265-272.

Johnston, R.E. 1977. Sex pheromones in golden hamsters, in "Chemical signals in vertebrates" (D. Müller-Schwarze and M.M. Mozell, eds.), pp. 225-251, Plenum Press, New York.

Macrides, F., Johnson, P.A. and Schneider, S.P. 1977. Responses of the male golden hamster to vaginal secretion and dimethyl disulfide: Attraction versus sexual behavior. Behav. Biol. 20: 377-386.

Michael, R.P., Bonsall, R.W. and Zumpe, D. 1976. Evidence for chemical communication in primates, Vitamins and Hormones 34: 137-186.

O'Connell, R.J., Singer, A.G., Macrides, F., Pfaffmann, C. and Agosta W.C. 1978. Responses of the male golden hamster to mixtures of odorants identified from vaginal discharge, Behav. Biol. 24: 244-255.

O'Connell, R.J., Singer, A.G., Pfaffmann, C. and Agosta, W.C. 1979. Pheromones of hamster vaginal discharge: Attraction to femtogram amounts of dimethyl disulfide and to mixtures of volatile components, J. Chem. Ecol. 5: 575-585.

Singer, A.G., Agosta, W.C., O'Connell, R.J., Pfaffmann, C., Bowen, D.V. and Field, F.H. 1976. Dimethyl disulfide: An attractant pheromone in hamster vaginal secretion, Science 191: 948-950.

Westphal, U. 1969. Assay and properties of corticosteroid-binding globulin and other steroid-binding serum proteins, Methods Enzymol. 15: 761-770.

CHEMICAL STUDIES OF THE PRIMER MOUSE PHEROMONES

M. Novotny, J.W. Jorgenson, M. Carmack, S.R. Wilson

Chemistry Department, Indiana University, Bloomington, IN 47405

E.A. Boyse, K. Yamazaki

Memorial Sloan-Kettering Cancer Center, New York, NY 10021

M. Wilson, W. Beamer, and W.K. Whitten[1]

The Jackson Laboratory, Bar Harbor, ME 04609

INTRODUCTION

Primer Pheromone Effects in Mice

Several primer pheromone effects in laboratory mice are now well known. These include the Whitten, Bruce, Vandenbergh, and Lee-Boot effects. The first three phenomena are associated with the effect of male mouse urine on females. In the Whitten effect, male urine promotes a regular four to five day estrous cycle in female mice (Whitten, 1956, 1958; Marsden and Bronson, 1964). In the Bruce effect, urine from a "strange" male mouse will cause failure of implantation and pregnancy in a female recently mated with another male (Bruce, 1960; Bruce and Parrott, 1960; Parkes and Bruce, 1962; Dominic, 1966). The Vandenbergh effect concerns the acceleration in the onset of puberty and first estrus in pre-pubertal female mice exposed to urine of males (Vandenbergh, 1967, 1969; Colby and Vandenbergh, 1974). The Lee-Boot effect is a phenomenon associated with confined groups of female mice, in which females become anestrous or develop spontaneous pseudopregnancies (reviewed by Parkes and Bruce, 1961).

Similar effects have been demonstrated in colonies of deer-mice (Bronson and Eleftheriou, 1963; Bronson and Marsden, 1964) and wild house mice (Chipman and Fox, 1966), indicating that these effects are potentially operative in wild mouse populations and are not necessarily a laboratory curiosity resulting from extensive inbreeding. However, among inbred strains there is variability in the ability of males to produce pheromones and females to respond (Hoppe, 1975).

It is reasonable that the three primer effects associated with male urine are actually mediated by the same chemical messengers. The work of Bronson and Desjardins (1974) demonstrates the rapid and pronounced effect on levels of lutenizing hormone and estradiol occurring in prepubertal female mice exposed to males. Similar hormonal responses in anestrous females might cause renewal of estrous cycles (Whitten Effect), while increasing estradiol levels in mated females might prevent implantation of the blasto-cysts (Bruce Effect). Thus the particular effect observed might simply depend on the condition of the female, the chemical messen-ger remaining the same (Bronson, 1971; Marchlewska-Koj, 1977).

It is unclear as to whether the urinary pheromone itself is volatile or not. Initial work with a wind tunnel by Whitten et al. (1968) suggested that the Whitten effect pheromone is volatile and airborne. Hoppe (1975) found that pregnancy blocking activity existed in a condensate obtained by bubbling nitrogen through urine and condensing the vapors in a dry ice acetone trap. However, work by Vandenbergh et al. (1976) seems to indicate that the pheromone mediating the puberty accelerating effect is of higher molecular weight and hence non-volatile. Our own work using a vapor-trapping system (operated for 48 hours) also failed to demonstrate a vola-tile component of urine capable of producing the puberty accelera-ting effect. In the experiments which tend to favor a volatile pheromone, it is not clear if transport of urine substances within aerosols can be ruled out. In view of the high sensitivity of female mice to male urine (pheromonal activity demonstrable in urine diluted 1,000-fold, personal communication from W.K. Whitten) transport of pheromone within aerosols may be important. This question of volatility will be considered again later.

Mating Preference in Laboratory Mice

Recent evidence shows that a male mouse caged with two gene-tically different estrous females will show a preference in mating with a particular female (Yamazaki, et al., 1976). This preference is controlled by genes present in both males and females in the major histocompatibility complex (H-2 region in mice). The females apparently produce a signal under control of an H-2 gene(s), while

males possess differential responsivity, also controlled by an
H-2 gene(s) (Yamazaki, et al., 1978; Yamaguchi et al., 1978;
Andrews and Boyse, 1978). It is suspected that the female's
signal may be volatile in nature and of urinary origin. The general
trend of this mating preference may be toward heterozygosity of the
H-2 region in a population, a trend which would confer a wider range
of immune responsiveness to the population (Yamazaki et al., 1976).

CHEMICAL INVESTIGATIONS OF MOUSE URINE

Partial Purification of Puberty-Accelerating Pheromone

Vandenbergh has demonstrated that the puberty-accelerating
pheromone is associated with the "pre-albumin" proteins present
in urine (Vandenbergh et al., 1975). Marchlewska-Koj (1977) also
demonstrated Whitten and Bruce Effect activities associated with
this material. These proteins have a molecular weight of approximately 17,500, are synthesized in the liver, and their synthesis
is regulated by sex hormones (Finlayson et al., 1974).

Using exhaustive ultrafiltration, Vandenbergh was able to remove puberty-accelerating activity from this protein material
(Vandenbergh, personal communication). It was found that material
passing through the smallest membranes (mol. wt. < 1000) did show
activity. Vandenbergh chromatographed this filtrate on Sephadex
G-15 and obtained a particular active fraction, which according
to its retention behavior on the gel should have a molecular
weight of about 860 (Vandenbergh et al., 1976).

We have repeated and extended this isolation work through
further chromatography on DEAE-Sephadex anion-exchanger and with
high-voltage paper electrophoresis.

The upper chromatogram of Figure 1 shows Sephadex G-15
chromatography of whole urine from male BALB/cWt mice. Two
fractions were cut and subjected to recycling chromatography
shown in the lower two chromatograms of Figure 1. Recycling
chromatography is a technique enhancing the separation of materials
by multiple passes through the chromatographic column (Chuang and
Johnson, 1973). This process yielded five fractions (1A, 1B, 1C,
2A and 2B) which were lyophilized and dissolved in pH 7.0 phosphate buffer. Fractions were reconstituted in a volume of buffer
corresponding to one-half the volume of the original urine used
in the isolation. The fractions were then used in a puberty
acceleration bioassay. This bioassay consisted of spraying 0.1 ml

of solution into the female's nose at 2:30 p.m. and 4:00 p.m. on
the first day of the assay. The females were then sacrificed and
uteri were weighed 48 hours after the first application of urine.
Females were the offspring of a cross between SWR males and SJL
females, and were between 23 and 28 days in age and 14 and 15 grams
in weight (Bioassays performed by M. Wilson and W.K. Whitten,
Jackson Laboratory, Bar Harbor, Maine). The results of the
bioassay are reported in Table 1, with fraction 1A showing the
most activity.

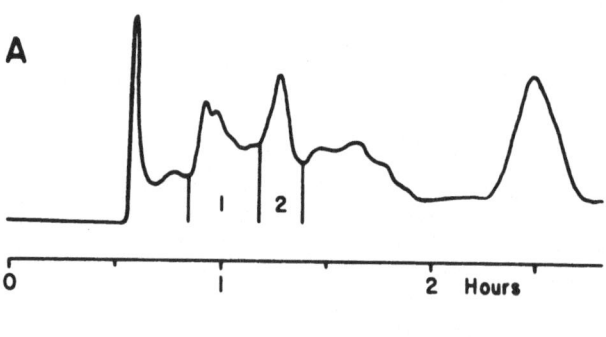

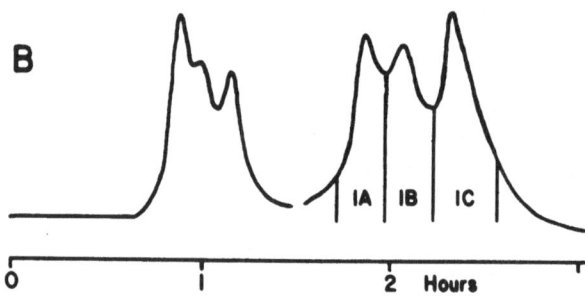

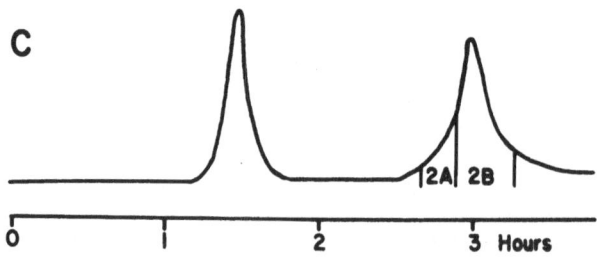

Figure 1.   Sephadex G–15 chromatography of whole male urine, BALB
             cWt (A). Recycle chromatography of fraction (A) 1 (B) and
             fraction 2 (C). Column dimensions 1.5 x 60 cm. Mobile
             phase 0.1 F ammonium bicarbonate, pH = 7.8. Detection:
             adsorption at 280 nm.

Table 1. Mean uterine weight for mice treated with urine fractions.

Fraction Number	N	Uterine Weight (mg) $\bar{x}$	$\pm$ S.E.
1A	5	51.2	8.9
1B	6	22.3	1.0
1C	6	28.2	7.5
2A	7	20.2	1.0
2B	6	29.1	5.8
saline	8	26.5	5.1
3	10	38.6	8.2
4	10	20.1	1.3
5	10	20.5	1.7
saline	10	19.1	0.8
6	14	35.7	7.3
7	14	20.4	1.4
8	12	27.4	4.3
9	10	24.1	0.9
10	10	21.2	1.9
saline	10	19.2	1.2
urine/10	13	49.4	5.2
11	11	24.1	2.0
12	12	34.7	4.4
13	12	43.3	9.4
14	12	23.0	1.3
15	11	21.1	1.0
12 + 13	12	52.7	7.9
urine/10	10	59.6	9.2
carboxy-peptidase	15	27.8	4.6
control	15	31.9	6.2
saline	15	19.9	0.8
urine/10	15	49.7	6.8

(urine/10 = urine diluted 10-fold, saline = buffer control)

Since we became aware that the higher molecular weight protein
material in urine has activity, we were concerned that the activity
found in fraction 1A simply represented carry-over of activity from
the earlier-eluting protein material (the first peak in the upper
chromatogram of figure 1). The Sephadex G-15 separation was repeated in order to test for activity in fractions eluting before
fraction 1A. Figure 2 shows this separation where three fractions
(3, 4 and 5) eluting before 1A were collected. The bioassay data
are shown in Table 1. The activity of the high molecular weight
protein peak (fraction 3) is seen in accordance with Vandenbergh's
results. However, there is minimal activity in fractions 4 and 5,
indicating that fraction 1A is a uniquely active fraction of lower
molecular weight.

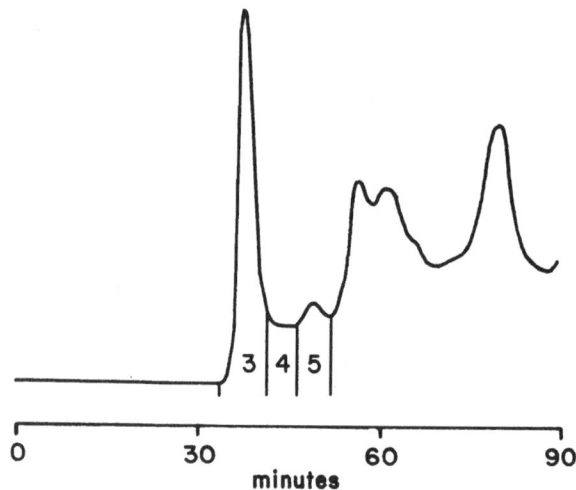

Figure 2.   Sephadex G-15 chromatography of whole male urine.   Con-
            ditions as in Figure 1.

Fraction 1A was further subjected to chromatography on DEAE-
Sephadex anion-exchanger. The upper chromatogram of Figure 3 shows
that fraction 1A is composed of a variety of compounds, with at
least one (fraction 10) being highly acidic, eluting only at low
pH. Most activity was found in fraction 6, as indicated in Table
1. The lower chromatogram of Figure 3 shows further cuts (fractions
11, 12, 13, 14 and 15) taken in the region of fraction 6. Of these,
fractions 12 and 13 show the most activity. However, it is interesting to see that by combining fractions 12 and 13 the highest
activity was obtained (this might suggest the presence of two
active components).

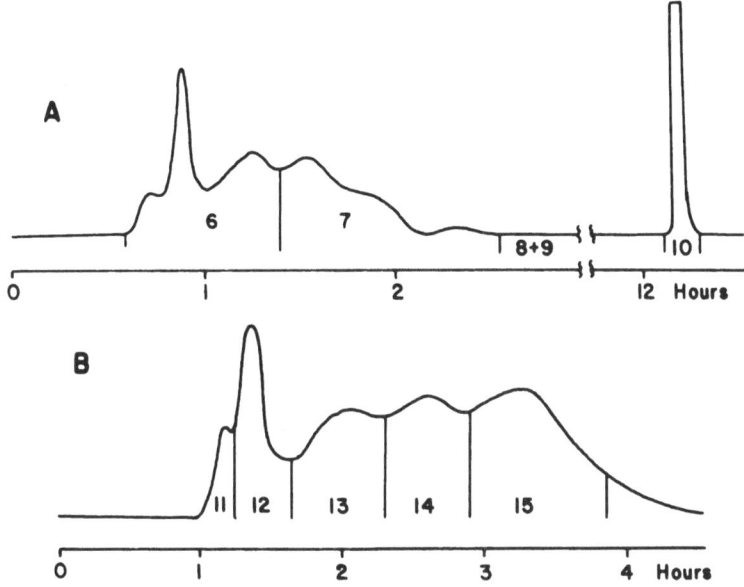

Figure 3. DEAE-Sephadex A-25 chromatography of fraction 1A.
Mobile phase gradient from 0.5 F ammonium acetate (pH
7.0) to 1.0 F acetic acid, step gradient to 1.0 F phos-
phoric acid to elute fraction 10 (A). Mobile phase
0.5 F ammonium acetate, isocratic (B). Column 2.5 x
20 cm. Detection: absorption at 280 nm.

High-voltage paper electrophoresis was performed on fractions
11 through 15. Separated bands were detected with fluorescamine, a
fluorogenic reagent specific for primary amines. In this manner,
fractions 12 and 13 were seen to be a mixture of several substances
(only those with primary amino-groups were detected), although in
each fraction a single spot was clearly prominent.

In view of Vandenbergh's suggestion that the pheromone is a
peptide (Vandenbergh et al., 1976), an experiment was performed to
test this possibility. Fractions 12 and 13 were combined,
lyophilized and subjected to treatment with carboxypeptidase-Y, a
non-specific peptidase from yeast. The material was incubated
with 0.1 mg of the enzyme in one ml of pH 5.5 pyridine-acetate
buffer (0.1 F) at 35°C for two hours. The material was then re-
chromatographed on Sephadex G-15 to remove the enzyme and the
pyridine buffer. A control sample was subjected to the same treatment with the exception that the enzyme was omitted. Bioassay
data are reported in Table 1, and indicate no significant effect
of the enzyme. As with any reaction of this type, the rate of
reaction is low as the substrate concentration becomes low. Thus
the enzyme will be unable to remove all traces of a peptide. If
the pheromone is a peptide and is active at low concentrations,

then the enzyme will most likely not be able to significantly
diminish the pheromonal activity.

A point relevant to the question of the pheromone's volatility
is our observation that following repeated purification and lyophilization, the fractions found active continue to possess an odor
reminiscent of the whole mouse urine. We have also found that
following extensive purging of whole urine with helium for a period
of 48 hours, the urine retains most of its strong odor. It may be
that the actual active chemical(s) is a volatile molecule bound to
a larger "carrier" molecule. Possible advantages of a scheme like
this could include: slow, consistent release of the pheromone from
the carrier; control of synthesis and/or excretion of the pheromone
by the carrier; protection of a labile pheromone from decomposition.
The possibility of a pheromone-carrier system needs to be considered
in order to avoid over-simplification of interpretation of chemical
and bioassay data. It may prove that a single application of an
active volatile material to the nose of a female will fail to produce a response, and that a more long-term exposure is necessary
(such as what might be obtained by application of carrier-pheromone
with subsequent slow-release of the active substance).

Investigations of Mouse Urine Volatiles

Representative samples of the volatile components of urine
can be obtained by modification of established head-space sampling
procedures (Zlatkis et al., 1973). These volatiles may then be
very effectively separated and identified by capillary gas chromatography and combined gas chromatography/mass spectrometry. In
this way differences between the volatile components of males, females, castrates, etc., may be observed and chemically identified

Figure 4 shows three chromatograms, the upper from normal male
mouse urine, the middle from diestrous females, and the lower from
estrus females (all mice C57BL/6). The most noticeable feature in
the comparison is the general greater intensity of peaks in the male
urine, corresponding well to the noticeably stronger odor of male
urine. Peaks B and D in the male chromatogram correspond to 2-
isopropyl-4,5-dihydro-1,3-thiazole and 2-sec-butyl-4,5-dihydro-
1,3-thiazole, respectively (Novotny et al., 1976; Liebich et al.,
1977). These compounds appear unique to the male, but have shown
no activity in puberty acceleration bioassays. Peaks F and G also
appear unique to males of this strain (they are not present in
BALB/cWt mice) but remain chemically unidentified. Peak C
corresponds to a compound which is present in much higher levels
in male urine than female. Its molecular weight is 154 and its
formula is $C_9H_{14}O_2$, but its structure remains unsolved. Comparison
of the lower two chromatograms reveals three peaks (A, E and H) which
are increased five- to ten-fold in females during estrus. Peak A
is 2-heptanone while E and H are unidentified.

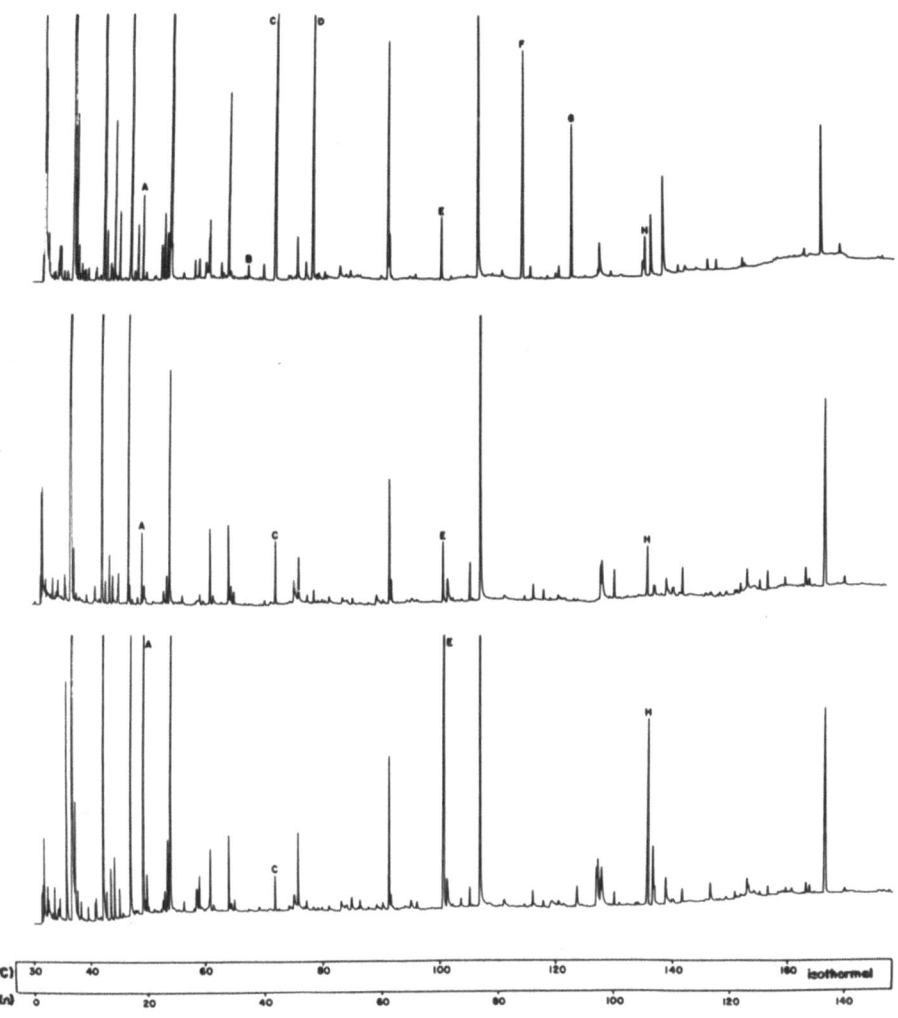

Figure 4. Gas chromatography of urine volatiles from C57BL/6 mice.
Normal males (upper), diestrus females (middle), and
estrus females (lower). Column: 60 meter x 0.25 mm
I.D. glass capillary coated with Ucon 50-HB-2000.
Detection: flame ionization detector.

Figure 5 shows three chromatograms from BALB/cWt mice, the
upper from intact males, the middle from castrate males and the
bottom from castrate males following ten days of testosterone
treatment (0.25 mg testosterone propionate per mouse, every other
day, I.M. injections in rear leg). Here, testosterone treatment
of castrates is seen to produce a generalized increase in the
volatile components. All of the peaks lettered are seen to increase
following testosterone treatment. Of these, 2-heptanone (A);
3,4-dimethyl-3-penten -2-one (tentative, B); 2-sec-butyl-4,5-

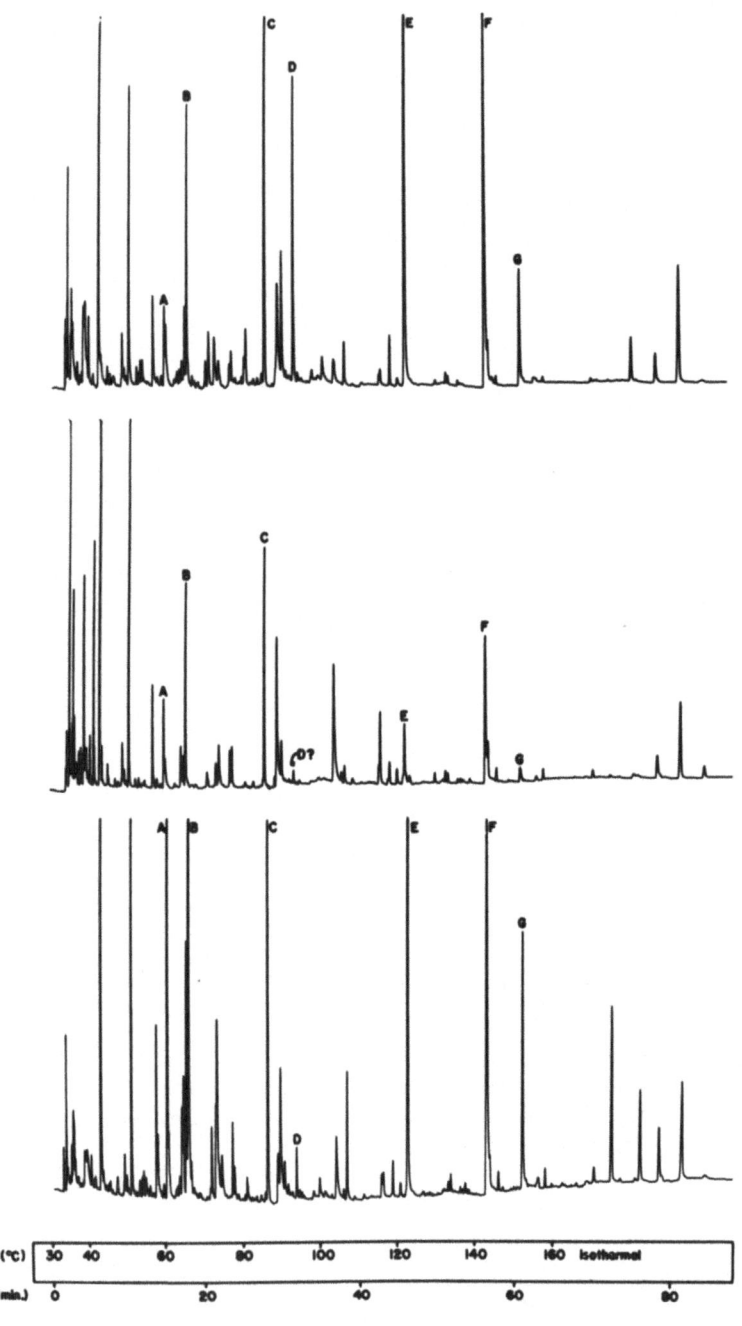

Figure 5. Gas chromatography of urine volatiles from BALB/cWt mice.
Normal males (upper), castrate males (middle) and
testosterone treated castrate males (lower). Column as
in Figure 4.

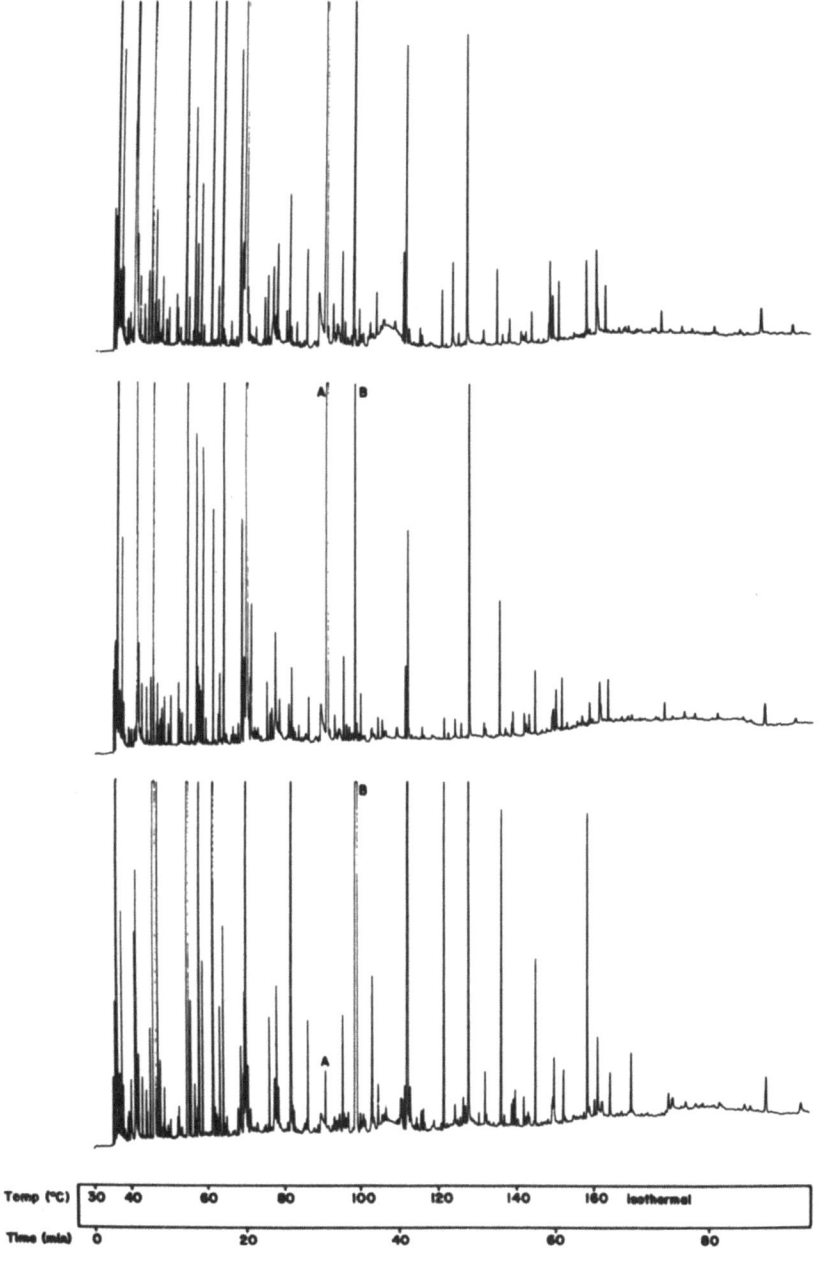

Figure 6. Gas chromatography of urine volatiles for male mice. AKR-H-2^b (upper), AKR-H-2^k (middle) and C57BL/6-H-2^k (lower). Column as in Figure 4.

dihydro-1,3-thiazole (D); a toluidine (E); phenol (F); and a cresol
(G) have been identified. Peak C, also under testosterone control, is
the unsolved molecular weight 154 compound previously mentioned.
The effect of testosterone on urine volatiles of castrates is
interesting since it parallels the effect of testosterone on pheromone production. Understanding of the effect of testosterone on
volatiles may provide a clue as to the nature of the pheromone.

Figure 6 compares the urine volatiles from three strains of
male mice. The first is from AKR-H-2^b, the second from AKR-H-2^k
and the third from C57BL/6-H-2^k.

The first two strains are genetically identical except for the
H-2 region. The second and third strains are genetically different,
but are identical in the H-2 region. The first two chromatograms
are very similar, and both are dissimilar to the third. Of particular interest are peaks A (mol wt. = 154) and B (2-sec-butyl-
4,5-dihydro-1,3-thiazole) since they are already implicated to be
under hormonal regulation. AKR mice tend to produce relatively
more of A while C57BL/6 produce much more of B.

No reliable differences in the volatiles associated with the
H-2 genes have as yet appeared. The general appearance of the
volatiles seems for the most part to be under the control of other
genes. Although the mice in this comparison were on identical
diets, it would be informative to know in general which of the
compounds are more genetically fixed vs which are more dietary
controlled. At this time there is no known correlation between
the mating preferences that are exhibited by mice and the profiles
of volatile components of mouse urine.

The volatile components of mouse urine presently have only a
tenuous relation to known pheromonal effects. However, the variety
of unique and unusual compounds (many of which are in various
stages of chemical identification) which are seen to be under
hormonal and genetic control presents an interesting study in
itself. Elucidating the biochemical and possible behavioral aspects of these compounds may prove fruitful in our understanding
of chemical communication in mice.

ACKNOWLEDGEMENTS

This work was supported in part by Grant No. GM-23668 from
the National Institute of General Medical Sciences, U.S. Public
Health Service. We are grateful to Drs. J.C. Cooke and R. Milberg,
School of Chemical Sciences, University of Illinois at Champaign-
Urbana for recording some high-resolution mass spectra.

REFERENCES

Andrews, P.W. and Boyse, E.A. 1978. Mapping of an H-2-linked gene that influences mating preference in mice, Immunogenetics 6: 265.

Bronson, F.H. 1971. Rodent pheromones, Biol. Reprod. 4:344.

Bronson, F.H. and Desjardins, C. 1974. Relationships between scent marking by male mice and the pheromone-induced secretion of the gonadotropic and ovarian hormones that accompany puberty in female mice, in "Reproductive Behavior" (E. Montogna and W.A. Sadler, ed.), pp. 157-177, Plenum Press, New York.

Bronson, F.H. and Eleftheriou, B.E. 1963. Influence of strange males on implantation in the deermouse, Gen. Comp. Endocrin. 3:515.

Bronson, F.H. and Marsden, H.M. 1964. Male-induced synchrony of estrus in deermice, Gen. Comp. Endocrin. 4:634.

Bruce, H.M. 1960. A block to pregnancy caused by proximity of strange males, J. Reprod. Fertil. 1:96.

Bruce, H.M. and Parrott, D.M.V. 1960. Role of olfactory sense in pregnancy block by strange males, Science 131:1526.

Chipman, R.K. and Fox, K.A. 1966. Oestrus synchronization and pregnancy blocking in wild house mice (Mus musculus). J. Reprod. Fertil. 12:233.

Chuang, J.Y. and Johnson, J.F. 1973. Exclusion processes, in "An Introduction to Separation Science" (B.L. Karger, L.R. Snyder, and C. Horvath, ed.), p. 454, John Wiley and Sons, New York.

Colby, D.R. and Vandenbergh, J.G. 1974. Regulatory effects of urinary pheromones on puberty in the mouse, Biol. Reprod. 11: 268.

Dominic, C.J. 1966. Observations on the reproductive pheromones of mice: I. Source, J. Reprod. Fertil. 11:407.

Finlayson, J.S., Potter, M., Shinnick, C.S. and Smithies, O. 1974. Components of the major urinary protein complex of inbred mice: Determination of NH_2-terminal sequences and comparison with homologous components from wild mice, Biochem. Gen. 11:325.

Hoppe, P.C. 1975. Genetic and endocrine studies of the pregnancy-blocking pheromone of mice, J. Reprod. Fertil. 45:109.

Liebich, H.M., Zlatkis, A., Bertsch, W., VanDahm, R. and Whitten, W.K. 1977. Indentification of dihydrothiazoles in urine of male mice. Biomed. Mass Spec. 4:69.

Marchlewska-Koj, A. 1977. Pregnancy block elicited by urinary proteins of male mice, Biol. Reprod. 17:729.

Marsden, H.M. and Bronson, F.H. 1964. Estrous synchrony in mice: alteration by exposure to male urine, Science 144:1469.

Novotny, M., Jorgenson, J.W., Carmack, M., Wilson, S. and Whitten, W.K. 1976. "A search for the pheromones of the male laboratory mouse" (presented at the 10th International Symposium on the Chemistry of Natural Products, Dunedin, New Zealand, August, 1976).

Parkes, A.S. and Bruce, H.M.   1961.   Olfactory stimuli in mammalian
    reproduction, Science 134:1049.
Parkes, A.S. and Bruce, H.M.   1962.   Pregnancy-block in female mice
    placed in boxes soiled by males, J. Reprod. Fertil. 4:303.
Vandenbergh, J.G.   1967.   Effect of the presence of a male on the
    sexual maturation of female mice, Endocrinology 81:345.
Vandenbergh, J.G.   1969.   Male odor accelerates female sexual
    maturation in mice, Endocrinology 84:658.
Vandenbergh, J.G., Whitsett,J.M. and Lombardi, J.R.   1975.   Partial
    isolation of a pheromone accelerating puberty in female mice,
    J. Reprod. Fertil. 43:515.
Vandenbergh, J.G., Finlayson, J.S.,Dobrogosz, W.J., Dills, S.S.
    and Kost, T.A.   1976.   Chromatographic separation of puberty-
    accelerating pheromone from male mouse urine, Biol. Reprod.
    15:260.
Whitten, W.K.   1956.   Modifications of the oestrus cycle of the
    mouse by external stimuli associated with the male, J. Endocrin.
    13:399.
Whitten, W.K.   1958.   Modifications of the oestrus cycle of the
    mouse by external stimuli associated with the male: changes
    in the oestrus cycle determined by vaginal smears, J. Endo-
    crin. 17:307.
Whitten, W.K., Bronson, F.H. and Greenstein, J.A.   1968.   Estrus
    inducing pheromone of male mice: transport by movement of air.
    Science 161:584.
Yamaguchi, M., Yamazaki, K. and Boyse, E.A.   1978.   Mating prefer-
    ence tests with the recombinant congenic strain BALB.HTG,
    Immunogenetics 6:265.
Yamazaki, K., Boyse, E.A., Mike, V., Thaler, H.T., Mathieson, B.J.,
    Abbott, J. Boyse, J., Zayas, Z.A. and Thomas, L.   1976.
    Control of mating preferences in mice by genes in the major
    histocompatibility complex, J. Exp. Med. 144:1324.
Yamazaki, K., Yamaguchi, M., Andres, P.W., Peake, B. and Boyse, E.A.
    1978.   Mating preferences of F_2 segregants of crosses between
    MHC-congenic mouse strains, Immunogenetics 6:253.
Zlatkis, A., Lichtenstein, H.A. and Tishbee, A.   1973.   Concentra-
    tion and analysis of trace volatile organics in gases and
    biological fluids with a new solid adsorbent, Chromatographia
    6:67.

VARIATION IN THE LEVELS OF SOME COMPONENTS OF THE VOLATILE
FRACTION OF URINE FROM CAPTIVE RED FOXES (<u>VULPES</u> <u>VULPES</u>) AND
ITS RELATIONSHIPS TO THE STATE OF THE ANIMAL

S. Bailey, P.J. Bunyan and Jane M.J. Page

Ministry of Agriculture, Fisheries and Food
Agricultural Science Service, Tolworth Laboratory,
Hook Rise South, Tolworth, Surbiton, Surrey   KT6 7NF

INTRODUCTION

The use of natural compounds to affect the behavior of mammals
may offer a specific method to aid the control of pest or disease
carrying species without adversely affecting non-target species.
The red fox (<u>Vulpes</u> <u>vulpes</u>) is the main reservoir and vector of
rabies in Europe and contingency plans for its control in the
British Isles have been formulated (Lloyd, 1977). During discussions on these plans, the use of synthetic scent substances for
control purposes was considered, and this gave rise to the work
described in this paper.

The red fox possesses several possible sources of scent marking compounds, the anal sac secretion, the supercaudal gland secretion and the urine. Albone et al. (1974) were already investigating the composition of the anal sac secretion and considering
similar work on the supercaudal gland so that a study of fox urine
volatiles to isolate and identify the components seemed to hold
the most promise, and it was started in 1975. Recently Jorgensen
et al. (1978) have published findings on some of the volatile
constituents of urine samples from wild foxes.

The initial work in our study was carried out using urine
samples taken from the bladders of trapped foxes but these samples
arrived irregularly, could not be repeated since the fox was killed
before sampling occurred and, since they had not been voided
naturally, may have differed from normal urine. It was therefore

decided that the preliminary study would concentrate on the
seasonal changes in excreted urine from captive male and female
foxes. These animals were obtained just after weaning in the wild,
but no attempt was made to acclimatize them to human contact.

EXPERIMENTAL

Housing of Animals

     In spring 1976, a pair of wild fox cubs approximately eight
weeks old were taken from the parents. These and subsequent animals have been housed in separate runs two meters square by seven
meters long. A concrete bunker is provided as a den and a
commercially available dog holding unit for urine collection
(Figure 1). The unit, manufactured by Forth-Tech Services Ltd.,
Dalkeith, Midlothian, Scotland, consists of a cage one meter cube
raised from the floor by half a meter. Beneath the floor grid,
it is fitted with a nylon screen and sloping tray to funnel the
urine into a glass bottle surrounded by ice. Each cage is fitted
with a door at either end to allow food to be inserted and allow
access by the fox. Both doors are closed in order to contain the
fox for urine collection. The cages are housed in a wooden building for weather protection and separated from each other by a two
meter gap.

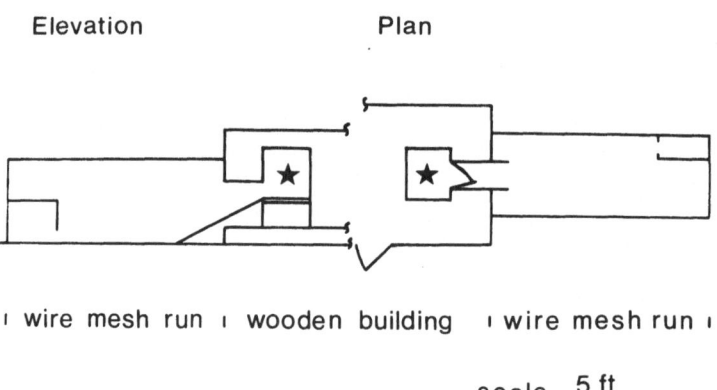

Figure 1. Plan/elevation of pair of symmetrical enclosures used
           to house male and female foxes.

The animals have been fed a diet of constant proportions of rabbit, rat and Japanese quail to remove possible changes due to different food intake. The foxes are always fed in the metabolic cages to combat cage-shyness and as a result little difficulty has been encountered in shutting the foxes in when necessary for urine collection. Urine collection, into a chemically cleaned bottle, is normally carried out at weekly intervals although frequency has been increased during the breeding season and decreased during midsummer. The normal procedure for urine collection has been to contain the fox overnight for a period of seventeen hours. The bottles containing the samples are then sealed with all glass stoppers and removed from the ice bath directly to a deep freeze at -20°C where they remain until immediately before analysis.

In spring 1978 a second pair of cubs was obtained to replace the original pair in order to confirm the results obtained to date.

Collection of Volatiles from Urine

After gentle thawing and mixing, urine (0.5 ml) is pipetted into the flat bottomed flask (40 ml) of the glass apparatus described by Zlatkis et al. (1973) and 4M phosphate buffer pH 6.5 (0.5 ml) is added. Cold water (ca. 12°C) is passed through the condenser and the tube (15 cm x 4 mm I.D.) containing preconditioned Tenax GC (Phase-Sep, 35/60 Mesh, 10 cm length between quartz wool plugs) fitted to the central outlet. Nitrogen gas at a flow rate of 37 ml/min is passed over the magnetically-stirred urine/buffer mixture and through the Tenax for 1 hr with the flask immersed in a boiling water bath.

The Tenax trap is then placed into a tube oven and a stainless steel capillary trap (1 mm x 0.75 mm I.D.), coated with silicone oil SF-96-2000 plus Igepal CO 880, attached to the end. The capillary trap is placed in liquid nitrogen and a flow of nitrogen of 30 ml/min established through the Tenax and capillary. The oven is heated quickly to 270°C and maintained at this temperature for 45 min.

Gas-liquid Chromatographic Analysis of Volatiles

In general, the gas-liquid chromatography has been carried out using a Pye GCD chromatograph fitted with a flame-ionization detector and an open tubular stainless steel column (100 m x 0.75 mm I.D.). The capillary column is coiled by the method of Verbeek and Maarse (1973) and coated by the dynamic method using a solution of 15 percent SF-96-2000 (Phase-Sep) plus 0.75 percent Igepal CO 880 (Phase-Sep) in chloroform (similar column to Robinson

et al. (1973)). After coating, the column is conditioned by
slowly raising the temperature to 190°C over 48 hr and maintained
at 190°C for 16 hr.

At the start of analysis, the capillary trap immersed in
liquid nitrogen is quickly detached from the Tenax tube and connected
as a precolumn in the Pye GCD chromatograph. The liquid nitrogen
reservoir is removed and the chromatograph temperature programmed
to a run of 30 min at 60°C followed by 2°C/min rise to 170°C.
Nitrogen carrier gas flow rate through the column is 3 ml/min,
and the flame ionization detector is used in the manufacturer's
recommended manner.

A small number of samples have been analyzed using an identical
column fitted in a Pye 104 chromatograph equipped with a Tracor
flame photometric detector in the sulfur mode with simultaneous
flame-ionization detection. Other samples have been analyzed by
coupled Gas Chromatography - Mass Spectrometry (GCMS) using an
identical capillary column with direct entry and high speed pumping
to a Kratos MS30 Double Beam Mass Spectrometer equipped with a
Kratos DS50 Data System. The carrier gas is helium at a flow rate
of 3 ml/min with source temperature 200°C and electron-impact
ionization.

Manipulation of Data

The chromatograph was fitted initially with a Pye DP80
integrator and since November 1978 with a Varian CDS III integrator.
The results from the integrator are transferred manually to punched
tape for input to a Hewlett Packard 9830A computer system programmed to calculate the area of each peak as a percentage of the
total area and to plot percentage height against retention time.
Since a small number of very large peaks are consistently obtained,
the plot has been modified to display values up to ten percent
on the y axis. Peaks in excess of ten percent are plotted as
broken lines to ten percent and the actual value printed at this
point (Figure 2). This process of expressing concentrations as a
percentage of total volatiles compensates for changes in relative
concentration between different samples of urine and allows minor
and major changes to be inspected more conveniently.

Assessment of Hormone Treatment on Foxes

Before removing the first pair of animals from the experimental regime in the summer of 1978, a small investigation was
undertaken to determine whether manipulation of the sexual status
affected the volatile profile. An attempt was made to induce
oestrus in the female by two sub-cutaneous injections of
oestradiol benzoate (2.5 mg per injection) with a seven day

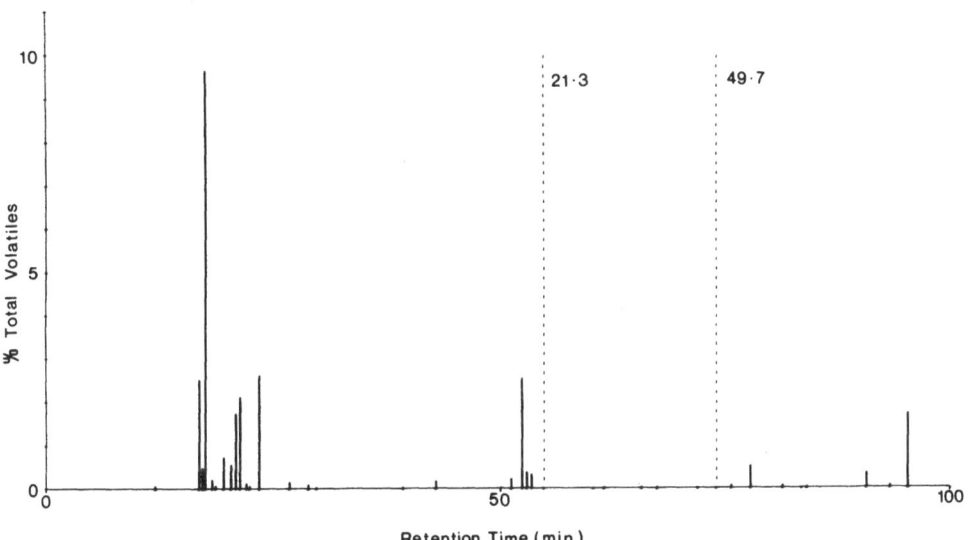

Figure 2. Computer plotted profile of volatiles from male fox urine collected 17 August 1978 showing components represented as percentage values of total volatiles. (Plot shows 21.3% Δ^3-isopentenyl methyl sulfide and 49.7% acetophenone).

interval. The male fox was castrated and following a recovery
period injected with two doses of testosterone propionate (20 mg
per injection) with a seven day interval between injections.
Urine samples were collected at frequent intervals from both animals before and for fourteen days following the first injection.

RESULTS AND DISCUSSION

A preliminary examination of the chromatograms or the raw data
from the integrator suggested that the pattern of volatiles obtained
from the urine of the adult male fox was more complex than that from
the juvenile although there was some evidence that this was a concentration dependent effect. With the female the patterns were more
similar. A typical chromatogram is shown in Figure 3. Direct comparisons are however difficult and the computer program previously
described was devised in an attempt to bring the data into a standard form. When the data is plotted in this mode, the urinary
profiles in the adult and juvenile are very similar. Thus there
would appear to be a difference in the overall concentration of
volatiles with the juvenile producing a much more dilute urine.

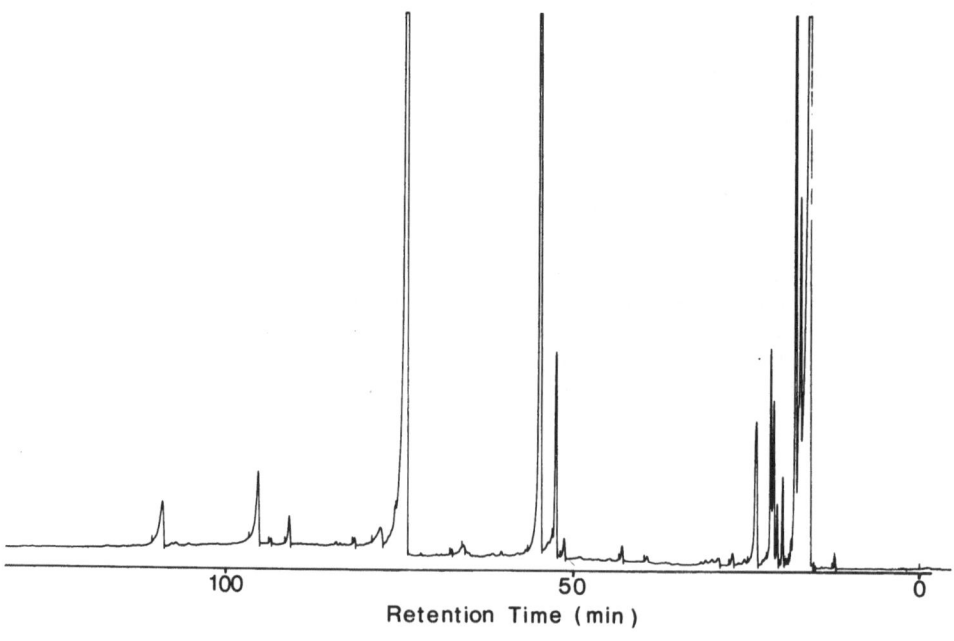

Figure 3.   Gas chromatographic trace of typical summer volatile
            pattern from male fox urine.

The use of the computerized format has allowed an assessment to be made of changes throughout the year. All the chromatograms are dominated by two major peaks. These peaks correspond to the Δ^3-isopentenyl methyl sulfide and acetophenone reported by Jorgenson et al. (1978) and we have confirmed their identity by GCMS. Similarly we have confirmed the presence of 2-phenylethyl methyl sulfide, 4-heptanone, benzaldehyde and 2-methylquinoline as reported by these authors. In contrast, 2-methylquinoline previously reported to be present in the male we have found also in the female urine. We have not so far found 6-methyl-Δ^5-hepten-2-one or geranyl acetone in any urine.

Jorgenson et al. (1978) carried out their investigations by collecting snow containing urine from sites marked by various foxes of unknown age or status living free. They were unable to follow the volatile pattern throughout the year and assess seasonal changes. In contrast, we have been able to follow the production of volatiles in animals of known status throughout the year and have demonstrated a seasonal variation in the components of male fox urine (Figure 4). The first, 4-heptanone shows a relative increase in concentration over the period December to April with a maximum in mid-February (Figure 5).

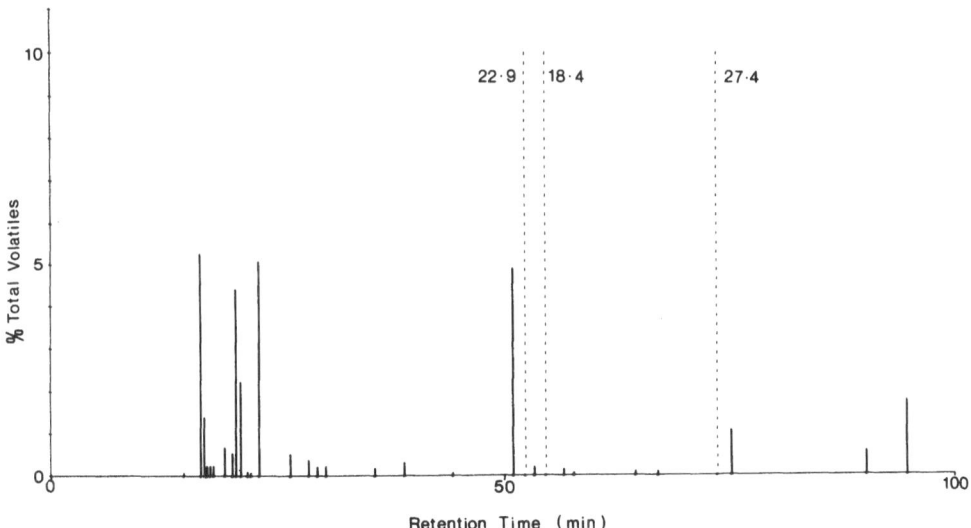

Figure 4. Computer plotted profile of volatiles from male fox urine collected 19 February 1979 showing seasonally variable components. (Peak area 22.9% is previously unreported component).

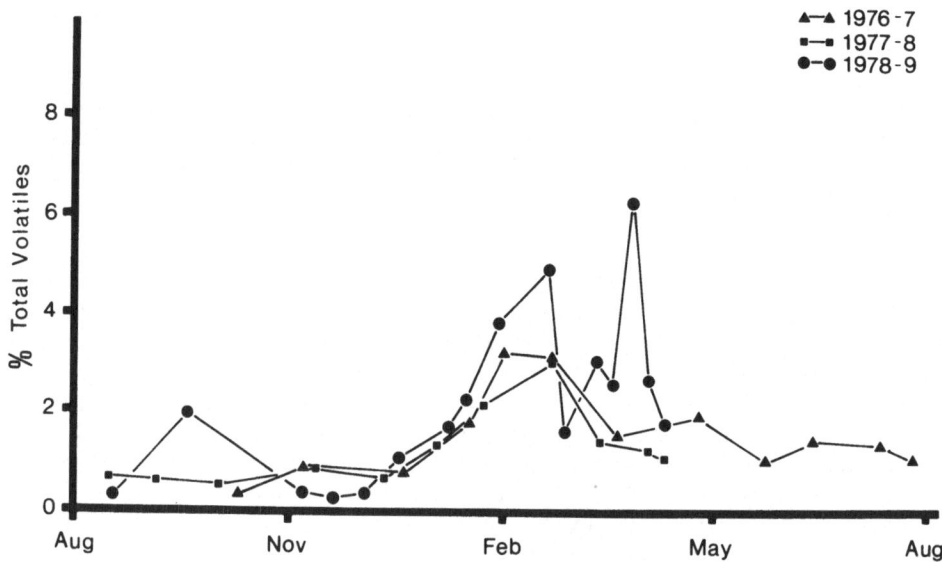

Figure 5.   Annual variation of 4-heptanone over three year sampling
             involving two male foxes.

The second component to show a seasonal change has not been
previously reported in urine. This compound has a retention time
of 0.95 relative to Δ^3-isopentenyl methyl sulfide on our SF 96
column. Again the cycle shows a maximum relative concentration in
February. Figure 6 shows the changes observed over three years in
two different dog foxes. Analysis using the Tracor flame photometric sulfur detector with simultaneous flame ionization detection
showed that the compound contains sulfur. Combined GCMS indicated
a molecular weight of 118 corresponding to an empirical formula
$C_6H_{14}S$. Interpretation of the mass spectrum suggested that the
component is 3-methylbutyl methyl sulfide $(CH_3)_2CHCH_2CH_2SCH_3$. The
component was synthesized by reacting metallic sodium with methane
thiol and adding 3-methylbutyl chloride to the product thus: -

$$2Na + 2HSCH_3 \longrightarrow 2NaSCH_3 + H_2$$

$$NaSCH_3 + (CH_3)_2CHCH_2CH_2Cl \longrightarrow (CH_3)_2CHCH_2CH_2SCH_3 + NaCl$$

The synthetic product was found by glc to have the same retention time as the natural product and comparison of the mass
spectra under identical conditions is shown in Figure 7. 3-methyl-

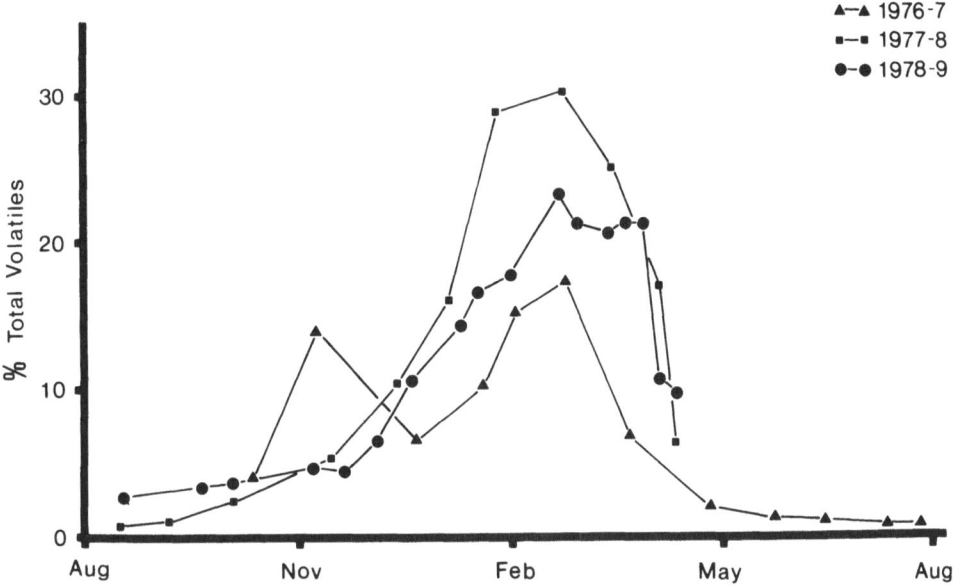

Figure 6.   Annual variation of "unknown" over three year sampling
            involving two male foxes.

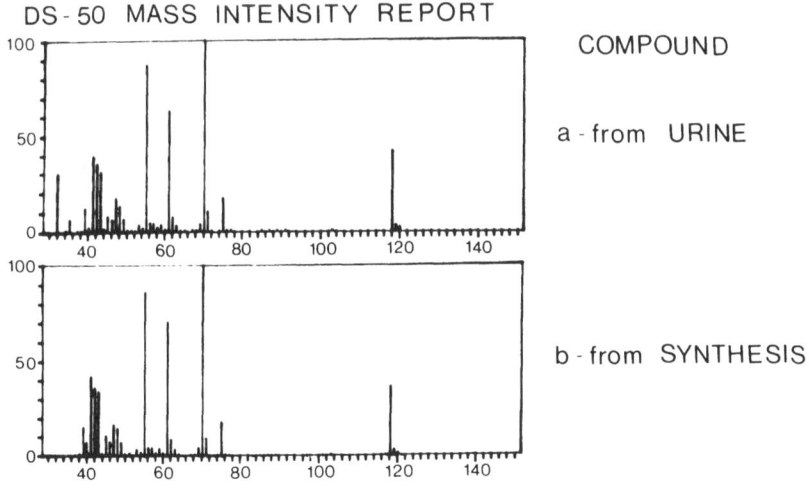

Figure 7.   Mass spectra of synthetic and natural 3-methylbutyl
            methyl sulfide recorded under identical GCMS conditions.

butyl methyl sulfide is also present in female fox urine and its
production during the breeding season is subject to considerable
variation although the maximum values observed are less than 20
percent of those in the male.

The analysis using the Tracor detector also showed the presence of a number of other sulfur containing compounds, some of
which show considerably enhanced sulfur response when compared to
the flame ionization response. However, only the 3-methylbutyl
methyl sulfide and its unsaturated analogue are present in large
amounts (Figure 8). The minor sulfur containing components remain
to be identified.

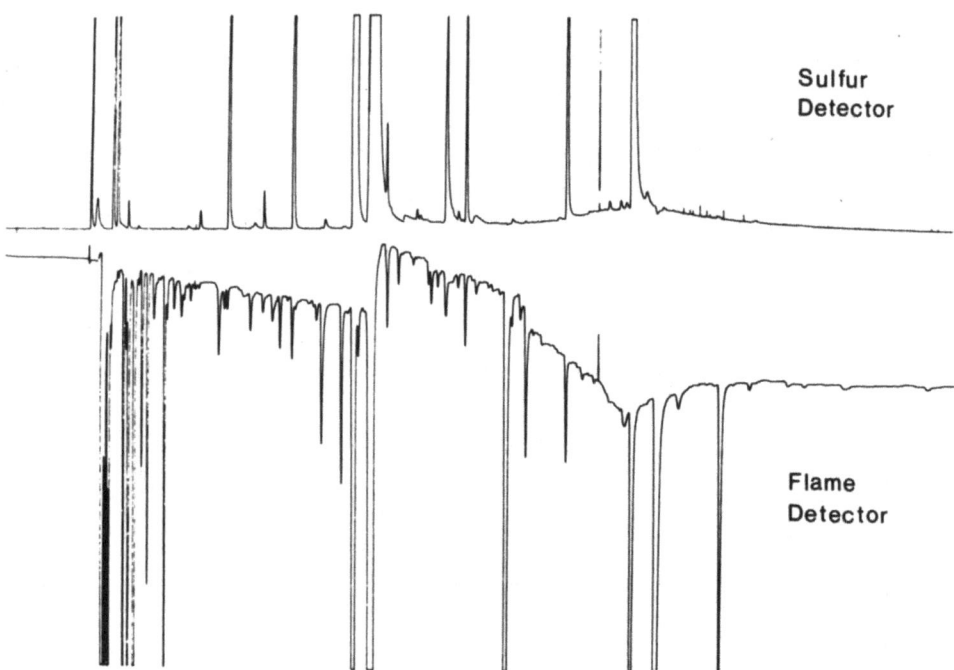

Figure 8.   Gas chromatographic trace using simultaneous flame
            ionization and sulfur specific detection of volatiles
            from male fox urine collected 30 October 1977.

The volatile profiles obtained from the male fox after castration showed no significant differences from those obtained from the intact animal at a corresponding time of year. Injection of testosterone propionate after a recovery period following the castration appeared to give an increase in the production of 3-methylbutyl methyl sulfide although the amounts were not highly significant especially since levels might have been expected to begin to rise at the time of year. Similarly injection of oestradiol benzoate into the female appeared to give a rise in excretion of 3-methylbutyl methyl sulfide three days after injection but again the levels were not highly significant. No other changes in volatile profile were observed from the female. It may be significant that on each of the nights following the injection of the female fox with oestradiol benzoate the urinary profile from the intact male caged nearby was observed on subsequent analysis to show an increase in the concentration of 3-methylbutyl methyl sulfide. These experiments need to be repeated under more carefully controlled conditions using many more samples of urine in order to obtain sufficient data to enable meaningful statistical analysis to be applied. They probably also need to be repeated at different times in the summer when the natural level of 3-methylbutyl methyl sulfide is at its lowest.

At present, the significance of the production and seasonal changes in concentration of 3-methylbutyl methyl sulfide are not known, but it is significant that in South-East England the percentage frequency of pregnant vixens reaches one hundred during the first week in February (Lloyd and Englund, 1973). Curve fitting techniques applied to the data from three years sampling from male foxes indicates that exponential equations in the positive and negative directions fit the data better than polynomial equations. In particular, the data from the first male in its second year shows a highly significant fit to the exponential curve (Figure 9). It could be that the compound is produced exponentially to reach a maximum at the time when females are receptive and could have biological importance as an olfactory signalling mechanism. However, the presence of 3-methylbutyl methyl sulfide should perhaps also be considered in conjunction with that of Δ^3-isopentenyl methyl sulfide. It is possible that production of the two compounds is linked. They may be a by-product of similar biochemical or physiological processes or may arise from totally different pathways. Jorgenson et al. (1978) suggest that Δ^3-isopentenyl methyl sulfide may be produced from isopentenyl pyrophosphate. It is possible that 3-methylbutyl methyl sulfide could be produced following enzymic reduction of isopentenyl pyrophosphate or of Δ^3-isopentenyl methyl sulfide. This conversion could be dependent on the reductive potential of the animal which in turn may reflect its hormonal status. However, it is possible that 3-methylbutyl methyl sulfide may arise from an isovaleric acid derivative. The biochemical basis for the production of these compounds remains to be investigated.

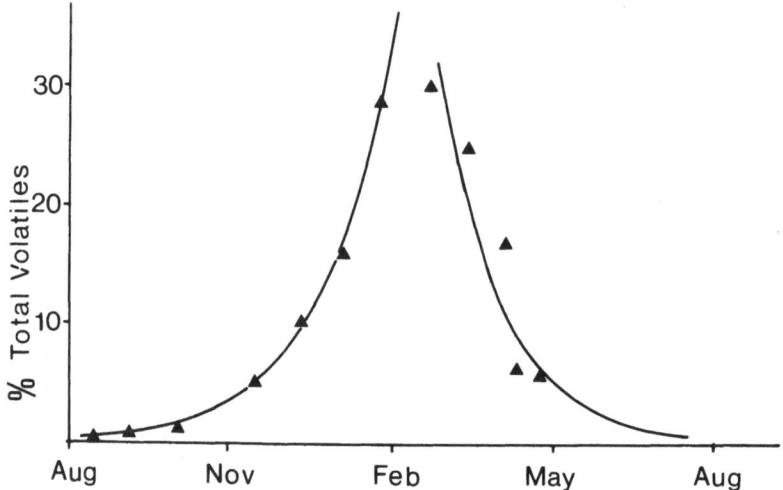

Figure 9. Computed exponential curves fitted to data on 3-
           methylbutyl methyl sulfide from analytical findings
           on urine volatiles from second year male fox.

     Whether 3-methylbutyl methyl sulfide is a signalling compound,
or only an excreted waste product indicating a sex linked physiological change, now remains to be biologically tested. Meanwhile
attempts to isolate and identify other minor components continue
in order to ascertain whether some combination of major and minor
components is necessary for chemical communication.

ACKNOWLEDGEMENTS

     The authors wish to thank R.J.C. Page and Miss L.M. Lelliot for
animal handling, P.J. Garland, L.W. Huson and R.R. Page for
technical assistance and H.G. Lloyd for helpful discussions.

REFERENCES

Albone, E.S., Eglinton, G., Walker, J.M. and Ware, G.C. 1974.
     The anal sac secretion of the red fox, Vulpes vulpes; its
     chemistry and microbiology. A comparison with the anal sac
     secretion of the lion, Panthera leo. Life Sci. 14: 387.
Jorgenson, J.W., Novotny, M., Carmack, M., Copland, G.B.,Wilson,
     S.R., Katona, S. and Whitten, W.K. 1978. Chemical scent
     constituents in the urine of the red fox (Vulpes vulpes L.)
     during the winter season. Science 199: 796.

Lloyd, H.G. 1977. Wildlife rabies: prospects for Britain, In Rabies - the Facts (C. Kaplan, Ed.), pp. 91-103, Oxford University Press, Oxford.

Lloyd, H.G. and Englund, J. 1973. The reproductive cycle of the red fox in Europe. J. Reprod. Fert. Suppl. 19: 119.

Robinson, A.B., Partridge, D.,Turner, M., Teranishi, R. and Pauling, L. 1973. An apparatus for the quantitative analysis of volatile compounds in urine. J. Chromatogr. 85: 19.

Verbeek, A.M. and Maarse, H. 1973. The honeycomb spool: a modified method for winding capillary columns. J. Chromatogr. 84: 389.

Zlatkis, A., Lichtenstein, H.A. and Tishbee, A. 1973. Concentration and analysis of trace volatile organics in gases and biological fluids with a new solid adsorbant. Chromatographia 6: 67.

CHEMICAL SIGNALS ASSOCIATED WITH AQUATIC PREDATION SITES

Daniel Rittschof

Department of Physiology
University of California
Los Angeles, CA

Predation sites were simulated and observed in the gulf intertidal near the Edward Ball Marine Laboratory, Sopchoppy, Florida. Bivalve molluscs, brachyuran crabs and several gastropod species were tested as prey. In the case of the gastropod sites, experiments were performed to determine the approximate size and source of the signal molecules. The sites were visited by Fundulus similis, Callinactes sapidus, Melongena corona, Clibanarius vittatus, and Pagurus longicarpus. All attendant species were attracted via chemical signals released passively from the prey. Attendants were attracted equally as well to gastropod sites at which only small molecules could act as signals. Site visitation was dependent upon the prey species, and attendant species appeared at characteristic times after site initiation. This temporal pattern was to a large extent independent of the size of the predation site. F. similis, C. sapidus, and M. corona were attracted to small molecules from the flesh of bivalves, gastropods and brachyurans in an apparently non-specific manner. F. similis and C. sapidus visited the sites almost immediately after initiation. They continued to visit at increasingly less frequent intervals for about six hours. M. corona visited until several days after initiation. All three species fed upon the prey. P. longicarpus and C. vittatus were attracted only to gastropod flesh sites and in a species-specific manner. Although the hermit crab species were attracted to small molecules released from the flesh of gastropods, visual and tactile secondary cues appeared to be important. Neither species of hermit crab remained at a site unless there was a shell present. Neither species fed upon the prey, but came and waited for the shell to be free for occupancy. Shell-fit measures of the general populations of C. vittatus and P. longicarpus, when compared to the fit of the attendants at predation sites, indicated

that predation site attendants had a poorer shell fit than the
general population. Both species of hermit crabs could detect
a newly released shell even after occupancy by another crab.
Shell exchange behavior was rapid and without preliminary exploration in the immediate vicinity of a predation site. P. longicarpus
was attracted to Polinices duplicatus sites, while C. vittatus was
preferentially attracted to Busycon and M. corona sites. Continuous in situ treatment of gastropod tissue at a predation site with
a serine protease resulted in an increase of attendants over control with no loss of a specificity. These data support the hypothesis that a small amino acid is responsible for feeding attraction and that peptides or a different mix of amino acids may be
acting as signal molecules for the attraction of shell utilizing
species. It is postulated that naturally occurring serine pro-
teases are acting upon the protein substrate of the wounded or
dead prey gastropod to release proteolytic cleavage products which
are characteristic of the prey involved. The controlled release
proteolytic mechanism would explain the latent, extended attractive period of the predation site to the shell utilizing species.
Potential peptide signal molecules were partially characterized by
electrophoretic techniques.

The author would like to thank William Herrnkind, William
Lingberg, Brian Hazlett, Tara Missick and the staff of the Edward
Ball Marine Laboratory for assistance with this work.

THE ROLE OF CHEMORECEPTION IN SOME SOCIAL AND SYNCHRONOUS BEHAVIOR OF THE CRAYFISH, PROCAMBARUS CLARKII

Mary Ellen Harris

Biology Department
University of Dubuque
Dubuque, IA 52001

The effects of chemical stimuli on some behavior in the crayfish were studied. There were five test series, utilizing two experimental systems. The first system consisted of placing crayfish in standing water from a known source and monitoring general behavior. A box with six impulse buttons was attached to a strip chart recorder. This allowed the observer to record six different behavioral parameters on data rolls. The second system was composed of a flowing water apparatus. A stimulus animal was placed in an upper level tank which was supplied with freshwater inflow. Water then flowed into a lower level tank which contained a receiving experimental animal. Reactions of one animal to another active animal's water could be determined. Two strip chart recorders were used to monitor behavior of both animals simultaneously. The first test series contained three control situations and three experimental situations in the standing water experimental system. These controls proved important in establishing a linear activity decline with time. Non-parametric regression analyses were used to define the experimental effects. The final four test series had an initial control situation with subsequent experimental conditions in the flow-through system. Rank-order non-parametric statistics were used to determine experimental effects.

Individual recognition was tested by first establishing dominant and submissive status within twenty-five pairs of crayfish, Procambarus clarkii. Juveniles were used to prevent sexual cues. Each animal was then exposed to standing water of the other pair member and to water from a communal tank. There was no behavioral change towards water of the other pair member. However, all juveniles responded to the mixed chemical stimuli from the communal

tank water by significantly increasing activity (p < .001). Further, dominant crayfish were more active than submissives upon exposure to communal tank water. Juvenile individual recognition was
then tested in the flow-through system. No reactions to water
flowing from known pair members occurred. Submissive crayfish
significantly decreased resting (p < .05) upon exposure to water
flowing from unknown animals. Thus, chemical individual recognition was shown. Rather, animals responded to unknown chemical
stimuli from individuals and a mixed environment.

Synchronous activity was tested in the flow-through system.
It was thought that twenty crayfish pairs might coordinate activity patterns through chemical communication. It was found that
activity was governed by other than chemical stimuli. Synchronous
feeding was discovered to be mediated by a pheromone. Feeding
responses and activity levels for twenty adults and juveniles
significantly increased (p < .05) upon exposure to water flowing
from animals feeding on a non-stimulatory substance. Juveniles
reacted more than adults. It is theorized that aggregation around
a food source may be beneficial to juveniles. Although aggressiveness is developed, juveniles could easily gather without doing
each other great damage because the claws are not fully grown.

Responses to water were analyzed in the flow-through system.
Twenty adult crayfish in the lower level tank significantly suppressed activity levels (p <.01) upon exposure to flowing water
from their own holding chambers. This activity decrease was
markedly greater (p < .0005) than the decrease shown by animals
exposed to water flowing from unknown individuals. Chemical
recognition of "home" water may function in population spacing,
with identified foraging areas providing for adequate food. Crayfish seem content to remain in such a known area.

Chemoreception has an important role in the behavior of
Procambarus clarkii, as seen from the results of these five test
series.

BEHAVIOURAL AND ELECTROPHYSIOLOGICAL EFFECTS OF NATURAL CHEMICAL STIMULI IN THE GOLDFISH (CARASSIUS AURATUS)

H.P. Zippel, R. Voigt, and D. Hager

Physiologisches Institut
Universität Göttingen
Humboldtallee 7
D 3400 Göttingen, West Germany

Although the biological importance of chemical signals in the olfactory and gustatory systems of aquatic vertebrates would appear to be well established, comparative behavioural and electrophysiological investigations of this phenomenon are few and far between.

Over the last decade, experiments in our laboratory have demonstrated a correlation between training behaviour (Zippel, H.P.: Z. vergl. Physiol. 69, 54-78, 1970) and neuronal responses recorded from single neurons in the olfactory bulb (Zippel, H.P., Breipohl, W.: in Olfaction and Taste V., D.A. Denton and J.P. Coghlan, eds.; 163-167, 1975) during application of synthetic odours (e.g., amyl acetate, coumarin, phenol). An extension of this work showed that, in comparison with synthetic stimuli, natural odours (Tubifex-food-extract, skin extract) not only caused a larger number of neurons to respond but also induced more pronounced and longer-lasting effects. The hypothesis that these findings merely reflected the action of low molecular weight compounds containing amino acids could not be confirmed: the application of synthetic amino acids produced effects comparable with those of synthetic odours (Zippel, H.P., Schumann, R., Tiedemann, W.: Pflügers Arch. 362, R. 46, 1976).

On the basis of the above findings, the following questions were the subject of our most recent experiments:

(i) What are the behavioural olfactory thresholds of Tubifex extract and its fractionated high and low molecular weight components?

(ii)   Is there any evidence of a smell/taste distinction in the behaviour of the animals to the above stimuli?

(iii)   Are the behavioural findings reflected in the activity of single neurons in the olfactory bulb?

The behavioural data show that (on the basis of 1g wet weight) crude Tubifex extracts elicit positive reactions in intact animals down to a concentration of 10^{-7} (odour threshold).   The frequency and persistence of behaviour to the test stimuli undergoes an almost continuous decrease from high (10^{-3}) to low concentrations. In bulbectomized animals, the threshold concentration is increased to 10^{-5} (taste threshold).   The action of the non-dialysate (high molecular), the dialysate (low molecular), or a mixture of both fractions is not essentially different from that of the crude extract.

The electrophysiological findings are in accordance with the behavioural data: generally, the number of responding neurons and the intensity and duration of the effects decrease progressively from high to low concentrations.   Although the majority of cells conform to this pattern, a small number of neurons react to only one (or two) of the applied stimuli.

In conjunction with earlier findings, the above data indicate that the behavioural correlate is recordable in the olfactory system at the level of the second neuron.   Whether the pronounced effects caused by complex natural stimuli are attributable to inputs from the olfactory receptors or to reverberatory circuits in the olfactory bulb itself, or whether they are mainly influenced by descending pathways from central nervous structures, cannot be resolved at present since recordings have only been made in the intact olfactory system.   The hypothesis that the long-lasting effects, particularly following application of natural stimuli, might be derived from higher centres warrants further investigation.

OLFACTORY USE IN FOOD-LOCATION AT SEA BY TUBENOSED PELAGIC BIRDS

Larry V. Hutchison and Bernice M. Wenzel

Department of Physiology and Brain Research Institute
UCLA School of Medicine
Los Angeles, CA 90024

For at least the last hundred years, conjecture has existed
that procellariiform birds (albatrosses, fulmars, shearwaters,
and petrels) might use olfaction in locating food and in return to
their nesting sites. Anatomical studies showed that the olfactory
mucosa is comparatively more extensive in tubenosed birds.
Measurements have shown that they have larger olfactory bulb/
cerebral hemisphere ratios than most other birds for which measurements are available. Our controlled experiments conducted in the
Pacific Ocean consistently support the conclusion that procellariiforms detect and are attracted by food-related odors. Fish oils,
squid homogenate, vegetable oil, bacon grease, mineral oil, and
motor oil were the odor stimuli. These stimuli were presented as
surface slicks, slicks in clear plastic enclosures floating on
the ocean, and saturated wicks on free-floating rafts. Significantly greater proportions of procellariiforms than nonprocellariiforms (gulls, terns, pelicans, etc.) were sighted within 10 meters
of fish oils, squid homogenate, vegetable oil or bacon fat and
they approached from downwind significantly more often. Only
procellariiforms regularly landed and fed on slicks of edible oils.
Neither the plastic enclosure nor the wick attracted procellariiforms unless baited with food-related stimuli. Small volumes of
major complex fractions of tuna oil prepared from gas chromatography
were presented in hexane via the saturated wick method and were
found as effective in attracting procellariiforms from downwind as
the unrefined substances. Sea water, a dry wick, and hexane served
as control conditions. A behavior pattern unique to procellariiforms, which we termed foraging flight, was characteristic of these
bird's approaches to sources of food-related odors. In conjunction
with the experimental results, the foraging flight patterns was
indicated as a behavioral manifestation of the ability to follow

an air-borne odor to its source. When puffed cereal was presented,
with or without oily slicks, nonprocellariiforms approached from
all directions in significantly greater proportions than when cereal
was absent. With cereal alone, they were significantly more
numerous than procellariiforms.

Our field work on procellariiforms is conducted as an extension of our continuing neurophysiological study of the olfactory
system in avian forms. We have collected considerable data on the
central projections of the olfactory bulb in the pigeon, a species
with only moderate olfactory endowment. Recently, anatomical and
electrophysiological study of the olfactory pathway of the Northern
fulmar (Fulmarus glacialis) was begun. Electrical stimulation
(8-12 V pulses, single or train, 0.5-1.0 ms duration, 0.1-1.0 Hz)
of one bundle of olfactory nerve twigs resulted in evoked responses
in both olfactory bulbs and in bulbar projection sites in the forebrain, as well as in significant modification of spontaneous unit
activity ($\pm$ 20% change in rate). Responsive forebrain sites were
homologous to those receiving olfactory projections in the pigeon.
Varying concentrations of certain odor stimuli combined with
ambient air blown directly into the nares by a syringe produced
significant changes from pre-odor spontaneous firing rates in
several bulbar cells. Single cells showed either enhancement or
inhibition depending upon qualitative differences among stimuli.
Unit activity was consistently inhibited by natural food odors
(cod liver oil and krill homogenate) but was not significantly
modified by ambient air and control solvent. (Supported by a grant
from the National Geographic Society to B.M. Wenzel and NINCDS
Postdoctoral Fellowship NS 05896 to L.V. Hutchison).

PARTIAL ISOLATION OF PREGNANCY BLOCK PHEROMONE IN MICE

Anna Marchlewska-Koj

Department of Genetics, Institute of Zoology
Jagiellonian University
Kracow, Poland

Implantation failure in the early stage of mouse pregnancy after exposure to urine of alien males is a well documented phenomenon.

The present report describes the results of studies on the identification of the priming pheromone eliciting the pregnancy block effect. CBA/kw females mated to CBA/km males were used for biological tests. The morning when the vaginal plug was found was designated as the first day of pregnancy. The female was transferred to a clean cage and treated with the test solutions during the second and third day of pregnancy. The solution was given in drops on the oro-nasal groove. Every morning vaginal smears were obtained, and the first day when the smear contained only squamous epithelial cells was regarded as the oestrus. The appearance of blood in vaginal smear between the 10th and 12th day was considered as indicating pregnancy.

The priming pheromone was assayed in outbred male urine. The urine was always dialyzed against phosphate-buffered physiological saline. Chemical analyses and biological activity were tested in the non-dialyzable fraction in the following way:

1. The proteins were salted out with solid ammonium sulphate, dissolved in distilled water and tested. Eighty per cent of females exposed to the protein fraction terminated pregnancy.

2. The non-dialyzable fraction was separated by column chromatography on Sephadex G-75 into three different fractions as measured by optical density at 280 nm. Only the peptide fraction was biologically active, and the Bruce effect was evoked in 40% of tested females.

3. The peptide fraction was further analyzed by using thin-layer electrophoresis on Sephadex G-25. By staining with ninhydrin, eight peptide fractions were identified. The block of pregnancy was observed after exposure of mice to six fractions with the following efficiency: after fraction 4 - 100%; fraction 5 - 71.4%; fraction 3 - 57.1%; fraction 1 and 7 - 28.6% and fraction 2 - 14.3%.

For further identification of the chemical nature of the pheromone, males were injected with 1 µCi of ^{3}H-testosterone. Their urine was subjected to chromatography on a Sephadex G-75 column, and the radioactivity was determined in a liquid scintillation spectrometer. Most of the radioactivity was located in the peptide fraction.

THE ROLE OF THE TARSAL GLANDS IN THE OLFACTORY COMMUNICATION OF THE ONTARIO MOOSE

A.B. Bubenik[1], M. Dombalagian[2], J.W. Wheeler[2] and O.Williams[1]

[1]Ministry of Natural Resources, Maple, Ontario
[2]Department of Chemistry, Howard University, Washington, D.C.

From behavioral studies of moose and elkhounds, the authors have concluded that the tarsal gland secretion (TGS) is the primary olfactory signal in searching out con-specifics. For this study TGS alcoholic extracts of 49 males and 49 females collected between October--December were analyzed by gas chromatography--mass spectroscopy. In 28 males (57%) and 17 females (34.7%) o-cresol was found as the major volatile, accompanied by smaller amounts of higher homologs. The o-cresol content varied with sex, age and date from 0-170% of a C_{16} ethyl ester used as an internal standard. Male and female moose from the same locality showed low (or no) cresol content for calves and one-year olds. Glands containing o-cresol were found in females from the same area while males having these "active" glands were scattered in time and location. Since samples containing o-cresol did not elicit any sexual response from moose, we suggest that o-cresol acts only as a "conspicuous informer", attracting the animals attention and leading it to other more definitive olfactory cues. The higher homologs in the phenolic fraction were also identified, but did not elicit any sexual response. Controls from other parts of the moose body, urine extracts, extracts of Norwegian moose tarsal glands, and the variation with sex, age and condition of the animal in Ontario moose indicate that the cresols are tarsal gland components and not urinary contaminants. Further studies of the role of the tarsal gland are in progress.

DIOSMIC RESPONSES TO SCENT-SIGNALS IN <u>LEMUR</u> <u>CATTA</u>

C.S. Evans

Dept. of Biological Sciences
Glasgow College of Technology
Scotland

SUMMARY

Behavioral and structural aspects of the genital scent-signalling system were examined in the social diurnal lemur <u>L. catta</u>. Among the prosimian primates the Malagasy lemurs are found in complex societies (Jolly, 1967) which not only retain, but have elaborated on, the level of olfactory signalling which arose from nocturnality (Charles-Dominique, 1978). The selective incorporation of specialized exocrine glandular secretions into the elaborate (dimorphic) behaviors of <u>Lemur catta</u> were suggested as a likely source of chemical signal complexity (Evans and Goy, 1968). Evidence on function exists for only some of the glands of a few non-primate mammals, (e.g. Rasa, 1973). A wider interpretation of the relative importance of olfactory cues (mainly genital tract secretions) in sexual behavior (Keverne, 1976; Goldfoot et al., 1978) requires primate olfactory studies to be extended to the "lower" but macrosmatic members, e.g., Haplorhines (Epple, 1974). Studies of the ring-tailed lemur have examined responses to products of the homologous and heterologous secretory areas, the brachial, ante-brachial, the scrotal and labial skin glands, using the choice behavior of adult males when presented with scent from the possible contributory sources to object marks. Male/male recognition may involve the dimorphic glands, possibly for both intra- and inter-troop interactions (Mertl, 1977).

Males show consistent individual response levels for investigative and marking behaviors. These could clearly be scored in association with a given stimulus, such as a scent-bearing dowel or swab; oriented responses only were analyzed. Intact female-scent marks (genital secretions) elicit-preferential responding

by males, although there is a lack of selective deposition of marks
by females (Bailey and Evans, 1979; Evans, 1979). Presented with
contrasting pairs of stimuli on or near "neutral" (familiar) marking sites males divided their attention equally between unlike-sex
genital scent-pairs but spent significantly more time at the labial
source of a within-sex pair (Evans, 1979). Evidently female marks
disperse mainly labial secretions which may not provide immediate
reproductive information but do contain attractant components. The
common secretion source tested - the labial sebaceous/apocrine
field (histology will be reported elsewhere) - elicited a similar
level of sampling (sniffing + licking + Flehmen) in two comparison
series. Male lemurs respond to, but do not discriminate between,
male/female (external) genital stimulus pairs; similarly there is
a lack of consistent preference by male tamarins (Epple, 1974) to
male vs. female marks. These findings may reflect inadequate
stimulus differential (e.g., similar sources) rather than a lack
of discriminative ability. Diestrus vaginal secretions failed to
sustain interest although approached at (low) but consistent rate;
however, their attractiveness may change to signal receptivity.
This suggests the identity of the source responsible in part for
the four-fold increase in <u>direct</u> ♂♂ sniffing/licking of the vulval
area during vaginal estrous (Evans and Goy, 1968, Tb3). Further
(hormonal) manipulation of the secretion sources so far identified
is indicated, while estrous urine/vaginal secretions remain to be
investigated. Diestrous lemur urine does not contain behaviorally
detectable attractants (Bailey and Evans, in prep. 1979) unlike
that of tamarins which contains gender cues (Epple, 1978).

The Flehmen response is most readily given following
nasal contact-investigation (c.f. Verberne, 1976) at
genital marks and to the scent-gland component of these, at least
by males. Further studies of the determinants of this response
are in progress. At present, the lack of discrimination between
male/female scents emphasizes the need to examine the contribution
of Flehmen to chemosensory functions. In contrast to the Flehmen
response to urine in ungulates, Flehmen in lemurs seems to be a
discriminator between genital and other skin gland secretions.
Flehmen is proposed as part of the sampling mechanism for the
exchange of low- or non-volatile compounds within the VNO (Bailey,
1978). This interpretation is supported by the following findings:
the presence of a distinct AOB in <u>Lemur</u> (Stephan, 1965), and
histological evidence of both a complete neuroepithelium and
vascular surround within the VNO (Evans, unpubl. data); these
being more extensive than in the mouse lemur (Schilling, 1970).
This development of the VN system suggests adequate chemoreceptive
potential with good fluid-exchange efficiency, requirements also
of the "vascular pump" mechanism (Meredith, 1978, this vol.).
Hence the secondary (VN) system may well allow the detection of
social and other sources of environmental cues (c.f. Bertmar, 1975).

These findings and related studies of ♂ + ♀ discriminations of masculine and feminine scent-sources (Bailey and Evans, in prep.), if considered in parallel with reproductive and field studies, may outline the role of chemosensory involvement in complex social interactions.

Direct functional studies of the dual olfactory system, particularly the dynamics of Jacobson's Organ, in a suitable diosmic primate are indicated.

REFERENCES

Bailey, K. 1978. Flehmen in the ring-tailed lemur (L. catta). Behav. 65: 309-319.

Bailey, K. and Evans, C.S. 1979. Chemosensory discrimination of conspecific stimuli in Lemur catta L. In prep.

Bertmar, G. 1975. Ecochemical studies on reindeer, In Olfaction and Taste V., 231-233. D.A. Denton and J.P. Coghlan, Eds. Acad. Pr.

Charles-Dominique, P. 1978. Solitary and gregarious prosimians In Rec. Adv. Primatol. V. III, 137-150.

Epple, G. 1974. Olfactory communication in S. American primates. Ann. N.Y. Acad. Sci. 237, 261-278.

Epple, G. 1978. Studies on the nature of the chemical signals in scent marks and urine of Saguinus fuscicollis. J. Chem. Ecol. 4, 383-394.

Evans, C.S. and Goy, R.W. 1968. Social behavior and reproductive cycles in captive ring-tailed lemurs (Lemur catta L.). J. Zool. 156: 181-197.

Evans, C.S. 1979. Olfactory responses to heterosexual genital secretions by male and female ring-tailed lemurs. In prep.

Goldfoot, D., et al. 1978. Anosmia in male rhesus monkeys. Science 199, 1095-1096.

Jolly, A. 1967. Estrous synchrony in wild Lemur catta. In Altmann, S.A., Ed., Social Communication in Primates.

Keverne, E.B. 1976. Sex receptivity and attractiveness in monkeys. Adv. Stud. Behav. 7, 155-200.

Meredith, M. 1978. Vomeronasal/accessory olfactory system: stimulus access and unit response. In Olfaction and Taste VI: 200 (abs.) LeMagnen J. and MacLeod, P. Eds. IRL.

Mertl, A.S. 1977. Habituation to territorial scent marks in the field by Lemur catta. Behav. Biol. 21. 500-507.

Rasa, O.A.E. 1973. Marking behavior and its social significance in the African Dwarf Mongoose. Z. Tierpsychol. 32: 293-318.

Schilling, A. 1970. L'organ de Jacobson du lemurien Malgache
    <u>Microcebus murinus</u>. Mem. mus. Nat.d'Hist. Nat Paris 61 (A)
    203-280.
Stephan, H, 1965. Accessory olfactory bulb in insectivores and
    primates. Acta. Anat. 62, 215-253.
Verberne, G. 1976. Chemocommunication among domestic cats,
    mediated by olfactory and vomeronasal senses ii., Z.
    Tierpsychol. 42, 113-128.

STUDIES ON CHEMICAL COMMUNICATION IN SOME AFRICAN BOVIDS

R.C. Bigalke, P.A. Novellie, Dept. of Nature Conservation
Maritha le Roux, Department of Chemistry
University of Stellenbosch
Stellenbosch 7600, South Africa

Bontebok. Responses of a captive male bontebok Damaliscus dorcas dorcas to secretions of interdigital and preorbital glands were tested by presenting them at a food trough.

Interdigital secretion was offered on gauze swabs in steel containers and, in subsequent tests, in an airstream. The airstream was also used to present each of the major components, (Z)-6-dodecen-4-olide and (Z)-5-undecen-2-one (Burger et al., 1976). Mean length of feeding bouts was not significantly affected by any of these olfactory stimuli. Behavioural responses were confined to irregular brief bouts of sniffing at scent sources, which also occurred during some control runs, and a few shows of aggression. These, involving horning, glandular weaving, sniffing at and defaecating on the dung pile, were all performed fairly regularly outside the test situation so that it is uncertain whether the experiments elicited them.

Preorbital exudate also had no significant influence on feeding time. However, in 10 presentations, 11 bouts of sniffing at the scent port were recorded. There was clearly much greater interest in this secretion.

The limited response to interdigital secretion and its components, and the fact that chemical composition shows very little individual variation, suggests that it may be a species-specific scent. Spread about an inhabited area it would merely indicate use by conspecifics and would not elicit particular reactions.

Preorbital marking is performed mainly by terrestrial males in the rutting season. In the field David (1973) saw no behaviour patterns suggesting recognition of these scent marks, but our tests indicate that they elicit interest, at least in a male. No repellant effect was apparent although this might perhaps have been expected. Territorial advertising does however involve a great deal of visual display and invaders are also driven off. Olfactory cues might therefore be expected to be relatively unimportant. It is suggested that preorbital scent marks may play a role as familiar features of the territory holder's environment - perhaps mainly in the early stages of occupancy ("self-reassurance") - and that in some situations marking may be merely an expression of aggression.

Grysbok. Responses to preorbital secretions were tested in captive grysbok Raphicerus melanotis, a small solitary antelope inhabiting dense shrub. The major chemical components of the exudate have been identified as formiates but have not yet been individually separated.

Twigs marked with preorbital secretion placed next to food troughs failed to affect food intake significantly even when tested in a strange environment or when the secretion came from a totally strange male. Marked twigs were also placed next to Pinus seedlings, which were eaten within twenty four hours. Seedlings treated with a cowdung and lime repellant mixture, were untouched.

Since olfaction appears to be the principal form of communication and male grysbok mark actively with their preorbital glands, the apparent absence of a repellant effect by preorbital secretion is puzzling. Perhaps marking serves rather to saturate the territory with the scent of the occupant ("self-reassurance"), and to inform neighbours of his occupancy. However, the tests have not excluded an area repellant effect.

Springbok. No evidence has been found of springbok (Antidorcas marsupialis) marking with the preorbital glands which occur in both sexes. In this the species resembles Gazella granti (Leuthold, 1977) and differs from all other gazelles studied to date, which mark with the glands. The well developed dorsal gland or rump patch has been shown to secrete isoprenoid and terpenoid hydrocarbons and ketones (Burger et al., 1978); terpenoids have proved rare in mammalian gland secretions to date. The putative alerting function of the scent, which is presumed to be released when the rump patch is extended in alarming situations has yet to be tested.

REFERENCES

Burger, B.V., M. Le Roux, C.F. Garbers, H.S.C. Spies, R.C. Bigalke, K.G.R. Pachler, P.L. Wessels, V. Christ and K.H. Maurer. 1976: Ketones from the pedal gland of the bontebok (Damaliscus dorcas dorcas). Z. Naturforsch. 31c: 21-28.

Burger, B.V., M. Le Roux, H.S.C. Spies, V. Truter and R.C. Bigalke.
    1978: Mammalian pheromone studies - III. (E.E.)-7, 11, 15-
    trimethyl-3-methelenehexadeca-1, 6, 10, 14-tetraene, a new
    diterpene analogue of α-farnesene from the dorsal gland of the
    springbok, Antidorcas marsupialis. Tetrahedron Letters 52: 5221-
    5221-5224.
David, J.H.M. 1973: The behaviour of the bontebok, Damaliscus
    dorcas dorcas (Pallas 1776) with special reference to terri-
    torial behaviour. Z. Tierspychol. 33: 38-107.
Leuthold, W. 1977: African ungulates, a comparative review of
    their ethology and behavioural ecology. Springer, New York.